WASHINGTON POST BESTSELLER

WALL STREET JOURNAL BESTSELLER

LOS ANGELES TIMES BESTSELLER

AN INDIE NEXT LIST SELECTION

NAMED A BEST BOOK OF 2017 BY:

Smithsonian.com	*Business Insider*
"Science Friday"	Barnes & Noble
Brainpickings	BookBrowse
The Christian Science Monitor	Chicago Public Library

"Astonishing. . . . Mundy . . . joins authors such as Margot
Lee Shetterly and Nathalia Holt in giving the women behind
great twentieth-century scientific endeavors their due."

—*NATURE*

"Mesmerizing. . . . There are thrilling moments in the story
when the women succeed in breaking codes . . .
a much-needed update to the canon of World War II literature."

—*THE MISSOURIAN*

"I cannot overstate the importance of this book;
Mundy has rescued a piece of forgotten history, and given
these American heroes the recognition they deserve."

—NATHALIA HOLT, *New York Times* bestselling
author of *Rise of the Rocket Girls*

"The book not only shines a light on a hidden chapter of American
history, it also tells the kind of story of courage and
determination that makes you want to work harder and be better."

—*THE DENVER POST*

"PRODIGIOUSLY RESEARCHED AND ENGROSSING." —*THE NEW YORK TIMES*

Praise for

CODE GIRLS

"An absorbing portrait of not only these marvelous, brilliant, hard-working women, but of the era just before, during, and after WWII in the United States. It was intriguing to read an account of what it was like to live in the country during a time when every citizen contributed to the war effort in very tangible ways."　　　　　—Book Browse

"*Code Girls*...finally gives due to the courageous women who worked in the wartime intelligence community."　　　　　—Smithsonian.com

"Mundy's book is expansive and precise. It's anecdotal enough to make it an entertaining read for the layperson, and there's plenty of technical detail to interest the crypto-nerd."　　　　　—*Houston Chronicle*

"Like *Hidden Figures*, this well-crafted book reveals a remarkable slice of unacknowledged U.S. history...Captivating."
　　　　　—*The Christian Science Monitor*

"Meticulously researched...By combining the personal and procedural, Mundy makes the women's experiences vivid and their successes deeply felt."　　　　　—*The New Mexican*

"Mundy unveils the untold story of a very important part of American history that otherwise would have been kept secret."　　　　　—*Miami Herald*

"A well-researched, compellingly written, crucial addition to the literature of American involvement in World War II."
　　　　　—*Kirkus* (starred review)

"A very engaging read on an important topic, a welcome reminder that not all the allied codebreaking efforts occurred at Britain's Bletchley Park. While the current social and work environment for women in general has changed dramatically and markedly improved since World War II, it is well to remember that progress on this front has been uneven, to say the least. *Code Girls* pays tribute to an unsung group of patriotic Americans who, more than seven decades later, are just now receiving their due." —*Studies in Intelligence*, Central Intelligence Agency

"Indispensable and fascinating history. Highly recommended for all readers." —*Library Journal* (starred review)

"Mundy's fascinating book suggests that [the Code Girls'] influence did play a role in defining modern Washington and challenging gender roles—changes that still matter seventy-five years later."

—*Washingtonian*

"Salvaging this essential piece of American military history from certain obscurity, Mundy's painstaking and dedicated research produces an eye-opening glimpse into a crucial aspect of U.S. military operations and pays overdue homage to neglected heroines of WWII... Captivating portraits of women of sacrifice, initiative, and dedication."

—*Booklist* (starred review)

"Women who helped bring victory achieve visibility, at last, in this history." —*Military Times*

"Mundy tells these remarkable women's individual stories, intertwined with the broad historical narrative of how military intelligence evolved during this time. In a clear, braided narrative, she reveals how these code breakers broke both codes and new ground—and why it's high time they were recognized for their achievements."

—Smithsonian.com

"A superbly researched and stirringly written social history of a pivotal chapter in the struggle for women's rights, told through the powerful and poignant stories of the individuals involved. In exploring the vast, obscure, and makeshift offices of wartime Washington where these women performed seemingly impossible deeds, Mundy has discovered a birthplace of modern America."

—Glenn Frankel, Pulitzer Prize–winning author of *High Noon*

"A riveting account of the thousands of young coeds who flooded into Washington to help America win World War II. Liza Mundy has written a thrilling page-turner that illuminates the patriotism, rivalry, and sexism of the code-breakers' world."

—Lynn Povich, author of *The Good Girls Revolt*

"An extraordinary book by an extraordinary author. Liza Mundy's portraits of World War II code breakers are so skillfully and vividly drawn that I felt as if I were right there with them—mastering ciphers, outwitting the Japanese army, sinking ships, breaking hearts, and even accidentally insulting Eleanor Roosevelt. I am an evangelist for this book: You must read it."

—Karen Abbott, *New York Times* bestselling
author of *Sin in the Second City* and
Liar, Temptress, Soldier, Spy

"*Code Girls* is not just a great slice of history—one that would have been lost to us without Liza's storytelling and the work of some heroic archivists—but a story relevant to every discussion we have now about America's security agencies and how they came to be. I am delighted readers will finally know about these pioneering women and their incredible contributions to America."

—Gayle Tzemach Lemmon, senior fellow at the Council on
Foreign Relations and *New York Times* bestselling author of
The Dressmaker of Khair Khana and *Ashley's War*

CODE GIRLS

The UNTOLD STORY *of*
the AMERICAN WOMEN
CODE BREAKERS OF WORLD WAR II

LIZA MUNDY

NEW YORK BOSTON

Hachette Books
Hachette Book Group
1290 Avenue of the Americas, New York, NY 10104
hachettebooks.com
twitter.com/hachettebooks

Originally published in hardcover and ebook by Hachette Books in October 2017.

First Trade Edition: October 2018

Hachette Books is a division of Hachette Book Group, Inc. The Hachette Books name and logo are trademarks of Hachette Book Group, Inc.

The publisher is not responsible for websites (or their content) that are not owned by the publisher.

The Hachette Speakers Bureau provides a wide range of authors for speaking events. To find out more, go to www.hachettespeakersbureau.com or call (866) 376-6591.

Library of Congress Cataloging-in-Publication Data

Names: Mundy, Liza, 1960– author.
Title: Code girls : the untold story of the American women code breakers of
World War II / Liza Mundy.
Description: First edition. | New York : Hachette Books, 2017. | Includes bibliographical
references and index.
Identifiers: LCCN 2017020069 | ISBN 9780316352536 (hardcover) | ISBN 9780316439893
(large print) | ISBN 9781478922704 (audio book) | ISBN 9781478922711 (audio download) |
ISBN 9780316352550 (ebook)
Subjects: LCSH: World War, 1939–1945—Cryptography. | World War, 1939–1945—
Participation, Female. | Cryptographers—United States—History—20th century. |
Cryptography—United States—History—20th century.
Classification: LCC D810.C88 M86 2017 | DDC 940.54/86730922—dc23
LC record available at https://lccn.loc.gov/2017020069

ISBNs: 978-0-316-35254-3 (trade pbk.), 978-0-316-35255-0 (ebook)

Printed in the United States of America

LSC-H

10 9 8 7 6 5 4 3 2 1

To all these women,
and to Margaret Talbot

I'm in some kind of hush, hush business. Somewhere in Wash. D.C.
If I say anything I'll get hung for sure. I guess I signed my life away.
But I don't mind it.

—Jaenn Magdalene Coz,
writing to her mother in 1945

Contents

PART III
The Tide Turns

Author's Note

In researching and writing this book over several years, I drew from three large archival collections of documents produced by the U.S. Army and U.S. Navy code-breaking units during and after the war. Most were classified for many decades, and now can be found at the National Archives at College Park, Maryland. The collections run to hundreds of boxes and include thousands of memos, internal histories, reports, minutes, and personnel rosters, citing everything from lists of merchant ships sunk, to explanations of how certain codes and ciphers were broken, to names and addresses of newly arrived code breakers, to captured codebooks. I filed Mandatory Declassification Review requests with the National Security Agency, resulting in the recent declassification of more material, including some fifteen oral histories conducted by NSA staff over the years with women code breakers, as well as volumes of a multi-part history of wartime Arlington Hall. (Somewhat astonishingly, other parts of that history remain classified.) I located some forty more oral histories, as well as scrapbooks and rosters, at the Library of Congress and other archives. I consulted scholarly articles and the many books on code breaking and the war.

I interviewed more than twenty surviving code breakers, located in various ways. A few had contacted NSA, or their family members had. I placed notices on websites. I obtained rosters and consulted databases to find contact information. In other cases, friends and acquaintances provided names, or, often, one woman would lead me to another. I also obtained civilian and military personnel records that are publicly available in the National Archives personnel records facility in St. Louis,

Missouri. These were supplemented by high school and college year-books, scrapbooks, recruiting pamphlets, newspapers, personal letters, and the very good alumnae records that many colleges maintain. In some cases, of course, I had to trust the women's memory, but a surprising amount of what they recollected could be confirmed with archival records. In just a few instances, however, archives proved insufficient. I wish, for example, that I could include more information on Arlington Hall's African American unit, but very few records of that unit seem to exist.

I have included dialogue only when it was related to me, or recited in an oral history, by someone who was present. I use maiden names and other terms of the time, except in the epilogue, acknowledgments, and notes.

THE SECRET LETTERS

December 7, 1941

The planes looked like distant pinpoints at first, and few who saw them took them seriously even up to the moment they dropped their payloads. An Army private, training at a radar station on the northern tip of Oahu, spotted a blip on his screen suggesting that a large formation of planes was headed for Hawaii, but when he pointed it out to his instructor and they called their superior, he told them not to worry. The blip—he assumed—was just a group of American bombers, B-17 Flying Fortresses, arriving from California. A Navy commander, peering out his office window, saw a plane going into a dive and figured it must be a reckless American pilot. "Get that fellow's number," he told his junior officer. "I want to report him." Then the officer saw a dark shape fall out of the plane and whistle downward.

And now, just minutes before eight a.m., the planes erupted into full view, all of them, streaking in and filling the mild sky like a swift-moving thundercloud: nearly two hundred fighters and bombers, flown by Japan's best pilots. On the underside of their wings glowed the round red insignia of the rising sun. Finally, the people looking at them understood.

Below the planes lay Pearl Harbor's Battleship Row, a line of American warships tied up on mooring quays in the blue Hawaiian waters, placid and unprotected—no barrage balloons, no torpedo nets. Almost one hundred vessels in all, more than half of the entire U.S. Pacific Fleet, dotted the harbor. In nearby airfields, American planes sat arrayed on the ground, wingtip to wingtip, clustered invitingly, fat targets.

The screaming tangle of enemy planes—a second wave arrived an hour after the first—dropped bombs as well as torpedoes modified to navigate Pearl Harbor's shallow waters. One of the bombs found the USS *Arizona*, whose band stood on deck preparing for the morning flag raising. The bomb pierced the battleship's forward deck, setting off a cache of gunpowder and creating a giant fireball. The ship—hit over and over—rose out of the water, cracked, and sank. Other bombs and torpedoes found the *California*, the *Oklahoma*, the *West Virginia*, the *Tennessee*, the *Nevada*, the *Maryland*, and the *Pennsylvania*, flagship of the Pacific Fleet. Diving, peeling off, coming back and back again, the Japanese planes struck destroyers and cruisers as well as buildings. Three battleships settled to the harbor bottom, another capsized, and more than two thousand men were killed, many still asleep. Nearly half of the men who died were on the *Arizona*, among them twenty-three pairs of brothers.

The planes on the airfields were virtually obliterated.

On the mainland, telephone switchboards lit up. Operators plugged calls as fast as they could. It was early afternoon on the East Coast and news of the Pearl Harbor attack raced through the country, traveling by radio, in special editions of newspapers, by people running along the street, crying out. Broadcasts and concerts were interrupted, the Sunday calm shattered. Congress declared war on Japan the next day. Germany—Japan's ally—declared war on the United States three days later. Men flooded recruiting stations in the weeks that followed. Every American felt affected by the tragedy and by the abrupt entry of the United States into a global, two-ocean war.

War had been coming to America for more than a year. Even so, once it arrived the fact of total war was astonishing, unthinkable, as were the events that caused it. The first unthinkable thing was that Japan—seeking a decisive blow that would destroy the American fleet and end the Pacific War almost before it started—would attack unprovoked and without warning. But it was equally unthinkable that America's own planners had been caught so unawares. Despite years of tensions with Japan over its aggression in China and around the Pacific, despite the fact that President Franklin D. Roosevelt had frozen Japan's assets, despite awareness in much of the Navy that something was going to happen somewhere in the Pacific, America's leaders had not seen Pearl Harbor coming.

The attack set in motion a lasting controversy. How could the United States have been taken by surprise? Congressional hearings would be held, fingers pointed, scapegoats identified. Conspiracy theories would be floated. Careers would be ended and reputations ruined. Chaos prevailed as the war establishment suffered upheaval along with instant expansion—what would be called, today, scaling up.

America could no longer be blind and deaf to enemy intentions. A failure on the scale of Pearl Harbor must not happen again. The country was fighting a global war against adversaries who had been preparing for years, if not decades. Intelligence was more important than ever, yet intelligence was hard to come by. Emerging from two decades of disarmament and isolationism, America had a clubby Navy with a disorganized intelligence apparatus; a small skeleton Army; no freestanding Air Force; and—as hard as this may be to believe, in this era of proliferating and overlapping spy agencies—barely any spies abroad. Building an overseas spy network would take time.

In the present—and for the foreseeable future—a first-rate codebreaking operation was needed to crack enemy message systems. Foreign diplomats; political leaders; German submarine captains; Pacific island lookouts; weathermen; skippers of rice ships; airmen in the heat of combat; even companies and banks—if anybody was saying anything, anywhere in the world, America wanted to know about it.

And so the secret letters began going out.

* * *

Some had already been issued. Months before the attack on Pearl Harbor, the U.S. Navy was beginning to realize that unprecedented action would be needed to address the nation's intelligence deficit. Thus, a handful of letters materialized in college mailboxes as early as November 1941. Ann

White, a senior at Wellesley College in Massachusetts, received hers on a fall afternoon not long after leaving an exiled poet's lecture on Spanish romanticism.

The letter was waiting when she returned to her dormitory for lunch. Opening it, she was astonished to see that it had been sent by Helen Dodson, a professor in Wellesley's Astronomy Department. Miss Dodson was inviting her to a private interview in the observatory. Ann, a German major, had the sinking feeling she might be required to take an astronomy course in order to graduate. But a few days later, when Ann made her way along Wellesley's Meadow Path and entered the observatory, a low domed building secluded on a hill far from the center of campus, she found that Helen Dodson had only two questions to ask her.

Did Ann White like crossword puzzles, and was she engaged to be married?

Elizabeth Colby, a Wellesley math major, received the same unexpected summons. So did Nan Westcott, a botany major; Edith Uhe (psychology); Gloria Bosetti (Italian); Blanche DePuy (Spanish); Bea Norton (history); and Ann White's good friend Louise Wilde, an English major. In all, more than twenty Wellesley seniors received a secret invitation and gave the same replies. Yes, they liked crossword puzzles, and no, they were not on the brink of marriage.

Anne Barus received her own letter during the fall of her senior year at Smith College. A history major, she was head of the International Relations Club, and had been accepted into a prestigious internship in Washington, D.C. It was a rare opportunity for a woman—for anybody—and she was looking forward to exposure to a range of government work. But when she found herself invited to a clandestine meeting in Smith's science building, together with a group of mystified classmates, she quickly put her own plans aside.

At Bryn Mawr, Mount Holyoke, Barnard, Radcliffe, the letters went out, throughout the fall and into the terrible winter of early 1942, as undergraduates began rolling bandages and sewing blackout curtains, taking first aid courses, learning to do plane spotting, sending bundles to Britain. Meat became scarce and dorm rooms grew cold from lack of fuel. The schools were members of the Seven Sisters, and had been founded in the nineteenth century to educate women at a time when many leading colleges—Harvard, Yale, Princeton, Dartmouth—would not admit them. On many of these campuses, the wartime menace felt particularly close. In the cold waters of the North Atlantic, Navy men and merchant seamen were running the gauntlet of German U-boats— enemy submarines, which often traveled in groups called wolf packs, preying on convoys transporting food and supplies to beleaguered England. At Wellesley, twenty miles from Boston, lights were doused to hide the ships in Boston Harbor, and the students learned to find their way around by flashlight.

At the time these schools were founded, many considered higher education to be poorly suited for girls. Now the views had changed. Educated women were wanted. Urgently.

The students were called to secret meetings where they learned that the U.S. Navy was inviting them to embark on a field called "cryptanalysis," a word, it was soon made clear to them, they were never to utter outside the confines of the gatherings. They were being offered a training course in code breaking and, if they passed, would proceed to Washington after graduation, to take jobs with the Navy as civilians. Sworn to secrecy, the women were forbidden from telling anybody what they were doing: not their friends, not their parents, not their family, not their roommates. They were not to let news of their training leak into campus newspapers or disclose it in a letter, not even to their enlisted brother or boyfriend. If pressed, they could say they were studying communications: the routing of ordinary naval messages.

In their introductory meetings, the chosen women were issued manila envelopes containing a brief introduction to the arcane history of codes and ciphers, along with numbered problem sets and strips of paper with the letters of the alphabet printed on them. They were to complete the problem sets every week and turn them in. They might help one another, working in groups of two and three. Each week, professors such as Helen Dodson, selected by the Navy, would lead the students through the material. Their answers would be sent to Washington and graded. The meetings functioned as a kind of proctored correspondence course. Speed was of the essence, and often the professors were little more than a chapter ahead of the students in learning the material.

And so the young women did their strange new homework. They learned which letters of the English language occur with the greatest frequency; which letters often travel together in pairs, like *s* and *t*; which travel in triplets, like *est* and *ing* and *ive*, or in packs of four, like *tion*. They studied terms like "route transposition" and "cipher alphabets" and "polyalphabetic substitution cipher." They mastered the Vigenère square, a method of disguising letters using a tabular method dating back to the Renaissance. They learned about things called the Playfair and Wheatstone ciphers. They pulled strips of paper through holes cut in cardboard. They strung quilts across their rooms so that roommates who had not been invited to take the secret course could not see what they were up to. They hid homework under desk blotters. They did not use the term "code breaking" outside the confines of the weekly meetings, not even to friends taking the same course.

The summons spread beyond the Northeast. Goucher, a four-year women's college in Baltimore, Maryland, was known for the caliber of its science departments. Goucher's dean, Dorothy Stimson, a noted authority on Copernicus, happened to be a cousin of Secretary of War Henry Stimson. After Pearl Harbor, the war secretary put in a quiet word, asking for some of Dean Stimson's best senior girls. At Goucher, it was an English professor—Ola Winslow, awarded a Pulitzer Prize for her biography of the American theologian Jonathan Edwards—who was selected to teach the secret course, once a week, in a locked room at the top of Goucher Hall, together with a Navy officer.

Goucher was located in the heart of urban Baltimore. The U.S. Naval Academy was thirty-two miles away in Annapolis, and "Goucher girls," as they were called, traveled there often for dates and dances.

One of the most well-liked students in the Goucher class of 1942 was Frances Steen, a biology major and the granddaughter of a shipping captain who ferried grain between the United States and his native Norway, which now was under Nazi occupation, its king compelled to leave the country under fire. Her father ran a grain warehouse at the Baltimore dock. Her brother, Egil, had graduated from the Naval Academy, and by the time Fran got her own secret letter, Egil Steen was on North Atlantic convoy duty. The Steen family was doing everything they could to support the war effort. Her mother was saving grease from bacon, giving away pots and pans to be made into tanks and guns.

Now, it appeared, there was something else the Steen family could contribute to preserve their son's safety: Fran.

As war engulfed the nation, the summonses continued to go out. Even after Pearl Harbor's shock receded, secret letters were sent again, in 1942, 1943, and 1944, as code breaking proved crucial to disrupting enemy operations and saving Allied lives. At Vassar, nestled in the hills of Poughkeepsie, New York, Edith Reynolds received a letter inviting her to appear in a room in the library at nine thirty a.m. on a Saturday. Edith was barely twenty. She had skipped two grades in elementary school and entered Vassar when she was just sixteen years old.

The letter invited Edith to a room in the college library, where she stood, dazzled, as a hulking Navy captain walked in, covered top to toe, it seemed to her, in the most magnificent gold braid. "Your country needs you, young ladies," he told Edith and a few chosen classmates.

By the time Edith got her summons, German U-boats had attacked shipping up and down the Atlantic coast. On the New Jersey shore, where her family spent summers, bits of shipwreck would wash up and they could hear guns booming. It did not seem out of the question that Japan would invade the U.S. mainland—Alaska, even California—or that America would come under fascist domination.

The U.S. Army, meanwhile, needed its own cadre of code breakers and set out to recruit apt young women. At first, the Army approached some of the same colleges the Navy did, prompting angry memos from top Navy brass, bitterly upset that the Army was "cutting in" to try to get their girls. Like the Navy, the Army wanted women who were college educated—ideally women who had pursued a rigorous liberal arts education that encompassed foreign languages as well as science and math. In the United States, in the 1940s, there was really only one job consistently available to a woman with such a fine education: schoolteacher.

And so—while the starchy, white-gloved Navy officers targeted the fancy women's colleges of the Northeastern Seaboard—the U.S. Army sent recruiters to teaching colleges, far humbler institutions, throughout the South and Midwest. At Indiana State Teachers College in Indiana, Pennsylvania, Dorothy Ramale was taking high-level math classes in the hope of becoming a math teacher. Dorothy had grown up in rural Pennsylvania—a tiny place called Cochran's Mills, known, if it was known at all, for being the birthplace of Nellie Bly. The middle of three girls, Dorothy used to sit on the porch of their playhouse and dream about the world beyond. It was Dorothy's ambition to see every continent on earth. As a child, her only contact with the wider world came from rare events like the time when Amelia Earhart flew over a local cemetery to salute a relative who was buried there, and Dorothy, along with other schoolchildren, waved as the legendary aviatrix passed overhead.

Dorothy's father had supported his family through the Depression by farming and doing maintenance at a church cemetery. People sometimes asked if he regretted not having a son to help him, and he was quick to retort that his three girls were as good as any boy. Dorothy helped him with stacking hay and other outdoor chores, sometimes crawling down into a freshly dug grave to retrieve a tool he needed. At college, she often was the only girl in her trigonometry classes. Math was not a subject women were encouraged to study, and certain parts of the country had no female math teachers at all. Dorothy knew the odds, but math was her passion.

During her senior year, Dorothy stayed up late one night in February and went out to watch as her male classmates were put on a bus and sent to Pittsburgh to begin their military service. She could hardly see them through her tears. Afterward, the campus felt lonely and awful. When the dean of women invited her in for a quiet talk, she instantly agreed to do what the dean was proposing.

But even that was not enough.

The Army needed more code breakers, and then more still. So it went looking beyond college campuses, for female schoolteachers interested in pursuing a new line of work. Such women were not hard to find. Teachers' pay was notoriously low, and classrooms often were enormous. The Army dispatched handsome officers to small towns, remote cities, and farm communities, where they stationed themselves in post offices, hotels, and other public places. Posters and newspaper ads promoted their appearances, seeking women open to a move to Washington to serve the war effort, women who could "keep their lips zipped."

* * *

And so it was that on a Saturday in September 1943, a young schoolteacher named Dot Braden approached a pair of recruiters standing behind a table in the soaring lobby of the Virginian Hotel, the finest lodging place in her hometown of Lynchburg, Virginia, and one of the grandest hotels in the state. The man behind the table was an Army officer, and the other recruiter was a woman dressed in civilian clothing. Dot herself was twenty-three years old, dark-haired, slight, adventurous, confident in her own abilities. She was a 1942 graduate of Randolph-Macon Woman's College, where she studied French, Latin, and physics, among other subjects. She had spent one year teaching at a public high school and wished never to repeat that experience. She was the eldest of four children and had two brothers serving in the Army. She needed to earn her own living and help support her mother.

Without knowing what she was applying for—the recruiters provided no concrete job description—Dot Braden filled out an application for a job with the War Department. Just a few weeks later, Dot found herself on a train rattling out of the tobacco-growing countryside of Virginia's Southside region, headed 180 miles north to Washington, D.C., with excitement in the pit of her stomach, very little money in her pocketbook, and not the faintest idea what she had been hired to do.

"Your Country Needs You, Young Ladies"

The U.S. military's decision to tap "high grade" young women—and the women's willingness to accept the summons—was a chief reason why America, in the aftermath of its entry into World War II, was able to build an effective code-breaking operation practically overnight. That millions of women served the war effort by rolling up their sleeves and donning trousers and jumpsuits to work in factories—the celebrated Rosie the Riveter, who helped build bombers and tanks and aircraft carriers—is well known. Far less well known is that more than ten thousand women traveled to Washington, D.C., to lend their minds and their hard-won educations to the war effort. The recruitment of these American women—and the fact that women were behind some of the most significant individual code-breaking triumphs of the war—was one of the best-kept secrets of the conflict. The military and strategic importance of their work was enormous.

During World War II, code breaking would come into its own as one of the most fruitful forms of intelligence that exists. Listening in on enemy conversations provides a verbatim, real-time way to know what that enemy is thinking and doing and arguing about and worrying over and planning. It provides information on strategy, troop movements, shipping itineraries, political alliances, battlefield casualties, pending attacks, and supply needs. The code breakers of World War II advanced what is known as signals intelligence—reading the coded transmissions of enemies, as well as (sometimes) of allies. They laid the groundwork for the now burgeoning field of cybersecurity, which entails protecting one's data, networks, and communications against enemy attack. They pioneered work that would lead to the modern computing industry. After the war, the U.S. Army and Navy code-breaking operations merged to become what is now the National Security Agency. It was women who helped found the field of clandestine eavesdropping—much bigger and more controversial now than it was then—and it was women in many cases who shaped the early culture of the NSA.

The women also played a central role in shortening the war. Code breaking was crucial to Allied success in defeating Japan, both at sea and during the bloody amphibious assaults on Pacific islands against a foe that was dug in, literally—the cave fighting toward the end of the war was terrible, as were kamikaze attacks and other suicide missions—and willing to fight to the death. And in the all-important Atlantic theater, U.S. and British penetration of the Nazi Enigma cipher that German admiral Karl Dönitz used to direct his U-boat commanders helped bring about the total elimination of the Nazi submarine threat.

The chain of events that led to the women's recruitment was a long one, but a signal moment occurred in September 1941, when U.S. Navy rear admiral Leigh Noyes wrote a letter to Ada Comstock, the president of Radcliffe College, the women's counterpart to Harvard. For more than a year the Navy had been quietly recruiting male intelligence officers from elite colleges and universities, and now it was embarking on the same experiment with women. Noyes wanted to know whether Comstock would identify a group of Radcliffe students to be trained in cryptanalysis. He confided that the Navy was looking for "bright, close-mouthed native students"—that is, high-achieving women who had the sense and ability to keep a secret and who had been born in the United States and were free of close ties with other nations.

"Evidence of a flair for languages or for mathematics could be advantageous," Noyes said, adding that "any intense sociological quirks would, of course, be undesirable." Without stating what such "quirks" might be, the admiral suggested that a handful of promising seniors could enroll in a training course the Navy had developed.

"In the event of total war," Noyes told her, "women will be needed for this work, and they can do it probably better than men."

Ada Comstock was happy to comply. "It interests me very much and I should like to take whatever steps would be thought serviceable," she promptly wrote to her friend Donald Menzel, an astronomy professor at Harvard who was serving as a point person for the broader naval recruiting effort. Astronomy is a mathematical science and a naval one—for centuries, navigation was done using the position of the sun and the stars—and many of the instructors who taught the secret course would come from the field.

Comstock also received another letter, this one from Laurance Safford, one of the U.S. Navy's few experienced code breakers, now responsible for building a much-enlarged code-breaking unit. Safford elaborated on the qualifications they wanted by spelling out the kind of young women the Navy did *not* want.

"We can have here no fifth columnists, nor those whose true allegiance may be to Moscow," Safford wrote. "Pacifists would be inappropriate. Equally so would be those from persecuted nations or races—Czechoslovakians, Poles, Jews, who might feel an inward compulsion to involve the United States in war." Ada Comstock duly put together a list of about forty Radcliffe seniors and young graduate students, with the idea that twenty or so might meet the standards. Disregarding the not-so-subtle anti-Semitism of the Navy's communications, she included the names of two Jewish women.

At the Navy's request, Comstock also approached leaders of other women's schools. These deans and presidents were devoted to the cause of educating women and eager to defend liberty and freedom of thought against fascism and totalitarian belief systems. They also were keen to develop career opportunities for their students. The leaders savvily perceived that war might open up fields—and spots in graduate schools—that up to now had been closed to women. Even before Comstock received the Navy's letter, many of the leaders had been strategizing over how they could provide what Virginia Gildersleeve, dean of Barnard College, called "trained brains" to a war effort that would depend on advances in science and math.

The women's college leaders met at Mount Holyoke on October 31 and November 1, 1941, with representatives from Barnard, Bryn Mawr, Vassar, Wellesley, Radcliffe, Smith, and Mount Holyoke attending. Comstock told them about the Navy's request and said Radcliffe would participate. She distributed some materials the Navy had developed: a "Guide for Instructors" and an "Introduction to Students." The idea was that selected students would take the course during the remainder of their senior year, then go to work for the Navy, in Washington, as civilians. The "Guide for Instructors" assured them that no prior experience was necessary and that they would receive a "gouge," or answers to the problems. The instructors would be given a few texts to jump-start their own education, including a work called *Treatise on Cryptography*, another titled *Notes on Communications Security*, and a pamphlet called *The Contributions of the Cryptographic Bureaus in the World War*—meaning World War I, the so-called war to end all wars.

The result was the wave of secret letters that appeared in college mailboxes in the fall of 1941, summoning surprised young women to secret meetings. Most were in the top 10 percent of their class, selected based on academic performance as well as character and loyalty and grit. (A memo from a Radcliffe administrator explaining why one young woman was not selected said that she lacked "gumption," had "perhaps been spoiled by money and a very domineering mother," and seemed unlikely to "develop a serious interest in the work or to stick to it.") The chosen women not only were cautioned against uttering the word "cryptanalysis" outside the confines of the classroom; they were not to say the words "intelligence" or "security" to any person outside their study group, lest they tip off the enemy. Pembroke, the women's college affiliated with

Brown University, soon found itself in hot water. A professor at Brown took the "bit in his teeth," as one angry Navy memo put it, and began bragging about the course. As a result, in February 1942, Brown and Pembroke were blacklisted from the program.

There was controversy over whether the course would be offered for academic credit. The Navy at first was against this, because it wanted young women who were motivated and independent. But John Redman, a high-ranking naval communications officer with a keen interest in the program, came to think that giving credit was a good idea. Being graded would make the students try harder, and it would ease the heavy workload the women already faced in order to graduate. Many of the colleges did offer the class for credit, without listing it in the catalog. For secrecy's sake, it went on the women's transcripts as a math course.

By March 1942, the women were well embarked and had turned in some of the problem sets. Lieutenant Commander Ralph S. Hayes wrote Menzel, saying that the students had performed so well that the Navy was hoping women's schools would contribute more students the following year, and from a wider array of institutions, including Wheaton and Connecticut Colleges. He added that, in the Navy's view, "it looks as if the demand for high grade women in this work would continue for some time."

By mid-April 1942, Donald Menzel reported that the Radcliffe women were shaping up ably. He knew this because he was teaching the Radcliffe contingent, and he felt very proud of their performance. "Miss McCormick, my star student, who incidentally is the only girl taking the Harvard Japanese course, has beat out all the men in an extremely large class." By mid-May, twenty-five Radcliffe women had been certified by the Civil Service Commission for naval work as civilians and were expected to start duties in June.

The Navy was a service that cared about status. It wanted women who were well connected socially, and there also seems to have been interest in knowing what the women looked like. The application asked that the women submit passport photos, some of which excited a bit of commentary. "I might point out that the passport photos will scarcely do justice to a number of the members of the course," enthused Harvard's Donald Menzel, saying that the women's "appearance is such that large-scale photographs would be a grace to any naval office."

Around the same time, another meeting was taking place. Twenty women's colleges sent representatives to the elegant Mayflower Hotel in Washington, D.C., where the U.S. Army was working to forge its own ties with institutions that schooled women. Already it was clear that educated women would be needed for the broader war effort. As the country coped with an acute labor shortage, the inspector general of the Department of Labor noted that adult civilians would not be sufficient to stock an economy bereft of its male workers. Students would be needed, and it made sense to start with the female ones. So the Army worked to tap its own network of women's colleges before the Navy could reach them; indeed, the Navy suspended its own efforts to set up training at Connecticut College when it learned that the Army had gotten there first.

Disparate as their backgrounds were, the women who answered these summonses—that of the Navy and that of the Army—had a handful of qualities in common. They were smart and resourceful, and they had strived to acquire as much schooling as circumstances would permit, at a time when women received little encouragement or reward for doing so. They were adept at math or science or foreign languages, often all three. They were dutiful and patriotic. They were adventurous and willing. And they did not expect any public credit for the clandestine work they were entering into.

The last fact was perhaps the most important. In the late spring of 1942, the first wave of women recruited by the U.S. Navy finished their secret courses and turned in their final problem sets. Those who had stuck with the course and answered enough problems correctly—less than half of those recruited—arrived to start their duties, working in the Navy's downtown Washington, D.C., headquarters, which quickly became so crowded that a few found themselves sitting on upturned wastebaskets.

The women were told that just because they were female, that did not mean they would not be shot if they told anybody what they were doing. They were not to think their sex might spare them the full consequences for treason in wartime. If they went out in public and were asked what they did, they were to say they emptied trash cans and sharpened pencils. Some would improvise their own answers, replying lightly that they sat on the laps of commanding officers. People readily believed them. For a young American woman, it was all too easy to convince an inquiring stranger that the work she did was menial, or that she existed as a plaything for the men she worked for.

"Almost everybody thought we were nothing but secretaries," one of the women would say years later.

During the most violent global conflict that humanity has ever known—a war that cost more money, damaged more property, and took more lives than any war before or since—these women formed the backbone of one of the most successful intelligence efforts in history, an effort that began before the Pearl Harbor attack and lasted until the Second World War's very end. In the manila packets they opened before arriving in Washington, the women were told that, up to now, the secret cryptanalytic work they had been selected to perform had been done by men.

"Whether women can take it over successfully," the Navy letter told them, "remains to be proved."

The letter added: "We believe you can do it."

* * *

The women recruits were entering an environment of large and clashing male egos. There was furious infighting between the U.S. Army and the U.S. Navy, to a degree that would have been comic if it weren't taking place in the middle of a war. For several decades the two services had built small and separate code-breaking operations, which were competitive to a point where it sometimes was not clear who the real enemy was. "Nobody cooperated with the Army, under pain of death," said naval code breaker Prescott Currier. This was an overstatement, but not by much. Part of the clash had to do with money; as they began expanding, both services were competing for appropriations. Part of it had to do with glory. Part of it had to do with jealousy. Part of it had to do with the fact that the military-industrial complex was being ushered into full flower. Any number of agencies were founded and nurtured during World War II, and all jostled for power and resources. The Federal Communications Commission and the Federal Bureau of Investigation and the new Office of Strategic Services (OSS)—forerunner of the CIA—all vied for a piece of the code-breaking action.

The British—who centralized their own operation—were appalled. One British liaison described the Americans at the war's outset as "just a lot of kids playing at 'office.'"

As that comment suggests, the Americans also had to contend and cooperate with England's own code-breaking venture, Bletchley Park, the storied British operation that employed "debs and dons": brilliant Oxford and Cambridge mathematicians and linguists—mostly men, but also some women—who labored in dim and chilly "huts" at a drab and ungainly estate some sixty miles outside London. In and around Bletchley Park there also were thousands of women, many from upper-class families, who operated "bombe" machines, which were built to crack the Enigma ciphers used not only by the German Navy, but also by the German Army, Air Force, and security services. In 1941, when America entered the war in earnest, the British had an older and more sophisticated code-breaking operation than did their Yankee allies.

But the American code-breaking operation ramped up quickly and became ever more crucial as the war progressed, growing larger than Bletchley Park. At the outset of the joint effort, the Allies decided that the British would lead code-breaking efforts in the European theater. The Americans had lead responsibility for the immense Pacific, with help from their allies. As the war went on, the United States' code-breaking operation also became central to the European conflict. The ranks of that operation became steadily more female, as men shipped out to the hot, dry sands of North Africa, to Italian mountain ranges and snowy

European forests, to the decks of Pacific aircraft carriers, to the beaches of Iwo Jima.

In such a competitive culture, it was easy for the women's contribution to be overlooked. The women took their secrecy oath seriously, and they came from a generation when women did not expect—or receive—credit for achievement in public life. They did not constitute the top brass, and they did not write the histories afterward, nor the first-person memoirs. And yet women were instrumental at every stage. They ran complex office machines that had been converted to code-breaking purposes. They built libraries and information sections that were vital sources of "collateral," the term for subsidiary material such as public speeches, shipping inventories, and lists of ship names and enemy commanders, which helped break messages and illuminate their content. They worked as translators. They pushed forward new fields such as "traffic analysis," which is a method of looking at the external features of a coded message—the stations where messages are being sent from, or to; sudden fluctuations in radio traffic; ominous silences; abrupt appearances of new stations—to learn about troop movements. Women were put in charge of "minor" systems—weather codes, for instance—that turned out to be crucial when major systems went dark and could not be read.

And a number of predominately female teams attacked and broke major code systems. Once broken, a code must be exploited and, often, rebroken, and women formed the great assembly line of workers who did this.

Women also tested America's own codes to make sure they were secure. They worked as radio intercept operators at global listening posts. The Navy did not permit its women to serve overseas—much as many wanted to—apart from a few who went to Hawaii, but the Army did admit its code-breaking women into the war theater. Some Army women would be sent to Australia and to Pacific islands such as New Guinea. Some would move with General Douglas MacArthur when he occupied Tokyo after the war. Other women helped create "dummy traffic": fake radio signals that helped fool the Germans into believing the D-Day invasion would take place in Norway or the Pas de Calais region of France—rather than on the beaches of Normandy.

These were the formative days of what is now called "information security," when countries were scrambling to develop secure communications at a time when technology was offering new ways to encipher and conceal. As in other nascent fields, like aeronautics, women were able to break in largely because the field of code breaking barely existed. It was not yet prestigious or known. There had not yet been put in place elaborate systems of regulating and credentialing—professional associations, graduate degrees, licenses, clubs, learned societies, accreditation— the kinds of barriers long used in other fields, like law and medicine, to keep women out.

Women also were put in a peculiar position by dint of being brought into the workforce to free up men for military service. "Release a man to fight" was the phrase of the day. As a result, men who had been doing office work sitting at desks were able to ship out to the violent war theater. This led to the women's being welcome, but also to their being resented, and resented in a very different way from the women who came into workplaces in the 1960s and 1970s.

It was a profound situation, psychologically. The women were brought in to free men to go forth and, potentially, die. Yet the work they were doing was intended to ensure that those men lived. They were trying to protect the very men whose lives their arrival put in danger. The women were recruited at a time when psychological testing was not yet widely used—to see who could cope with this sort of thing and who could not—and post-traumatic stress was not a recognized condition. All of the women had brothers and lovers and fiancés and friends serving in the war. A number of them broke messages that told the fate of their own brothers' ships and units. The work took its toll. Some of the women never recovered. Louise Pearsall, recruited to work as a mathematician on the Enigma code-breaking project, was one of those who suffered a breakdown afterward. "She was a total wreck," said her brother William.

That women were considered better suited for code-breaking work—as the letter that Rear Admiral Noyes sent to Ada Comstock suggested—wasn't a compliment. To the contrary. What this meant was that women were considered better equipped for boring work that required close attention to detail rather than leaps of genius. This was a widespread view in the 1940s. In the field of astronomy, women long had been employed as "computers," assigned to do lower-level calculations. This was seen as women's rightful domain: the careful repetitive work that got things started, so that the men could take over when things got interesting and hard. Men were seen as more brilliant than women, but more impatient and erratic. "It was generally believed that women were good at doing tedious work—and as I had discovered early on, the initial stages of cryptanalysis were very tedious, indeed," recalled Ann Caracristi, whose first job as a code breaker was sorting reams of intercepted traffic.

To a real extent, the same prejudice against women persists to this day. Even now, the disciplines that are hardest for women to break into—like math and laboratory and computer science—are the ones that are believed to depend on innate genius, a trait long, and wrongly, associated chiefly with men. The literature of code breaking has played into this myth. Stories about code-breaking exploits often focus on a titanic genius who enjoys a flash of inspiration that breaks the code and releases a stream of vital information. Narratives like these encourage the notion that genius springs isolated from the brain of a lone individual, inevitably male. The legendary names enshrined in the public history of code breaking are those of brilliant men like Alan Turing at Bletchley Park; Joe Rochefort, the American naval officer who helped mastermind the code breaking that led to victory in the Battle of Midway; or Meredith Gardner, the American cryptanalyst who recovered the meanings of Russian messages as part of a U.S. program, code-named Venona, that led to the exposure of Soviet spies. Their reputation is often burnished by eccentricity—Joe Rochefort is depicted as pacing his underground lair in Oahu, wearing a smoking jacket and slippers—or their tragic end, as when Alan Turing was persecuted for homosexuality and committed suicide.

But all along there have been female geniuses whose contributions are as important. It's just that far less attention has been paid to them, and often these women were denied the top spots that would have brought them more recognition.

But the genius narrative is overblown. Code breaking is far from a solitary endeavor, and in many ways it's the opposite of genius. Or, rather: Genius itself is often a collective phenomenon. Success in code breaking depends on flashes of inspiration, yes, but it also depends on the careful maintaining of files, so that a coded message that has just arrived can be compared to a similar message that came in six months ago. Code breaking during World War II was a gigantic team effort. The war's cryptanalytic achievements were what Frank Raven, a renowned naval code breaker from Yale who supervised a team of women, called "crew jobs." These units were like giant brains; the people working in them were a living, breathing, shared memory. Codes are broken not by solitary individuals but by groups of people trading pieces of things they have learned and noticed and collected, little glittering bits of numbers and other useful items they have stored up in their heads like magpies, things they remember while looking over one another's shoulders, pointing out patterns that turn out to be the key that unlocks the code.

One of the best code-breaking assets is a good memory, and the only thing better than one person with a good memory is a lot of people with good memories. Every step of the process—the division of enemy traffic into separate systems; the noting of scattered coincidences; the building up of indexes and files; the managing of vast quantities of information; the ability to pick out the signal from the noise—enabled the great intuitive leaps. The precursor work during the war was almost always done by women, and many of those intuitive leaps were made by women as well.

Precisely because they did not expect to be celebrated or even promoted, the women tended to be collegial. This was in marked contrast to the Navy men—especially—who were fighting for recognition in a hotly careerist service. "The women who gathered together in our world worked very hard. None of us had an attitude of having to succeed or outdo one another, except in trivial ways," recalled Ann Caracristi years later. "I mean, you wanted to be the first to solve a particular problem, or you wanted to be the first to get this recovery. But there was very little competition for, you know, for money, or anything of that nature, because everybody really assumed that when the war was over we would be leaving... The majority of the people considered it a temporary way of life."

* * *

The women were of a unique and overlooked generation. Many were born in 1920, the historic year when American women won the right to vote. Their early life was led in an atmosphere of broadening opportunity. Women made gains in some professions. The flapper era promised a loosening of restrictions on behavior and a growing awareness of female potential; the exploits of Amelia Earhart and other women aviators, such as Elinor Smith—the "Flying Flapper of Freeport," who at age seventeen flew under all four bridges across New York's East River—all pointed to new freedoms, as did the high-profile work of stunt-seeking women journalists like Nellie Bly.

When these women born in 1920 or thereabouts were children, however, the Great Depression hit. Opportunity stopped, and progress reversed. Many women were fired so that what jobs remained could be given to men. Families were destabilized, especially fathers, many of whom also were struggling with the traumatic effects of military service during World War I. Daughters were responsible for helping inside the house; this meant girls were keenly attuned to the emotional disarray of their households. It was not only poor families that were impacted. At the elite women's colleges from which the Navy recruited its code-breaking cohort, it is striking how many students were there on scholarship and how many had suffered trauma as children. Jeanne Hammond, a member of the class of '43 at Wellesley, was the daughter of a businessman who lost almost everything in the Depression. Hammond never forgot the night she was eating dinner at the kitchen table and her father came home after meeting with his broker and announced, shoulders sagging, that their money was gone. "Don't worry, you'll always have enough bread and milk," he said, and Hammond, a child, thought bread and milk was all she would be eating forever. She attended Wellesley on a scholarship and felt anxious about keeping her grades up so as not to lose it. She declined her "secret letter" because she was terrified of being subjected to military discipline; it felt too much like the strict and demoralized environment she had grown up in.

Thousands of women did say yes. For them, the chance to come to Washington, to work and live on their own or with friends, was a respite from the cares of home life. It got them out in the world. It provided a breathing period between the demands of being a daughter and the responsibilities of being a wife. Much of America in the 1930s and 1940s was still very rural. The women often came from farms and towns where the future stretched out, unchanging. After the war, many never returned. Their lives were irrevocably altered by the work they did.

Freedom of movement was one of the differences between the British code breakers and their American counterparts. Once British women arrived at Bletchley Park and the surrounding estates, they were obliged to stay where they were, apart from trips to London. Their experience was one of urgency but also, to an extent, confinement. The American women were able to clamber onto trains and buses and travel. Living in the fast-growing nation's capital, they rented rooms, scrambled for housing, shared beds—two women who worked different shifts and used the same bed called it "hot-bedding"—and settled where they could find a place. The capital was full of odd little boardinghouses. Some bedrooms had only curtains on their doorways. Some proprietors served collard greens and black-eyed peas to northerners who had never eaten southern food. From Washington they were able to jump on the train during a forty-eight-hour or seventy-two-hour leave and spend a weekend in New York or even Chicago.

Edith Reynolds, recruited out of Vassar, woke up her first morning in a street-level room in a Georgetown brownstone. She was wearing red pajamas, sitting up in bed, when she realized there were no curtains on the windows. A gentleman walking along the sidewalk looked inside, saw her, and tipped his hat. Life was like that, during the war.

In Washington, the women code breakers took buses and trolleys. They went to USO dances and to bars. They looked out for one another. One group agreed that if anybody ordered a vodka Collins when they were out at a bar together, that would be their signal that a stranger was showing too much curiosity about their work, and they were all to disperse to the ladies' room and then flee. They saw *Oklahoma!*, which premiered in March 1943 and presented a new vision of musical entertainment, one that was dark and included death.

The women learned a lot, in ways they had not expected. Suzanne Harpole, a Wellesley code breaker, returned from a short leave to find that her room in an oh-so-respectable downtown boardinghouse had been "rented" during her absence to a military officer and his mistress. Arriving home a bit early, she had to wait in the parlor until the clandestine couple was finished with it.

* * *

In 1942, only about 4 percent of American women had completed four years of college. In part this was because women were denied admission to so many places; even coeducational colleges capped the number of women they would admit, and often had a staff position called "dean of women," as if women were a special subset of the student body, not full members. Families were more likely to pay a son's tuition than a daughter's. For a woman, the financial rewards of a college education were so limited that a degree did not carry the same promise of future earnings that it did for a man, and many families did not consider it worthwhile for a daughter to attend. It was not as if a woman could be an architect, say, or an engineer. Or not many could. Professional graduate schools had few—sometimes no—spots for women.

Those who did go to college were unusually motivated. Some came from families who valued learning for its own sake. Other families viewed college as a way for a woman to be exposed to neighboring men's schools at dances and mixers and to ensure a marriage to a husband of good prospects. Any payoff derived from a woman's education would likely depend on the achievements of her spouse. Sometimes the women came from immigrant families—German, French, Italian—where having a daughter in college was a way of Americanizing the family as soon as possible. In some large first-generation families, to have an educated daughter was a way of competing with one's siblings in the ascent up the social ladder. It was a status symbol. Sometimes, a girl was so intellectually voracious that there was no way to keep her away from college. It was not easy being a smart girl in the 1940s. People thought you were annoying.

The four-year women's colleges were a mix of true cerebral inquiry, rank social climbing, and raw marital ambition. The messages the students received were mixed. The leaders of the Seven Sisters schools were committed to women's education, as were some southern leaders, including Meta Glass at Sweet Briar College in Virginia. Many women's colleges had rigorous programs in everything from zoology to classics. Students were pushed to excel, often by female teachers who had devoted their own lives to their disciplines: Greek, physics, Shakespeare. Perfection was encouraged.

And many of the students embraced these aspirations and high standards. At Wellesley, the Latin motto, *Non Ministrari sed Ministrare*, was an exhortation to good works and service that meant "Not to be ministered unto, but to minister." Some undergraduates preferred a proto-feminist rallying cry: "Not to be ministers' wives, but to be ministers."

But the pressure to marry was intense. At Wellesley, there was an entire page of the yearbook, the *Legenda*, devoted to a list of engagements (Mabel J. Belcher to Raymond J. Blair; Alathena P. Smith to Frederick C. Kasten) and naming women who had recently married (Ann S. Hamilton to Lieutenant Arthur H. James). The school in 1940 delivered a series of marriage lectures to its students, including one on "marriage as a career," another on "biological aspects of marriage," and others on "obstetrics" and "the care of the young child." School lore had it that the winner of the senior class hoop roll—performed with much hilarity in cap and gown—would be the "class's first bride," and the winner was awarded a bouquet in recognition.

What is interesting about this generation of women is that they did understand that at some point they might have to work for pay. Forged by the Depression, they knew they might have to support themselves, even on a teacher's salary, no matter how "good" a marriage they did or did not make. Some were sent to college with the idea that it would be ideal to meet a man, but their degree would permit them to "fall back" on teaching school. And some women went to college because they were, in fact, ambitious and planned to compete for the few spots in law or medical schools that were available to them.

Suddenly these women were wanted—for their minds. "Come at once; we could use you in Washington," was the message conveyed to Jeuel Bannister, a high school band director who had taken an Army course on cryptanalysis at Winthrop College, in Rock Hill, South Carolina.

In the 1940s, the American labor force was strictly segregated by gender. There were newspaper want ads that read "Male Help Wanted" and others that read "Female Help Wanted." For educated women, there was a tiny universe of jobs to be had, and these always paid less than men's jobs did. But it turned out that the very jobs women had been relegated to were often the ones best suited to code-breaking work. Schoolteaching—with the learning it required—was chief among these. Knowledge of Latin and Greek; a close study of literature and ancient texts; facility with foreign languages; the ability to read closely, to think, to make sense of a large amount of data: These skills were perfect.

But there were other women's jobs that turned out to be useful. Librarians were recruited to make sense of discarded tangles of coded messages. "Nothing had been filed. It was just a mess," said Jaenn Coz, one of a number of code-breaking librarians who came to work for the

Navy. "They sucked us out from all over the country." Secretaries were good at filing and record keeping and at shorthand, which is itself a very real kind of code. Running office machines—tabulator, keypunch—was a woman's occupation, and thousands were now needed to run the IBM machines that compared and overlapped multidigit code groups. Music majors were wanted; musical talent, which involves the ability to follow patterns, is an indicator of code-breaking prowess, so all that piano practicing that girls did paid off. Telephone switchboard operators were unintimidated by the most complex machines. In fact, the communications industry from its origins was one that had been considered suitable for women. Boys delivered telegrams, but women connected calls, in large part because women were considered more polite to callers.

Character also mattered. Here again, women's colleges were ideal. All the schools had codes of comportment—curfews, housemothers, chaperones, rules about not smoking in your room and not having men visit you in private and not having sex and not wearing trousers or shorts in public. All of this enabled the women to sail through the military's background checks. Bible colleges were even better; many of those graduates didn't drink.

To be sure, women had a strike against them in that they were considered bad at keeping secrets—women, everybody knew, were gossips, rumormongers, talkers. Then again, when it came to sexual behavior, women were seen as less of a security risk than men. Just before the United States entered the war, when the Army began recruiting privates to train as radio intercept operators, an internal memo raised a concern about the ramifications of entrusting young men to do top secret work. Youth, the memo noted, is "a time for sowing of wild oats and under the influence of women and liquor, much is said that the speaker would not dream of saying when uninfluenced." Women were thought to be less problematic, at least when it came to drinking and bragging.

It was a rare moment in American history—unprecedented—when educated women were not only wanted but competed for. Up to now, many college leaders had hesitated to encourage women to major in math or science, because jobs for women were so scarce. Soon after Pearl Harbor, however, companies like Hercules Powder started recruiting at places like Wellesley, looking for chemists. The Office of Strategic Services was avidly recruiting women, as was the FBI. The jobs landscape for female college graduates changed even just between 1941 and 1942. The men were gone but the war industry was complex and ongoing, and somebody had to staff it. The ballistics industry needed mathematicians to calculate weapon trajectories. Lever Brothers and Armstrong Cork also needed chemists. Grumman Aircraft Engineering Corporation needed "specifications men"—it used the same term for women—who could read blueprints. Raytheon needed engineers. Bethlehem Steel needed designers for armor plates. MIT needed female graduate students to run its analog computers. The military services were competing with the private sector and with one another. It is true that this was seen as a temporary state of affairs and also true that sexism persisted: Educators worried that they might encourage women to pursue math and science who would then be left high and dry. One electrical company asked for twenty female engineers from Goucher, with the added request, "Select beautiful ones for we don't want them on our hands after the war."

* * *

The Axis powers never mobilized their women to the extent that the Allies did. Japan and Germany were highly traditional cultures, and women were not pressed into wartime service in the same way, not for code breaking or other high-level purposes. There are of course many reasons why the Allies prevailed in World War II—the industrial might of the United States, the leadership of military commanders and statesmen, the stoicism of British citizens who endured years of bombing and deprivation, the resistance of the French and Norwegian undergrounds, the cunning and resourcefulness of spies, the heroism of citizens who helped and harbored Jewish neighbors, and the bravery and sacrifice of sailors and airmen and soldiers, including the millions of Russian soldiers who bore the brunt of military casualties and deaths.

But the employment of women also was one of these factors. It wasn't just that the women freed the men to fight, enabling General Dwight Eisenhower to load more men into landing craft at Normandy, or Admiral Chester Nimitz to staff more Pacific aircraft carriers. Women were more than placeholders for the men. Women were active war agents. Through their brainwork, the women had an impact on the fighting that went on. This is an important truth, and it is one that often has been overlooked.

On the eve of Pearl Harbor, the U.S. Army had 181 people working in its small, highly secret code-breaking office in downtown Washington. By 1945 nearly 8,000 people would be working stateside for the Army's massive code-breaking operation, at a much-expanded suburban Virginia venue called Arlington Hall, with another 2,500 serving in the field. Of the entire group, some 7,000 were women. This means that of the Army's 10,500-person-strong code-breaking force, nearly 70 percent was female. Similarly, at the war's outset the U.S. Navy had a few hundred code breakers, stationed mostly in Washington but also in Hawaii and the Philippines. By 1945, there were 5,000 naval code breakers stationed in Washington, and about the same number serving overseas. At least 80 percent of the Navy's domestic code breakers—some 4,000—were female. Thus, out of about 20,000 total American code breakers during the war, some 11,000 were women.

Many of the program's major successes did emerge at the end of the war, and the public could appreciate what had been accomplished in secret. Late in 1945 the *New York Times* published a letter that General George Marshall had written to Thomas Dewey during his 1944 presidential campaign against Roosevelt, laying out some of the victories the country owed its cryptanalytic forces and begging Dewey to keep them secret. Once the war was over, the letter was made public. In it, General Marshall pointed out that thanks to the country's cryptanalytic forces, "we possessed a wealth of information" regarding Japanese strategy. He revealed the hidden cause of certain famous naval victories and pointed out that "operations in the Pacific are largely guided by the information we obtain of Japanese deployments."

Also after the war, the Joint Committee on the Investigation of the Pearl Harbor Attack noted that Army/Navy signals intelligence was "some of the finest intelligence available in our history" and that it "contributed enormously to the defeat of the enemy, greatly shortening the war, and saving many thousands of lives." Major General Stephen Chamberlin, who served in the Pacific, announced that military intelligence, most of which came from code breaking, "saved us many thousands of lives" in the Pacific theater alone, "and shortened the war by no less than two years."

Members of Congress were quick to commend the code-breaking forces. "Their work saved thousands of precious lives," orated Representative Clarence Hancock of New York, speaking on the floor of the House on October 25, 1945.

> They are entitled to glory and national gratitude which they will never receive. We broke down the Japanese code almost at the beginning of the war, and we knew it at the finish of the war. Because of that knowledge we were able to intercept and destroy practically every supply ship and convoy that tried to reach the Philippines or any Pacific Island. We knew, for example, that shortly after MacArthur landed on Leyte a large convoy with 40,000 Japanese troops was dispatched to reinforce the Japanese forces there. They were met by our fleet and by our airplanes at sea and were totally destroyed.

He stated, "I believe that our cryptographers...in the war with Japan did as much to bring that war to a successful and early conclusion as any other group of men."

That more than half of these "cryptographers" were women was nowhere mentioned.

PART I

"In the Event of Total War Women Will Be Needed"

CHAPTER ONE

Twenty-Eight Acres of Girls

September 1942

Chatham is a bump in the road in the southern part of Virginia: a small, picturesque town of just over a thousand people, with well-tended Victorians and a downtown known for its handsome redbrick Greek Revival courthouse. Originally a center of commerce for farmers tending the miles of tobacco land surrounding it, Chatham is the seat of Pittsylvania County, named for William Pitt, first Earl of Chatham, a member of the English Parliament who sympathized with the American colonists and objected to unjust taxation. Though somewhat remote, the town has always been high-minded. In 1942, Chatham boasted both a fancy boarding school for girls and a private military academy for boys, educating the children of the Virginia and North Carolina gentry and would-be gentry.

Unfortunately for Dot Braden, neither of those tony academic establishments was where she had been engaged to teach. It was Dot's job to teach—or try to—at Chatham's ordinary public high school, a modest brick structure located a few blocks off the main street, and one that in 1942 was suffering from any number of maladies and hardships, chief among them a near-total personnel turnover and full-fledged chaos among its teaching staff. In this, Chatham High was not unusual: America had been at war for nearly a year, virtually all able-bodied men of military age had signed up to fight, and schools around the country were coping with chronic teacher shortages. Male teachers, never in great supply, had mostly vanished. So had many female ones: The war set off a national rush to the altar, and many female schoolteachers quit their own jobs to marry the men before they shipped out. The upshot of so much upheaval and personnel reallocation was that at the beginning of the 1942 school year, all of the teaching at Chatham High School had fallen, or so it felt, on twenty-two-year-old Dot.

Fresh out of college, Dot Braden had never taught before she took the job in Chatham. In her first week she found that she was now the high school's eleventh- and twelfth-grade English teacher, its first- and second-year French teacher, its ancient-history teacher, its civics teacher, its hygiene teacher, and its calisthenics teacher, assigned to enforce a new exercise program the government had put in place to encourage fitness among young people. The latter responsibility mostly entailed marching the senior girls back and forth from lunch. She had to suppress a laugh at being called "General Braden" during this exercise, but Dot—blue-eyed, brown-haired, five feet four inches tall, determined, forceful—did all she was asked. When the physics teacher departed, there was another panic and scramble: During one faculty meeting Dot unwisely mentioned that her graduation certificate qualified her to teach physics, and lo and behold, she became the advanced physics teacher as well. Five days a week, eight hours a day, Dot Braden ran from classroom to classroom, teaching, lecturing, grading, marching. For her pains she was paid $900 a year, or about $5 a day.

Dot was accustomed to hard work, but if anybody had asked her—and nobody did—she would have said that teaching anybody anything while America was at war in this way was impossible. The school's sophomore class (whose own ranks fell by half over the next two years, as even teenagers took full-time work to serve the war effort) later noted in their graduation yearbook, *The Chat*, that the 1942–43 year was "the most confusing" of their educational careers. "It was terrible," Dot recalled later. "I mean to tell you, they dumped everything on me."

The straw that broke the camel's back occurred at Christmas. Dot and another female teacher were boarding with the sheriff and his wife.

Halfway through the year her roommate also left to get married—just up and quit—and Dot inherited her English composition classes. This happened at about the same time that the restaurant where Dot was taking her meals closed down, a victim of worker shortages and the rationing of meat, coffee, butter, cheese, and sugar—most anything a person would want to eat or drink. The sheriff's housekeeper began preparing her meals and Dot ate alone in her room each night, exhausted.

Why she did not quit, she would have had a hard time explaining. Persistent and tenacious by nature, she was resolved to do her job come hell or high water. And she did her best: Dot's composition class beat the boarding school, Chatham Hall, in a themed essay contest—an unheard-of occurrence—and the mother of a girl in her physics class came to thank her for all she had done, saying that the girl planned to study physics in college. But when the year's final dismissal bell sounded, Dot Braden packed her skirts and sweater sets and saddle shoes and went home to Lynchburg, Virginia, some fifty miles away.

"I am never going back to that school," she told her mother. "I am not. It will kill me. I'm just through with teaching and all that."

The two women were standing together in the modest wood-frame Victorian that Dot's mother rented at 511 Federal Street in Lynchburg, at the top of a steep hill in a residential area not far from downtown. Dot's mother, Virginia—they called her Meemaw—was raising four children on her own, and Dot's income helped. There were few other good jobs available, even for a young woman as hardworking and well educated as Dot. Local classified ads seeking female labor mostly wanted telephone operators, waitresses, housemaids, and—always—schoolteachers. But Virginia Braden didn't want her oldest daughter doing work she hated, so they agreed Dot would look for something else.

Sometime after that conversation, Dot's mother came home and said that some government recruiters were set up at the Virginian Hotel, an imposing yellow-brick Classical Revival edifice that stood at the corner of Church and Eighth Streets in the center of town, about a half mile or so from their house. Word had it the recruiters were looking for schoolteachers. Her mother didn't know what the job was, exactly. She made it sound mysterious, maybe a little bit like spying. The job was in Washington, D.C., and it had something to do with the war.

Washington, D.C.! Dot Braden had never been to Washington, though the nation's capital was only a little more than three hours to the north. Like most people she knew, she had rarely left the state she lived in. She did not take vacations, except to visit family members, and seldom traveled. The only time she could remember leaving Virginia was when she and some friends had gone to West Point, in New York, for a dance with Army cadets. Apart from that, the bulk of her life had been spent in two places: Lynchburg, where she lived now, and Norton, a small coal-mining community that had the distinction of being the westernmost city in the state. The family had moved to Norton when Dot was a girl and her father got a job as a postal carrier for the railway, but the marriage had not worked out, and Dot's mother relocated with the children back to Lynchburg, where she had relatives. Dot, as a girl, had sat in her school classroom reading a book with an illustration showing a happy family gathered around a dinner table and felt ashamed when the teacher walked behind her and saw the tableau she felt transfixed by. Things like marital disharmony were not spoken about, even though many families were suffering under the hardships and dislocation of the Great Depression.

Lynchburg was not a big city, but it was bigger than Norton. Some forty thousand people lived there, and a number of railway lines—the Norfolk and Western, the Chesapeake and Ohio, the Southern—rattled into the city's train station. A hilly city, Lynchburg at one time had been the largest tobacco market in the South. It lay along the James River, and in its early days, tobacco had been transported in hogsheads to Richmond by slaves poling bateaux. Then the canals were built, and the railroads, and more industry came: a shoe factory, foundries, paper and flour and lumber mills. Dot's mother worked as a secretary at a factory that made work uniforms. Money was hard to come by. In some ways Dot hardly noticed the Depression, because it didn't seem different from ordinary life. She was born in 1920, and the stock market crashed when she was nine.

As the oldest of four siblings, Dot was what amounted to the assistant mother. She and the next-in-line sibling—a brother, nicknamed Bubba—often paired up, ruling the younger two with an iron fist and making sure chores got done. Dot's siblings called her Dissey—pronounced Dice-y—because one of them at some point had been too young to pronounce Dorothy. To her younger brother, who went by Teedy, Dot was an intimidating whirlwind of accomplishment, always busy with school and extracurricular activities.

It was clear to Teedy that his older sister possessed a highly analytic mind and a literary sensibility. There were not a lot of books in the Braden household, but the ones they did own, Dot read over and over. *Rebecca of Sunnybrook Farm*, for instance: She was given a copy one Christmas and probably read it twenty-five times. The family had a leather-bound volume of Keats and the collected works of Edgar Allan Poe, Dot's favorite author. As a schoolgirl, whenever she was assigned to do a presentation, she liked to act out a Poe poem or story—"Quoth the Raven, 'Nevermore'"—in an attempt to terrify her classmates. She had a histrionic streak and a tendency to drop words like "forthwith" and "epistle" into everyday conversation. She and her siblings also savored adult detective fiction, which they checked out of the public library. Her mother would come home and find them all inside reading, and cry, "I have to get you children out in the sun!"

Dot was very close to her mother, who was spirited and unsinkable. Virginia Braden had grown up without much in the way of financial advantage and was resolved that her children would have better opportunities. Virginia had foreordained that Dot would attend Randolph-Macon Woman's College, a four-year institution located in the heart of Lynchburg, some two and a half miles from their house, with a hundred acres of manicured campus and eighteen redbrick buildings overlooking the James River. Randolph-Macon was one of a number of private colleges established in the nineteenth century to educate young women in Virginia, a state in which no public university had been willing to fully admit them until 1918. Of the state's many well-regarded private women's colleges—including Sweet Briar, Hollins, Westhampton, and Mary Baldwin—Randolph-Macon was said to be the most rigorous and demanding.

Dot was happy to follow her mother's orders. Money presented a problem, however. A local businessmen's club awarded her a scholarship that covered tuition, but there was book money to pay as well, and they didn't have it. Dot shut herself in her closet and cried. A generous uncle came to the rescue: He would lend them the book money under the condition that Dot not tell his wife. During her college years Dot worked at a florist shop and earned extra money grading physics papers; while her well-to-do classmates made their debuts and went to football games in their free time, Dot was taking the streetcar between her home and her job and her college classes. At Randolph-Macon, she found she had an aptitude for languages. She had begun taking Latin and French in the seventh grade and continued both in college. Speaking French was a challenge—unlike some of her classmates, she had never been able to take a continental tour—but writing and reading it came as easy as sleeping.

* * *

Dot presented herself at the Virginian Hotel on September 4, 1943, a Saturday. War by now had taken over her city: All that summer, Lynchburg's morning and evening papers—the *News* and the *Daily Advance*—had been full of news about the state of the conflict in places like Rome and Sicily; about Allied bombing campaigns; about the condition of the German rail system and Hitler's European empire; about the bravery of "young American fliers" who "grin in the face of death"; about the newly important status of faraway locales like Rabaul, a Japanese stronghold in the South Pacific; and, closer to home, about local black markets and price ceilings; butter supplies; the coal situation; a cotton textile shortage; the issuing of extra shoe ration coupons for children, with their growing feet; and the recent arrival from Norfolk of the first oyster shipment, "those inevitable heralds of fall," and the lack of workers to shuck them. Display ads urged citizens to buy war stamps; food columns advised how to make wedding cakes in a time of hasty ceremonies and sugar rationing; a local store, Millner's, advertised "the perfect dress for your public life—when you want to look your best at committee meetings, charity drives and other patriotic activities."

In such an environment, the presence of war recruiters in a hotel lobby did not seem at all strange. Entering the doorway, Dot found the two recruiters—a man and a woman—standing behind a table in the lobby, beneath its glowing chandeliers and thirty-foot barrel-vaulted ceilings. The recruiters seemed very interested in Dot's facility with languages and asked her to fill out an application giving her name and address, education, and work experience. They told Dot to provide character references and to state whether she had family members serving in the military. Both brothers by now were in the service. Bubba—his real name was Boyd Jr.—was stationed at Scott Field in Illinois, where soldiers were being trained as radio operators and mechanics for the Army Air Force. Teedy—or John—was at Camp Fannin, a huge new Army training camp in Texas. The application included a line for enumerating clubs she belonged to, so Dot wrote down "National Honor Society" and "Quill and Scroll."

The recruiters told her they would be in touch. Dot made her way out onto the street and allowed herself to hope. The more she thought about it, the more she liked the idea of working in Washington and doing her part. Like every other American family, the Bradens were gung ho about helping the war effort—so much so that Teedy had asked Dot to write a letter volunteering their dog, Poochie, and they had received a polite letter from a War Dog Reception and Training Center, declining the dog's services. For Dot, working in the nation's capital would be a break from the life that had been lovingly planned for her, and that part—the unknown—was exciting.

Just a few weeks later, Dot's hopes were realized. A letter arrived at 511 Federal Street informing her that she was invited to work as a civilian for the U.S. Army Signal Intelligence Service, which was part of the

Signal Corps. She was expected to pay her own way to Washington, but she would be paid $1,620 a year, almost double what she had made teaching school.

On October 11, 1943, a cool Monday morning with the hint of autumn in the air, Virginia Braden came to the Lynchburg train station to see her daughter off. One of Dot's aunts came as well, and both of the older women were crying. Dot was too nervous for tears. She was carrying her raincoat and umbrella, along with two small hard-backed suitcases containing all the clothes she owned. The train was crowded. Gas was rationed, as were tires, and most people no longer drove long distances. Members of the military received seating preference, so civilians often had to stand, or sit on a suitcase in the aisle. This day, Dot was lucky and found a seat near a boy she knew from high school. He was on his way to start military training and asked where she was headed. "I'm going to Washington to take a government job," replied Dot proudly. When he asked her what the job was, Dot had to admit that she did not know. There had been something on the form about "cryptography," but she had no idea what that word meant.

* * *

Towns and trees and farms streaked by the window as Dot's train left Virginia's Southside and headed through the rolling Piedmont, the land flattening out as they drew closer to the capital. After several hours the train pulled into Washington's Union Station, where the unavoidable fact of world war—and the nation's resolve to win it—was everywhere apparent. As Dot stepped out of the train she could feel that the tempo here was more fast-paced and even more determined than it had been in her hometown: If the war had been present in Lynchburg, it was omnipresent in Washington, especially here in the city's Union terminal, where more than one hundred thousand travelers now alighted every day. The passengers were black and white, male and female: men in suits heading to and from cities around the country; servicemen in uniform, and servicewomen too; civilian women in hats and jackets or dresses with neatly tucked waists.

Passing through the ticket gate and into the concourse, Dot found herself dwarfed by the space, which was said to be big enough to hold the Washington Monument laid out flat. With its arched doorways, its marble and white granite, its elevated statuary depicting ancient legionnaires—shields added for modesty's sake, to hide their muscular legs and thighs from female view—Union Station was the most imposing public space she had ever stood in. Extra ticket windows had been added, platforms lengthened to accommodate the hordes of wartime travelers. There was a vast waiting room whose mahogany benches were always, now, full. The usual station amenities—newsstand, coin-operated lockers, drugstore, soda fountain—had been augmented by a servicemen's canteen. Above Dot's head was a big poster fluttering from the ceiling that declared, AMERICANS WILL ALWAYS FIGHT FOR LIBERTY. Amid the military flavor, though, there was a feminine cast: Female voices now announced the arrival of trains.

Clutching her belongings, Dot threaded her way through the crowd. She followed an overhead sign that said TAXICABS and made her way to the stand at one side of the station, where she hailed a cab for the first time in her life. She gave the driver the address she had been told to report to, then settled back in her seat with a sense of awe and nervous excitement. The cab drove her through the city's monumental core, familiar from schoolbooks though she had never actually seen it. Out the window Dot could glimpse the dome of the U.S. Capitol, the Washington Monument poking skyward. Soon the cab was skirting the western side of the National Mall and she could see the Lincoln Memorial, and, before her, the wide ribbon of the Arlington Memorial Bridge, a physical but also symbolic link that connected Abraham Lincoln's statue, and all that he stood for—emancipation, unity—with the defeated South in the form of Arlington House, the onetime home of Robert E. Lee, elegant and imposing on a Virginia hill. There was American history, old and new, stretching out on all sides as the cab carried her over the Potomac River and along a highway that skirted Arlington National Cemetery, with its green hills and stark white tablets. But now the river was behind her and they were plunging deeper into the Virginia suburbs. Where on earth were they going? The ride went on, and on and on, and Dot began to worry that she would not be able to pay the fare when they got wherever it was that they were headed.

Finally, the taxi pulled up in front of one of the strangest places Dot had seen. Set back from the thoroughfare, barely visible to pass-ersby, her destination was a compound, almost like a little city. Behind a screen of trees loomed a large school building, not unlike the private academies Dot was accustomed to: a four-story hall of creamy yellow brick, L-shaped, with a wide central drive and a high portico set off by six Ionic columns. Newer buildings were scattered about. Two steel mesh fences surrounded the compound, and each building had its own perim-eter fence. Dot's main impression, one that would stay with her for the rest of her life, was of wire: lots and lots of high, terrifying wire, beyond which high-ranking military officers were passing, and through which she, a twenty-three-year-old ex-schoolteacher armed with a raincoat, an umbrella, and two suitcases, was expected to make her way.

Dot paid the driver—it took almost all the money she had—and walked up to a gate. "I'm supposed to be here," she nervously asserted, giving her name and watching the guard pick up a telephone. She was directed to the main school building, where she opened the door and found, to her relief, that she was expected. She was at a place called Arlington Hall. Before the war, Arlington Hall had been a finishing school for girls: a two-year "junior college" complete with lily ponds and an indoor riding ring and flowering cherry trees lining a gracious cen-tral drive. Now it had been requisitioned by the U.S. War Department and transformed into a government operation whose purpose was not clear. But the presence of so many military officers impressed upon Dot that her new job must be even more important and serious than she had understood. The thought was intimidating. Inside the main building, the French doors and elegant moldings of the girls' school were intact, as was the gracious central staircase, but the furnishings consisted of no-nonsense chairs and desks. A self-assured civilian woman, no older than

Dot, possibly younger, seemed to be in charge of the whole place, as far as Dot could tell.

Other women were arriving, gathering in the hallway and looking as uncertain and travel grimed as Dot felt. When a good-size crowd had collected, the women were ushered into a kind of drawing room, where the self-assured young woman distributed printed copies of a loyalty oath. Dot signed it, swearing that she would "support and defend the Constitution of the United States against all enemies, foreign and domestic"; that she took this obligation "freely, without any mental reservation or purpose of evasion." She also signed a secrecy oath, swearing that she would not discuss her activities with anyone outside her official duties—not now, and not ever, and that to do so opened her up to prosecution under the Espionage Act. The whole exercise felt frightening; she was really in for it now. The self-assured young woman told Dot that was enough for the day. She could leave and come back tomorrow.

"There are buses here that will take you wherever you are staying," she said.

Dot looked blank. "Where I'm staying?"

"Aren't you staying somewhere?" the woman wanted to know.

"No, I'm not staying anywhere," Dot stammered, embarrassed. She had formed the notion—or had been led to believe—that the U.S. government would provide lodging in return for her wartime service. She had been mistaken. There was nobody with whom she could stay; she did not know a soul in the city of Washington or its suburbs.

The woman looked scornful but told Dot that there was a facility nearby where she could rent a room. Dot gathered her things and clambered onto a bus. In fifteen or twenty minutes she found herself standing on yet another campus. The buildings here, however, were neither old nor high ceilinged nor gracious; they were new and hastily constructed, ugly temporary structures arranged in rows and connected by freshly poured sidewalks. Giant shade trees were interspersed among them, but the effect—old arching foliage, new blocky buildings—was incongruous rather than pleasing. Dot made her way in the direction that other workers who had ridden the bus with her were now moving. Her companions were all female: The facility where she found herself had been built at the behest of First Lady Eleanor Roosevelt to accommodate the young women, like Dot, who were pouring into Washington to take jobs to serve the war effort. Past conflicts—the Civil War, the First World War—had brought female workers to the nation's capital, but this incoming wave was of another order of magnitude. Hundreds of women were arriving at Union Station every day, creating an acute housing shortage throughout the capital and the surrounding area.

Like Arlington Hall, this cluster of female-only dormitories was located in Arlington County, Virginia, a compact and nondescript suburb located across the river from Washington proper. Like that of the capital city, Arlington's population was in the midst of doubling, rapidly, as its southern portion became, with the construction of the Pentagon, a virtual military city. The women's dorms, wedged on a patch of land near the river, were intended to last for the war's duration, and they were so flimsy that whoever built them must have thought the duration would be short. A newspaper reporter described them as "gray and extremely temporary in appearance."

The dormitory complex had been finished just months before Dot arrived, constructed on land that had once been part of the family estate of Robert E. Lee's wife, Mary Anna Custis Lee, who was descended from First Lady Martha Washington. The federal government had come into control of much of this land after the Civil War, and until recently the U.S. Department of Agriculture maintained an "experimental farm" here, conducting horticultural research on crops and growing methods. Hence the name: Arlington Farms.

But there were other, informal names for the sprawling new women's residence. In Washington, Arlington Farms was fast becoming known as Girl Town.

Other locals had dubbed it "28 Acres of Girls."

Dot didn't know it, but the national media had already taken notice of what was happening in Washington—the influx of young women.

More than twenty-four thousand had entered government service in 1940 alone, when the Civil Service Commission began recruiting workers for the defense buildup, and tens of thousands more had arrived since then. The population of the city had swelled by more than two hundred thousand, served by innumerable boardinghouses. But nothing personified the city's changing demographics as much as Arlington Farms, a "duration residence for women" built to house seven thousand female war workers. Its denizens tended to be written about in condescending tones, as wide-eyed rubes from the heartland. "There's a new army on the Potomac," gushed *Good Housekeeping*, "the bright-eyed, fresh-faced young Americans who have poured into Washington from remote farms, sleepy little towns, and the confusion of cities, to work for the government in a time of national emergency."

The girls moved into Arlington Farms even before there was time for grass to grow. Some had come to work for Congress, some for federal agencies, some for the Pentagon, which was the new War Department headquarters. Many, like Dot, were employed at Arlington Hall. They were known as government girls, or g-girls for short. The complex was designed to make their life as easy as possible, so they could devote their full energies to helping win the war. At Arlington Farms, women could take meals in the cafeteria and send their clothes to be laundered or dry-cleaned. Maids cleaned their rooms weekly. There were pianos and snack bars, and little cubbyholes meant to resemble the "dating booths" of American drugstores. Each dormitory was named after an American state.

At the front desk, a clerk told Dot that there was a room available in Idaho Hall, but she would have to pay a month's rent—$24.50—in advance. She was appalled. Dot didn't have anything like that much money with her, and wouldn't until she got her first paycheck. Sheepish, she went into a phone booth and placed a long-distance call to her mother. "I've got to pay in advance," she told Virginia Braden.

"Well, I've never heard of anything like that in my whole life," came her mother's familiar voice.

"Well, Momma, that's what they say," she said. Dot felt guilty; she had taken the job to ease her mother's financial hardship, and here she was, exacerbating it.

"Well, I'll find the money and send it to you," her mother told her, and that was that.

Idaho Hall was a two-story building composed of prefabricated-looking squares and rectangles, containing single and double rooms, door after door lining long corridors. There were ten dorms in all, each housing about seven hundred women. Idaho Hall had a lobby and lounges where occupants could play bridge, dance, drink tea, or sit with the soldiers and sailors who were always coming to visit. ("I am not running an old maids' home," said the head of the place, William J. Bissell, defending the rather easy fraternization that went on.) There was a recreation room and a shop selling cosmetics and sundries, and a mail desk where women could collect letters from a vast warren of mailboxes. Women were everywhere—some of them young, about Dot's age, but some of them older, maybe as old as thirty or even forty. Dot found her way to Room I-106 and opened the door to find a tiny single furnished room with a bed, a desk, a mirror, an ashtray, two pillows, a chair, a wastebasket, and a window. Down the hall were a communal bathroom and showers, with sinks where residents could wash their clothes and their hair. Women were allowed to lounge in bathrobes or in bras and panties. Laundry had to be hung inside, not outdoors, so Arlington Farms would not look like a tenement. There were ironing boards, and a kitchenette on every floor. Murals by Works Progress Administration (WPA) artists adorned the walls, to inspire the women and prettify what was essentially a barracks. As she made her way down the hall, Dot could see that some occupants had put up curtains. She noticed quite a bit of flowered chintz.

That night was the first Dot had ever spent among strangers. In the morning she walked to a bus stop and waited with other young women who were clutching purses and wearing hats and shirtwaist dresses. The bus to Arlington Hall was unmarked, with no destination sign. Back at the fenced-in compound, Dot found herself again in a crowd of women, some of whom she recognized from the day before. None had any idea why they were here. Confusion reigned. As they milled about waiting for instructions, Dot chatted with a woman named Liz, who was from Durham, North Carolina, and who seemed a bit older than she was.

"I'm going to stick with you," Liz said. "You look like you know what you're doing."

Dot had to laugh, hearing that. She had been in Washington for twenty-four hours and still felt as confused as she had when she arrived. She was photographed from the front and the side holding a sign that said DOROTHY V. BRADEN and the number 7521. The photo was affixed to a badge, and the badge permitted entrance to certain parts of the compound. During several days of orientation, she received strict lectures about the need for absolute secrecy around the work she would be doing. She visited rooms and workspaces where the activity gave her an inkling of what "cryptography" entailed—and it entailed quite a bit. She went back to her single room each night with the dawning awareness that as bizarre and unlikely as it might sound, she, Dot Braden, ex-schoolteacher, graduate of Randolph-Macon Woman's College, had been hired to break enemy codes.

* * *

Dot was to spend the bulk of her time in Building B, a low two-story building built on sloping land behind the main schoolhouse. Like the dorm where she was staying, her new workplace was designed in a functional style that might be called "temporary wartime Washington." From above, Building B, like its nearby counterpart Building A, looked like a giant comb. A horizontal passageway at the rear formed the central body, with a dozen or so slim "wings" jutting out perpendicular to the body, forming the teeth. Inside, each wing had an aisle down the middle, and rooms branching out to either side.

Those rooms were crammed with people—almost all women—working with graph paper, cards, pencils, and sheets of paper. Some sat at desks, but most were working at tables. Dot had never seen so many women side by side, not even during exam time in the library at Randolph-Macon. No one person seemed to be in charge, and yet the women at the tables seemed to know what they were doing. They sat focused and intent. Some tables were piled with stacks of cards and papers. Others had thin strips of paper hung from lines, like drying pasta. Lined up against walls were boxes and file cabinets. The cabinets were wooden, metal ones having been sacrificed to the war effort. Big six-foot fans stood at the crosswalks where corridors intersected. The fans made so much noise and blew so much paper that people argued constantly over where they should be directed. It was not as quiet as a library, not as noisy as, say, a cafeteria; instead, there was a sort of constant low murmur.

It took Dot several days to learn her way around. At first, much of the complex looked alike, but she soon came to see that there were distinctions between what was going on in many of the spaces. The operation also included rooms where women worked at machines—tabulating machines, punch-card machines, strange sorts of typewriters. There were small machines, huge machines, noisy machines, machines hooked up to other machines through thick nests of cables. Dot didn't know this, but she had found her way into the largest clandestine message center in the world.

Everywhere women were attacking enemy messages pouring in via airmail, cable, and teletype. The messages originated in Nazi Germany, Japan, Italy, occupied Vichy France, Saudi Arabia, Argentina, even neutral countries like Switzerland, sent between top political leaders and military commanders. At the Allied listening stations where the messages were secretly intercepted, American operators further encrypted them using their own encoding machines. Once they got to Arlington Hall, the messages had to be stripped of the American encryption before they could be stripped of the underlying enemy encryption.

The whole thing was insanely complex.

Technically, Arlington Hall was a military base, known as Arlington

Hall Station. Dot could see that there were thousands of people working here, in an operation fashioned after an assembly line. The ranks included a small number of Army officers and enlisted men, and some male civilians including older professors and young men with disabilities—some severe, such as epilepsy—that disqualified them for military service. But by far the majority were female civilians like Dot, and most of those, like Dot, were ex-schoolteachers.

Already, teachers were turning out to be well suited for code-breaking work, for a number of reasons beyond their level of education. Around the time Dot was hired, a harried employee of Arlington Hall was working up a memo that would spell out the qualities that made for a good code breaker. These were hard to know. Administrators were finding to their chagrin that there often was not a correlation between a person's background and how well that person would do at breaking codes. Some PhDs were hopeless, and some high school dropouts were naturals. There was a stage actress who was working out wonderfully, as was a woman with little formal education who had been a star member of the American Cryptogram Association, a membership group for puzzle and cipher enthusiasts. Code breaking required literacy, numeracy, care, creativity, painstaking attention to detail, a good memory, and a willingness to hazard guesses. It required a tolerance for drudgery and a boundless reserve of energy and optimism. A reliable aptitude test had yet to be developed.

Thus far, Arlington Hall officials had found that problems based on "reasoning" and "word meanings" provided some insight into who might do well, but only in the sense that people with low scores would have "difficulty in following simple directions and understanding the simpler techniques." Those who scored high on arithmetic tests often did well.

Hobbies, especially artistic hobbies, were emerging as a good sign. "Those who have had some creative outside interest or hobby generally work out very well in comparison with those whose interests are movies or similar entertainment," the memo concluded.

Temperament mattered, and here, too, was where the schoolteaching advantage came in. Officials were finding the best code breaker was a "mature and dependable" person with a "clear, bright mind"—but someone "young enough to be alert, adaptable, able to make adjustments readily, willing to take supervision," and "able to withstand inconveniences of Washington." This description fit many schoolteachers, including Dot Braden, to a T.

There were a few truths emerging with regard to women. Married women were problematic, the memo observed, through no fault of their own but because they tended to move to follow their husbands. This was another reason schoolteachers like Dot were perfect: They were almost always unmarried. In America in the 1940s, three-quarters of local school boards (like telephone companies and other employers of female labor) had enacted a "marriage bar," which required that married women not be hired and that a teacher must resign when she did marry, in accordance with the prevailing belief that a wife's place was at home. By definition, then, many female schoolteachers were single.

Schoolteachers were smart, educated, accustomed to hard work, unused to high pay, simultaneously youthful and mature, and often unencumbered by children or husbands. In short: They were the perfect workers.

"The proverbial 'old maid' schoolteacher finds the adjustment hard from the complete 'ruler of the roost' to 'one among many,'" the memo noted. "However, many of our best workers come from this profession."

* * *

Before Dot Braden's arrival, the U.S. War Department had dispatched investigators to Chatham and Lynchburg. The investigators contacted her references and looked at police and birth records to see whether she was foreign-born; whether she had undesirable qualities like emotional instability, erratic behavior, poor work habits, or communist sympathies. The high constable of Lynchburg reported that he had known Dot for fifteen years and that she was of "above average intelligence and could be depended on in any position of trust." Clement French, a dean at

Randolph-Macon, called Dot a "very conscientious, hard-working girl whose efforts to help herself thru college showed her real stamina." The Lynchburg schools superintendent said she was a "fine young woman who has fought against difficulties, especially financial and succeeded well."

All agreed Dot had never been fired; she did not use intoxicating liquors; she had not had brushes with the law. Dr. A. A. Kern, head of Randolph-Macon's English Department, wrote that Dot "stood for the better things in college life."

Based on these inquiries, investigators had submitted a "loyalty and character report" on Dot Braden. The report noted that she had a good college record and was of "the Anglo-Saxon race and of normal appearance." She was found to be "native born," as were her parents. The investigators noted that her parents were separated—her father was living at the Lynchburg YMCA—but said that they were "loyal citizens of middle-class." The report concluded that Dot was "dependable and honest and of sober habits," that she was "single and boards with her mother in a desirable section" of Lynchburg, and concluded that there was no reason to question her loyalty.

* * *

It remained for Dot to be assigned a permanent code-breaking duty. At Arlington Hall there were no unimportant jobs, but some tasks were harder than others. There were typists and keypunchers, as well as people working on the codes themselves, with titles that included "starter" and "overlapper" and "reader."

Dot was put to work sorting messages, a common first assignment. A couple of days spent untangling intercepts showed that she was capable of recognizing the digits at the beginning of a message, which designated the station from which it had been sent and the system it was part of. She now was presented with a series of four-digit numbers and told to discern any pattern she might see. She had taken many tests in high school and college, had always done well, and faced the numbers before her with a reasonable degree of confidence. She was sitting at a big table with a group of newcomers, and a girl beside Dot started crying. Dot did not find the work easy; it felt like a complicated puzzle, but she must have done all right, as she presently was told that she was being moved on to the next level.

To say that Dot was "trained" would be an overstatement. Over the next few weeks she sat through lectures on the rudiments of code breaking and code making. She learned a bit about codes used by the Japanese Army, which controlled much of the Pacific Ocean, occupying captured islands and other territory. She absorbed the principles of how the Japanese Army was organized, and basics of the Japanese language as used in military communications: "enough so we could go at it," as she would later say. She took a test on the Japanese language, which she did not mind, having had so much language instruction already. She watched movies directed by Frank Capra called *Why We Fight* that aimed to inspire patriotism and build morale. Most of all, she sat through still more talks on the importance of security, secrecy, and silence, the result of which was that she felt perpetually terrified she would accidentally bring a piece of paper home or let the wrong word slip. She entered a constant state of monitoring her own behavior, on high alert even when she was not working at the compound.

Dot also was called for a one-on-one interview. The interviewer, a woman, asked her what languages she knew; what science and math courses she had taken; when she had graduated from high school and when from college. She was asked whether she had worked with radios and whether she liked the physics classes she had taken in college. The interviewer wanted to know what her hobbies were. "Books and bridge," Dot replied. The interviewer wrote that she was "attractive and well-dressed" as well as "intelligent" and "nice."

Based on this, her assignment was made, and it consisted of one word: crypt.

Specifically, Dot Braden was assigned to Department K of Section B-II of Arlington Hall. If these terms seemed vague, that was the idea—to preserve secrecy around the intelligence work she was doing. Up to now, Dot's life, like that of most Americans, had been circumscribed and upended by the war and the changes it wrought in every aspect of her life. Now she was in a position to impact the war's outcome. She did not realize it herself, but she had been assigned to one of the most urgent missions that Arlington Hall had undertaken: breaking the codes that were being used to direct merchant ships moving around distant Pacific islands. The ships were bringing vital supplies to Japanese Army troops. Dot would be cracking the messages that controlled—and foretold— their movements. Severing the enemy's lifeline of food, fuel, and other critical supplies would allow General MacArthur and Admiral Nimitz to push back against Japanese primacy in the Pacific Ocean, where tens of thousands of American men's lives were at stake and the balance of the war remained undecided.

Dot Braden would be sinking ships.

CHAPTER TWO

"This Is a Man's Size Job, but I Seem to Be Getting Away with It"

June 1916

When the U.S. Navy wrote its female recruits from the Seven Sisters schools in fall 1941, telling them the wartime work they were embarking on had been done up to then by men, this was not true—not remotely. Even before World War II, there were some very important female pioneers, and breakthroughs made by these crucial early women code breakers would have even greater significance once the war began.

In fact, to the extent that America had any code-breaking capability prior to World War II and the mass recruitment of women like Dot Braden, it was thanks in considerable part to this small but brilliant group of women. These were women who were curious and resourceful; who needed to earn their own living; and who were on the lookout for work that satisfied them intellectually as well as, you could almost say, spiritually. They almost always were dissatisfied schoolteachers eager to find some other venue for their intellect and talent. They often were fortunate to be mentored by or partnered with—or both—men who supported their goals and respected their minds. In science, there is something called a "jackpot effect," where a male scientist hires women in his lab early in the development of a certain field, and these women hire other talented women, and, as a result, the field ends up with an unusually high number of women. Something like this was at work in cryptanalysis. A few key women proved themselves gifted, early on; a few key men were willing to hire and encourage them; that early success led to more women being drawn in.

The other factor that led to women's involvement was the advent of code breaking for serious wartime uses. Like medicine, code breaking often makes advances during times of violent conflict, when life-and-death necessity becomes the mother of invention, technology drives innovation, and government funds are freed up. Military cryptanalysis certainly had been around before World War II—in the United States it tended to be an occasional affair, used during the Revolutionary and Civil Wars and then dismantled—and also was rather leisurely in the days when communications were too slow to affect combat in real time. Things picked up in World War I, when militaries started using radio to direct troops, ships, and—soon—aircraft. But despite its growing importance, cryptanalysis often was not a job that career military men wanted, at least not at the outset. Officers understood that it was better for their careers to spend a war in the theater, being shot at and commanding men, rather than sitting safely behind a desk. And so wartime, exactly when code breaking was most needed, was exactly when women were invited to pinch-hit.

It also helped that cryptanalysis in its formative stages was an occupation without fame or prestige, not yet a recognized or even a known field of endeavor. While it had existed in Europe for centuries, where furtive bureaus operated in the shadows to monitor diplomatic missives, cryptanalysis for quite some time—particularly in the United States—also tended to be an obscure and even slightly crackpot profession, more of a hobby or amateur calling. This lack of renown or regulation—the fact that it had not yet been established as a man's field, or even a field—created a wide crack through which women could enter. To do so, it helped to have a high tolerance for the clandestine and irregular, a lack of squeamishness about reading words intended for other people, and a willingness to embrace the unknown. A little bit of desperation was also not a bad thing.

* * *

All of these qualities were characteristic of Elizebeth Smith, who in 1916 was a restless midwesterner with a strong desire to get beyond the horizons of her own small known world and no field of endeavor, yet, that she had latched onto to her satisfaction. The youngest of nine children, she grew up in Indiana and as a young woman had hoped her father, a Quaker, might facilitate her admission to a renowned Quaker college like Swarthmore. Her father, however, was "uninterested in my going to college," as Elizebeth later put it, so "by my own efforts" she gained admittance to the College of Wooster in Ohio, borrowing the tuition money from her father, who charged her 6 percent interest. After two years she transferred to Hillsdale College in Michigan, majoring in English and studying Latin, Greek, and German. Elizebeth, whose mother chose an unusual spelling for her first name to ensure that her daughter was never called "Eliza," spent a year teaching and serving as principal at a small country school, and decided to seek a more "congenial way of earning my living." In the summer of 1916 she traveled to Chicago and stayed with friends on the South Side. Dispirited after breaking off an engagement with "a handsome young poet and musician," she had no clear idea what she wanted to do and knew only that it should not be "run of the mill."

A visit to a Chicago employment agency proved fruitless. But the agency did suggest she visit the Newberry Library, where there might be some sort of job involving a 1623 folio of one of Shakespeare's plays. Elizebeth was "stunned" to learn that such a thing as an original folio of Shakespeare's work existed, and she resolved to see it, regardless of whether there was a job attached. She took the L—Chicago's rapid transit—to the Newberry. Her first sight of a Shakespearean manuscript, she later wrote, prompted the kind of thrill that an archaeologist might feel at stumbling upon the tomb of a pharaoh. She was smitten and began chatting up a friendly librarian to see if there was in fact a position that might permit her to work around magnificent original documents like that.

As luck would have it, there was. It wasn't at the Newberry, but rather at the estate of a wealthy man named George Fabyan, who was looking for someone to "carry on research" on a literary project involving Sir Francis Bacon. He specifically wanted a woman who was "young, personable, attractive and a good talker." The librarian called up Fabyan then and there. He had an office in the city, and before long a limousine pulled up, "and in came this whirlwind, this storm, this huge man and his bellowing voice could be heard all over the library floor," Elizebeth later recalled. Her potential employer was a textile merchant whose family had made a fortune in cotton goods—a hyperactive, wild-eyed person of myriad scientific enthusiasms and no scientific training. Thanks to his wealth, Fabyan was able to indulge his many curiosities. He was incubating any number of so-called research projects at a place he called Riverbank Laboratories, a suburban "think tank" located on an estate in Geneva, Illinois.

At the library, George Fabyan asked Smith if she would be willing to go out to Riverbank and spend the night. She protested that she did not have a change of clothes, and he told her he'd lend her some. When she agreed, he swept her into the limo, which drove them to the Chicago and North Western railroad station. Before she knew it, Elizebeth was sitting on a commuter train wondering, "Where am I? Who am I? Where am I going?"

Smith was fascinated and slightly repelled by Fabyan, who was large, bearded, and unkempt. Despite having a reputation at college for "volubility," she feared he must think her a "demure little nobody" and resolved to correct that impression. Aware that she was in the company of a multimillionaire, she resolved to be well-spoken and proper, and her idea of well-spoken and proper seems to have been lifted out of a Gothic romance. When Fabyan asked, "What do you know?" she leaned her head against the train window, looked at him "quizzically" through half-lidded eyes, and replied in her best Jane Eyre: "That remains, sir, for you to find out." Fabyan roared with laughter; he seemed to consider it an ideal answer. Another car was waiting at their destination, and presently Elizebeth Smith found herself installed in a guest bedroom at Riverbank, where there was a full fruit bowl and a pair of men's pajamas. She proceeded downstairs to a formal dinner whose attendees included her new boss: Elizabeth Wells Gallup, another former schoolteacher, who was living at the Riverbank estate and was a crank of the first order.

Gallup belonged to an international cabal of similarly minded cranks, who subscribed to the notion that Sir Francis Bacon was the true author of the plays and sonnets of William Shakespeare. Bacon, an English statesman and philosopher who ran Queen Elizabeth I's printing press—among many occupations—was one of a number of Renaissance thinkers who dabbled in secret writing. In the sixteenth century Bacon had invented something called a "biliteral" cipher, with which—as he put it—it is possible to make anything signify anything. Bacon had shown that all you need are two characters or two symbols—*A* and *B*, say—to make any letter of the alphabet and spell out any word. For example, *AAAAA* can stand for *A*, *AAAAB* for *B*, *AAABA* for *C*, *AAABB* for *D*, and so forth. Communicating a full range of facts and ideas using only two symbols is slow and unwieldy, but it can be done. You could do the same with images—sun and moon, apples and oranges, men and women—or, if you happen to be a man who runs the queen's printing press, with thin printed letters and fat ones. It was the same binary principle on which a number of more modern systems were also built: Morse, with its dots and dashes, and digital computers, with their 0s and 1s, are also binary systems.

Gallup—elderly, aristocratic, mild-mannered, fanatical—was convinced that Bacon had used a biliteral cipher to thread a message confessing his authorship through the printed type of Shakespeare's First Folio. She had met Fabyan through a mutual acquaintance, and he had been instantly enamored of the earthshaking significance of her thesis. Gallup liked to inspect the typography of the folio with a magnifying glass, and she aimed to assemble a coterie of young women to apprentice with her and master her methods—bankrolled, of course, by Fabyan. For his part, her benefactor was fond of inviting reputable scholars to soirees at Riverbank and liked to offer lantern displays—the early version of a PowerPoint presentation—in an effort to impress them and persuade them of the Baconian thesis. It would be Elizebeth Smith's job to help with research and to deliver these lectures, serving as the public face and PR engine of the Baconian effort. Fabyan believed that debunking William Shakespeare's authorship would be the crowning intellectual achievement of the twentieth century and would make his own name famous for all time.

Elizebeth accepted, though the dark side of her employer soon announced itself. Fabyan insisted upon dictating what clothes she wore, compelling her to buy her hats and dresses at Marshall Field's, the high-end Chicago department store, where the offerings were more expensive than she could afford. The Riverbank estate itself was both pastoral and Mad Hatter eccentric: Fabyan and his wife, Nelle, had purchased some three hundred acres of Illinois landscape, where they engaged Frank Lloyd Wright to renovate the villa they lived in, and installed or rehabbed other dwellings, including the "Lodge," where Elizabeth Wells Gallup lived with her sister, Kate Wells. On the grounds Fabyan had installed a Dutch windmill, which he brought over piece by piece from Holland; a working lighthouse; a Roman-style bathing pool fed by spring water; a Japanese garden; a giant rope spider's web for recreational climbing; and something he called the Temple de Junk, in which he stored random things he found in the unclaimed packages that he liked to buy up: shoes, bottles, glass photography plates of nudes. He had what Elizebeth described as a "passion for furniture which swung on supports." The Fabyan marital bed was suspended from chains, as were divans and chairs in the drawing room of the villa. Outdoors were hammocks, and a hanging wicker chair installed near a fireplace that Fabyan kept lit, even in summer, to repel mosquitoes. He liked to sit in that chair, chain-smoking and poking the fire, surrounded by guests. If somebody said something he disagreed with, he would stand and, as he put it, give them hell. Fabyan, a loud man who cursed freely, called it his "hell chair."

Nelle Fabyan had passions of her own, including animal husbandry. On the grounds were a herd of prize cattle that was always being sent off to competitions; a bull imported from Scotland that was said to have cost $30,000; an open-air zoo; and a pet male chimpanzee named Patsy. The estate was divided by a thoroughfare, on one side of which were the living quarters and on the other, the research area. A river, the Fox, ran through it.

Fabyan had little formal education, and the resulting insecurity seems to have propelled him to, as Elizebeth put it, try to "break the back of the academic world" by proving mainstream scholars wrong in any number of ways. Pathologically inclined to self-aggrandize, Fabyan liked to call himself a colonel even though he wasn't one; it was an honorific bestowed upon him by the governor of Illinois. At the estate, he favored a costume of leather bootees and a Prince Albert riding habit with split tails on the cutaway coat, though he did not, in fact, ride. Otherwise careless of his appearance, on the train going to and from his Chicago office, he would light a match and burn off stray threads in his fraying cuffs.

His ambition in creating Riverbank Laboratories was to "wrest the secrets of nature" from not only literary manuscripts but acoustics and agriculture as well. Toward that end, Fabyan also had hired a number of young men. One of these was William Friedman, fresh from graduate work in genetics at Cornell, now engaged to conduct experiments in the Riverbank fields and gardens, sowing oats along some scheme Fabyan had read about that had to do with planting them during the dark of the moon. Friedman, the son of Jewish Russian émigrés, had studied agricultural science because it came with a scholarship, but he had other eclectic interests. He was living on the second floor of the windmill, where he had a studio and was doing experiments with fruit flies to test the Mendelian laws of heredity. A natty young polymath, he made a hobby of photography and before long was engaged to make enlargements of the folio pages, for inspection by Gallup and Smith.

In this singular setting, with its batty but open-minded atmosphere of inquiry, William Friedman and Elizebeth Smith soon sensed the absurdity of the Bacon theory. They realized that Gallup "dwelt only among those who agreed with her premise," as Elizebeth put it, and that nobody else seemed able to spot the typographic patterns she claimed to, which in truth were just the results of printers repairing and reusing old type. But they became drawn into the world of codes and ciphers. Fabyan had amassed a rare collection of books about cryptic writings, written over the centuries by individuals who trafficked in ways of making communications secret and unintelligible to others. If the Bacon theory was a dead end, the subject of cryptanalysis itself was fully legitimate and would prove increasingly vital, thanks in large part to Elizebeth and William.

Codes have been around for as long as civilization, maybe longer. Virtually as soon as humans developed the ability to speak and write, somebody somewhere felt the desire to say something to somebody else that could not be understood by others. The point of a coded message is to engage in intimate, often urgent communication with another person and to exclude others from reading or listening in. It is a system designed to enable communication and to prevent it.

Both aspects are important. A good code must be simple enough to be readily used by those privy to the system but tough enough that it can't be easily cracked by those who are not. Julius Caesar developed a cipher in which each letter was replaced by a letter three spaces ahead in the alphabet (*A* would be changed to *D*, *B* to *E*, and so forth), which met the ease-of-use requirement but did not satisfy the "toughness" standard. Mary, Queen of Scots, used coded missives to communicate with the faction that supported her claim to the English throne, which—unfortunately for her—were read by her cousin Elizabeth and led to her beheading. In medieval Europe, with its shifting alliances and palace intrigues, coded letters were an accepted convention, and so were quiet attempts to slice open diplomatic pouches and read them. Monks used codes, as did Charlemagne, the Inquisitor of Malta, the Vatican (enthusiastically and often), Islamic scholars, clandestine lovers. So did Egyptian rulers and Arab philosophers. The European Renaissance—with its flowering of printing and literature and a coming-together of mathematical and linguistic learning—led to a number of new cryptographic systems. Armchair philosophers amused themselves pursuing the "perfect cipher," fooling around with clever tables and boxes that provided ways to replace or redistribute the letters in a message, which could be sent as gibberish and reassembled at the other end. Some of these clever tables were not broken for centuries; trying to solve them became a Holmes-and-Moriarty contest among thinkers around the globe.

Many Renaissance cryptographers perceived (as counterparts in the Middle East had done before them) that the alphabet itself has underlying mathematical properties: six vowels, twenty consonants, some of these much more frequently used than others. Cipher systems often are created by juxtaposing or "sliding" two or more alphabets against each other so that *B* in one alphabet lines up with, say, *L* in the other, and *C* with *M*, *D* with *N*, et cetera. One of these, the Vigenère square, developed in the 1500s and named for the French diplomat Blaise de Vigenère, achieves this by creating a twenty-six-by-twenty-six-square letter table, in which twenty-six alphabets are stacked on top of one another, each alphabet beginning with a different letter, with a keyword telling which alphabet to select for each letter to be changed. America's great innovator Thomas Jefferson dabbled in secret systems, inventing a cipher wheel that could transform one letter into a new one; it was discovered in his personal papers more than a hundred years after his death.

During the American Revolutionary War, many code systems were employed not only by diplomats and statesmen but also by spies and traitors. Sometimes, of course, diplomats and statesmen *were* the spies and traitors. Jefferson and Ben Franklin at times used coded language, as did Benedict Arnold. During the American Civil War, the military began to experiment with codes and ciphers. Union commanders sent messages by means of a soldier working a little handheld disk. The disk changed each letter to a new one, and the new one would be transmitted by a soldier waving a big flag in the direction of the intended recipient, in a

signaling cipher called wigwag. Confederates used a cipher so complicated that they became confused by it. Sometimes they also would intercept Union messages, publish them in the newspaper, and invite readers
to submit solutions.

Elizebeth Smith and William Friedman were fascinated by this
history—and by each other. They took long bicycle rides, swam in the
Roman pool, drove through the countryside in a Stutz Bearcat. Theirs
was an early example of what sociologists nowadays call homogamy,
which is marriage between equals. Elizebeth found William to be sleek
and sophisticated; he found her vibrant, smart, and dynamic. The two
married in May 1917, less than a year after they met, and Elizebeth moved
into the windmill.

Despite the newly married pair's skepticism about Gallup, the soirees continued, though the topic expanded beyond Bacon to encompass
codes and ciphers and their solution. From time to time, Fabyan liked to
summon a University of Chicago English literature professor, an amateur cryptanalyst, John Manly, and pit him against Elizebeth Friedman.
Cryptanalysis at that time was a parlor game. The operation gained a certain cachet: Movie stars sometimes came to tour Riverbank. The Friedmans developed genuine expertise, attracted freelance assignments, and
won acclaim when they decoded a batch of correspondence that helped
expose a conspiracy between Hindu separatists and German agents trying to foment revolution against the British in India. Elizebeth was disappointed that William was the one who got to testify in the trials (during
which one defendant was shot dead); though they had done the work as
a team, she had to stay behind because, as she put it, "someone had to oil
the machinery at Riverbank."

War, though, began to transform the tenor of the operation. As
unlikely as the setting was, Fabyan's estate incubated the first serious cryptanalytic efforts of the U.S military. In 1916, Fabyan began to
sense that America's involvement in world war was in the offing. In his
voluble self-promotion he had devoted much energy to cultivating ties
with important people in Washington, often taking William Friedman

with him on visits. Domineering and massive, Fabyan was friends with another massive man, Joseph Mauborgne, an Army officer and radio enthusiast who had solved the notorious Playfair cipher—a table in which pairs of letters are substituted for other pairs, used as the primary cipher system of the British Army—during a six-month trip on a steamer. Now, thanks to Mauborgne's influence, the U.S. Army was taking a greater interest in code making as well as code breaking.

So were other agencies. As more and more messages began traveling by cable and radio, any number of government entities found themselves with communications they wanted to penetrate and puzzle out, as well as systems they needed to protect. Fabyan decided to make Riverbank the place where the U.S. government could outsource these cryptologic efforts. To their astonishment, William and Elizebeth—newly self-taught, both still in their twenties—found themselves running the shop. They proceeded to build a cipher department with as many as thirty employees, among them scientists and language majors, translators and stenographers. According to Elizebeth, the team began to perform "all code and cipher work for the government in Washington," receiving intercepts from the Army, the Navy, the State Department, the Justice Department, the postal service, and others. The staff studied all manner of correspondence—one message Elizebeth worked hard on turned out to be a Czechoslovakian love letter—and produced a series of books called the Riverbank Publications.

Meanwhile countries such as France and England were far ahead in maintaining real cryptanalytic bureaus, descended from the black chambers, as Europe's clandestine government code-breaking shops were called in their Renaissance heyday. It was in fact a message the British decoded—the Zimmermann telegram—that (along with Germany's declaration of unrestricted submarine warfare) helped bring the United States into World War I. German foreign minister Arthur Zimmermann sent an internal coded message to Germany's minister to Mexico, instructing him to offer the president of Mexico the territories of Texas, Arizona, and New Mexico if Mexico would ally with the German cause and invade its northern neighbor. The British broke the message in January 1917; the United States was appalled; America, as George Fabyan had foreseen, went to war.

* * *

Fabyan himself did not join the fighting, but he did enjoy the trappings of military life. At one point he had trenches dug on the Riverbank property, and—as America prepared an expeditionary force to be sent to the killing fields of France—he looked for ways to play a part. The U.S. military by now was beginning to declare some independence from its eccentric benefactor; when military intelligence officials asked Fabyan to move his cryptanalytic unit to D.C., he refused, and so the War Department began quietly to build its own small code-breaking bureau, bringing in none other than John Manly, the Chicago professor who had matched wits with Elizebeth, and Herbert O. Yardley, a former telegraph code clerk for the State Department. To retain his influence, Fabyan offered at his own expense to set up a training school at Riverbank, where military officers and others could take a crash course in code techniques. Before they departed for Europe, the trainees—seventy-one of them, plus William and Elizebeth and Fabyan and a few others—lined up for a panoramic photograph in front of the Aurora Hotel, where the trainees were staying. Each was told to look either to the side or straight ahead. They were creating a biliteral cipher spelling out "Knowledge is power," one of Francis Bacon's favorite phrases. Unfortunately they didn't have quite enough people, so they could spell out only "Knowledge is powe." There was also a typo in the sense that one man looked the wrong way.

Among those who visited Riverbank during this time was another female initiate: Genevieve Young Hitt, whose husband, an Army officer named Parker Hitt, had done pioneering cryptanalytic work decoding Mexican military and government communications that were picked up by American radio trucks operating on both sides of the border and sent to him at his post in Fort Sill, Oklahoma. Much like William Friedman, Parker Hitt had married an educated woman who shared his interests. Genevieve, the daughter of a doctor, had attended Saint Mary's Hall School in Texas, where she studied English, botany, chemistry, and astronomy and was evaluated by her principal as having "lady-like deportment, and Christian character." Her ladylike ways did not prevent her from helping her husband with his cryptanalytic eavesdropping. Genevieve also solved test problems for a manual that Parker authored, one of the first U.S. military cryptanalytic training handbooks, and mastered his "strip cipher" device, a means of lining up alphabets, which she demonstrated on her Riverbank visit.

When Parker Hitt left for Europe, Genevieve Hitt took over the code room at Fort Sam Houston in Texas. Her job entailed coding and decoding intelligence dispatches, maintaining control of codebooks, and breaking intercepted messages. Like Elizebeth Friedman, she found military brainwork to be refreshingly different from the idle and decorative life she had been brought up in. When she was sent to Washington to retrieve some secret material—requiring eight days round-trip on a train—she wrote to her mother-in-law that "at times I have to laugh. It is all so foreign to my training, to my family's old fashioned notions about what and where a woman's place in this world is, etc., yet none of these things seem to shock the family now. I suppose it is the War. I am afraid I will never be contented to sit down with out something to do, even when this war is over and we are all home again."

She noted with a bit of gloating, "Well, I got what I went after, and then some—and I can't help feeling a little puffed up about it," and continued by reflecting, "This is a man's size job, but I seem to be getting away with it, and I am going to see it through."

Parker Hitt, hearing about her work, wrote to congratulate her. "I am rather expecting to find you commanding Fort Sam Houston on my return...Good work, old girl!"

Like William Friedman, Parker Hitt was a champion of women and a believer in women's intellectual abilities as well as their bedrock stamina. In Europe, Parker Hitt was charged with overseeing battlefield communications for the Army's Signal Corps. The Americans, British, and French strung phone lines around Europe and needed telephone operators to connect the calls. Switchboard operation was women's work, and male soldiers refused to do it. French operators were not as adept as American ones, so the Signal Corps recruited U.S. switchboard operators who were bilingual in English and French and loaded them into ships bound for Europe. Known as the "Hello Girls," these were the first American women other than nurses to be sent by the U.S. military into harm's way. The officers whose calls they connected often prefaced their conversations by saying, "Thank Heaven you're here!" Parker Hitt pushed for the Hello Girls to be allowed to prove their competence and courage. They did so, remaining at their posts even when ordered to evacuate during bombing in Paris, and moving to the front lines, where they worked the switchboards during explosions and fires.

Elizebeth Friedman also wanted to serve her country. In 1917 she wrote the Navy asking to work in intelligence, but Fabyan, unbeknownst to the Friedmans, opened their mail and censored any letters that might weaken his grip on his star code-breaking team. For quite some time he also prevented the Army from contacting William, though eventually the U.S. Army managed to commission William as a first lieutenant. In May 1918, William, too, was sent to France, where he performed valuable service, developing codes for front-line use. He also studied German codes and began to immerse himself in the European tradition of cryptography, which is the term for code making, and cryptanalysis, the term he coined for code breaking. (The word "cryptology" embraces both.)

After the armistice, William Friedman had become one of the few people in the country who understood how to disguise military communications, something the United States knew it needed to get better at. The Army endeavored to hire both Friedmans, offering William a salary of $3,000 and Elizebeth a position at half that, $1,520. The Friedmans were eager to extricate themselves from Fabyan, whom Elizebeth regarded as a "vile creature." They accepted six-month contracts and in early 1921 gladly moved to Washington, where they attended the theater several times a week, found a house in the suburbs they loved, enjoyed the milder mid-Atlantic weather—a welcome change from Illinois—and worked on strengthening the Army's signaling systems. William Friedman was hired full-time and would stay with the Army Signal Corps for more than thirty years.

The interwar period was not an auspicious time for American code breaking, however. While other countries continued to run black chambers—during World War I, England's Royal Navy had a secret operation called Room 40, which later merged with Army intelligence and became the Government Code and Cypher School—U.S. military intelligence maintained only a tiny "cipher bureau," funded jointly by the War and State Departments. The operation was run by Herbert O. Yardley, the former telegraph clerk, who taught himself cryptanalysis and in 1919, at age thirty, set up shop in New York. He called it the Code Compiling Company and ran it out of a house at 141 East Thirty-Seventh Street. Yardley's employees were mostly women—foreign-language teachers plucked from the New York City public school system—who often were escorted to their job interviews by nervous parents wondering what their daughters would be doing, exactly, in an unmarked midtown brownstone.

Yardley was a genial and charismatic man of irregular habits who drank often, slept late, worked in his undershirt, and had an affair with an employee whom he later married. But he was effective. His triumph came when he broke a diplomatic code that gave the United States access to the Japanese negotiating position during the 1921–1922 Washington naval conference. During this uneasy interwar period, major governments were negotiating how much naval tonnage certain countries would be accorded. Yardley ascertained that the Japanese would accept less tonnage than they were publicly holding out for, a major intelligence coup and one the United States and Great Britain took advantage of. But when Herbert Hoover was elected U.S. president in 1928, Henry Stimson—Hoover's new secretary of state—was shocked to learn that Yardley's bureau was penetrating the private diplomatic missives of other countries. Stimson in 1929 shuttered the operation, cutting off State

Department funding and primly explaining that gentlemen do not read one another's mail—something European gentlemen did all the time, of course, and had been doing for hundreds of years.

Yardley, outraged and out of a job, in 1931 published a tell-all called *The American Black Chamber*, which became a bestseller in the United States and Japan. The U.S. Army managed to retain a shoestring code-breaking operation by moving the outfit to Washington, keeping mum about it, and putting William Friedman in charge. Already engaged in making codes for the Army, Friedman would now break them as well, heading up a unit called the Signal Intelligence Service. He inherited Yardley's files as well as a fierce contempt for his predecessor, never losing an opportunity to disparage Yardley in an official memo or history and ridicule his code-breaking abilities.

Elizebeth Friedman, having given birth to their first child, thought she now might stay home and peacefully write a book for children on the origin of the alphabet. But there were so few people who could do what she could—and so many entities who needed her skills—that this plan did not last. In 1924, Edward Beale McLean, publisher of the *Washington Post*, engaged the Friedmans to develop a code for his private use. They accepted with the idea that William would direct and Elizebeth, typical of wives of the time, would do most of the day-to-day work. It promised to be a cozy project involving joint work by the fireplace in the evenings; the problem came when McLean proved reluctant to pay. They abandoned the project, as Elizebeth put it, "weary of very wealthy men and their dealings in money matters."

* * *

If Elizebeth seemed consigned to second fiddle, she soon found herself carving out a niche far more high-profile than that of her husband. The Great War was over, but a new war was beginning: the war against alcohol and the criminals who sold it to a thirsty public. In 1919, the Anti-Saloon League succeeded in pushing through the Eighteenth Amendment. Prohibition outlawed the manufacture, transportation, and sale of alcohol, but—importantly—did not outlaw its consumption. This meant American citizens could drink alcohol if they could find a way to get it. This loophole created a tempting criminal opportunity. Foreign distillers partnered with American gangsters to ship contraband alcohol to U.S. shores. Elaborate maritime operations evolved, in which a ship carrying a big cargo of alcohol would station itself in international waters, out of reach of American law enforcement, and use coded radio messages to communicate with smaller boats, which would douse their lights and dart out to collect a shipment. It was called rum-running, and it was a wildly profitable endeavor, lucrative on the level of modern-day drug cartels. Criminals, by the way, are another constituency that very much like to use codes.

Against these crime syndicates, Elizebeth Friedman—slender, sleepy-eyed, adventurous, dashing—became the government's secret weapon. Her law enforcement career began in 1927 when the Coast Guard asked William Friedman to decipher the rumrunners' messages. He was busy with his Army work, so they turned to her. This, as Elizebeth wryly put it, became a pattern: "If we can't have William Friedman we will make use of his brains through his wife." The Department of Justice made her a "special agent," a flexible title that permitted her to work at home, where the Fried-mans now had two children. When her workload increased and she felt obliged to move into an office, she hired a housekeeper and a nurse. Variously employed by the Justice and Treasury Departments, the Customs Bureau, the Coast Guard, and other agencies—responsibility for enforcing Prohibition bounced around—she broke the rum-running messages, resulting in successful prosecutions in which she was called to testify as an expert witness in court. After Prohibition was repealed, she worked on other cases involving smuggling and organized crime, testifying against criminals dangerous enough that she sometimes needed government protection. Once when she was late getting home, William joked to their children that perhaps their mother had been "taken for a ride." At the Coast Guard, she trained men and built an antismuggling cryptologic unit.

Needless to say, the spectacle of a lady law enforcement code breaker proved irresistible to the news media. In the 1930s Elizebeth Friedman attracted articles with headlines like KEY WOMAN OF THE T-MEN, LOCAL MATRON DECODES CRYPTIC MESSAGE FOR TREASURY DEPARTMENT, and THE WOMAN ALL SPIES IN U.S. FEAR. Elizebeth felt the coverage was "lurid" and resented it. She noted that one article called her a "pretty middle aged woman," while another portrayed her as a "pretty young woman" in a frilly pink dress. Both depictions dismayed her. She also knew—after Yardley's *American Black Chamber* debacle—the harm that can come when cryptanalytic achievements are revealed. The publicity made colleagues jealous and created unease in a small intelligence community already wary of public notice.

She also ran into sexist condescension. She sometimes gave her husband old Coast Guard intercepts to use for training his own Army cryptanalysts. For her pains, some of her husband's trainees suspected William was secretly doing Elizebeth's work. "Our impression—and I think it was a mistaken one—was that much of her success was a result of Mr. Friedman's effort," one of his trainees, Solomon Kullback, later admitted. "We thought...that Mr. Friedman really was responsible in working with her on a lot of these problems." The newspapers, in contrast, liked to put it about that Elizebeth had trained William. But in fact—while they did play up their presence as Washington's premier cryptologic team, and liked to send cryptographic Christmas cards and hold dinner parties in which guests had to solve a cipher to move to the next course—the two could not always discuss their real work with each other, because both were deciphering secret material for different branches of a growing federal bureaucracy whose agencies were distrustful and often at odds.

* * *

The U.S. Navy, meanwhile, was developing its own female secret weapon, as part of a code-breaking operation that, true to the prevailing climate, was kept jealously separate from the Army or any other rival entity. Upon America's entry into World War I, the country had struggled to quickly enlarge its modest career Navy, and created a men's naval reserve that permitted civilian men to serve during wartime, often as specialists with expertise in areas such as math or science. Even this influx wasn't enough, however, and it occurred to Secretary of the Navy Josephus Daniels to wonder aloud whether there was any law "that says a yeoman must be a man." Remarkably, there was not. Nowhere in the Naval Reserve Act of 1916 did it say that a naval yeoman had to be male.

Thanks to that loophole, American women were permitted to enlist in the naval reserves during World War I, and the designation "Yeoman (F)" was created. The move was controversial, even shocking, to the public, but many more women hastened to enlist than the Navy had expected. To the women's disappointment, they were not allowed to serve on ships (nurses, who were in a different category, could do so) but mostly worked as clerks and stenographers, facilitating the towering stacks of paperwork that the naval bureaucracy generates—the original yeoman's work. During the first global conflict of the twentieth century, eleven thousand American women served as Yeoman (F)—also called yeomanettes.

Among these was Agnes Meyer, a brilliant young teacher who would become one of the great cryptanalysts of all time. Born in 1889 in Illinois, Meyer attended Otterbein College, and then Ohio State University, where she studied mathematics, music, physics, and foreign languages. For what it's worth, she was extremely beautiful, with long hair that she swept into a chignon, and an angular, chiseled face. She was heading up the Math Department at Amarillo High School, in Texas, when the United States declared war on Germany in 1917. Enlisting at age twenty-eight, one of the first women to do so, she quickly became a Chief Yeoman, the highest rating available to a woman. She started out as a stenographer but soon was assigned to the Navy's postal and censorship office, reviewing U.S. telegrams and letters to make sure they didn't contain security breaches. The Navy transferred her to its code and signal section at a time when the unit's purpose was protecting naval communications—encoding America's own messages. Like William

Friedman, Agnes Meyer got her start making codes, which is the best possible training for learning how to break them.

After the Great War ended, Agnes Meyer was discharged along with the other reservists. (Congress, ungratefully, amended the reservist law to make sure it included the word "male.") Her abilities were such that the Navy promptly hired her back as a civilian. As part of her duties, she was given the task of testing what were known as "nut jobs": machines marketed by inventors offering so-called foolproof enciphering systems.

* * *

Technically, there are two kinds of secret message systems. One kind is a code, in which an entire word or phrase is replaced by another word, a series of letters, or a string of numbers, known as a "code group." A code may be used for secrecy, but also for brevity and truncation. Shorthand is a code in precisely this way and so, often, is modern-day texting. Common phrases, even long ones, can be compressed into short code groups, making messages faster and—when using cable, as many people did in the early decades of the twentieth century—cheaper to send. Saving money has always been important to governments, so the compression advantage is a big deal. Cable companies typically charged by the word, so the fact that stock phrases like "your request of last month has been approved" could be boiled down to a code group, as could the names of places or people or units, meant governments could save a good bit of money when sending telegrams. In the War Department's "general address and signature" code that was employed in 1925, for example, the word "cavalry" was HUNUG, "Pursuit Squadron" was LYLIV, "Bombardment Squadron" was BEBAX, "Wagon Company" was DIGUF, "U.S. Naval Academy" was HOFOW, and "Fourth Division Air Service" was BABAZ. (Texting uses codes, like OMG and IMO, for much the same reason: brevity and, at times, concealment.) The best code is one in which code groups are randomly assigned, with no rhyme or reason that an enemy can discern. Codes are compiled and kept in codebooks, not unlike dictionaries, where the encoder can look up the word or phrase and the corresponding group that stands for it. But even random codes have an obvious vulnerability: Constant repetitious use of the same code groups in messages enables code breakers to tease out their meaning from context or position.

The other type of system is called a cipher, in which a single letter—or number—is replaced by another single letter or number. Ciphers can be created by scrambling letters, which is called transposition—turning the word "brain," for example, into "nirab." Or a cipher can be achieved by replacing individual units with other units, a method called substitution: By substituting X for b, T for r, V for a, O for i, and P for n, for example, brain becomes "XTVOP." For centuries, ciphers were created by hand, often by those clever Renaissance men who would line alphabets up against one another and create boxes and tables that gave a way to substitute one letter for another. But when radio and telegraph came along, messages could be sent much, much faster than a wigwag flag could do. Machines were needed that could encipher rapidly; and, because it became easier to spot simple patterns when so many messages were being sent and intercepted, more complicated ciphers were needed. People can make complex ciphers, but people make mistakes. Machines are less likely to do so. These machines created an early form of what would later be called encryption, which meant that people who broke them might be described as an early version of what would later be called hackers.

And that's what Agnes Meyer was. She hacked the nut jobs, broke enemy devices and machines that inventors were peddling to the U.S. Navy, uncovering their flaws and weaknesses. The proffered inventions included a supposedly invulnerable machine invented by Edward Hebern, a fly-by-night character who at one point had been imprisoned for horse thievery. Agnes easily broke an "unbreakable" message Hebern had put in a public advertisement. Impressed, Hebern lured Agnes away to help him develop a better one, a job she might have accepted because she was getting discouraged by her slim chances for promotion as a female civilian in a male military service. As it happened, Elizabeth Friedman was temporarily brought in by the Navy to substitute during Agnes's absence, establishing a rivalry between the two women. Elizabeth (who tended to downplay her own talents while touting those of her husband) scorned what she saw as Agnes's overweening ambition and her lack of loyalty to public service. She saw Agnes as "a person who thought only of furthering herself" and scoffed that she had "fallen for" the offer from Hebern. (Nothing if not frank, Elizabeth also thought Meyer's Navy boss, Laurance Safford, was a "nut.") In 1924, Hebern marketed his improved machine back to the Navy. William Friedman, summoned in to test it, managed to break it.

This cemented the rivalry. Agnes soon quit Hebern and returned to her civilian Navy post, and from then on, Agnes despised William Friedman. Part of her resentment stemmed from competitive instincts common among people who rely on their wits for a living (rivalry between code breakers is not unlike that in many university academic departments), but part stemmed from the fact that William Friedman was treated better by the Army than she was by the Navy.

"Friedman was always two, three [pay] grades ahead of her, and I think that her feeling that that was sexist was probably true," said Captain Thomas Dyer, a Navy cryptanalyst who was trained by Agnes. Dyer described her as "absolutely brilliant" and ventured that she "was fully" the equal of Friedman.

The group of people at this time who understood codes and ciphers made up a very small universe—interdependent, claustrophobic, jealous. Everybody knew everybody and had an opinion about what everyone else was (or was not) capable of. Like so many others, Agnes Meyer had done a stint at Riverbank; in 1920 George Fabyan wrote a complimentary letter to the Navy Department, saying, "We were very favorably impressed with the young lady," and adding that he'd be delighted to hire her if at any point the Navy wanted to release her. She also sojourned briefly with Herbert Yardley. In her own civilian post, Agnes would go on to train virtually all of the major male naval code breakers who became famous for their World War II exploits. "She not only trained most of the leading naval cryptanalysts of World War II," wrote one intelligence officer, Edwin Layton, "but they were all agreed that none exceeded her gifted accomplishments." Though she did not get any public credit, she laid much of the groundwork that made their exploits possible. She also helped design the Navy's first ciphering machine, a feat for which she and her co-designer later were awarded $15,000 by Congress. She married a Washington lawyer, Michael Driscoll; like the Friedmans, the Driscolls were a two-career couple a half century before two-career couples were a thing.

By the 1920s the American Navy was beginning to harbor its own code-breaking ambitions, sensing in Japan a future naval adversary. Japan had defeated Russia in 1905 and it clearly wanted to build a Pacific fleet to rival or surpass America's; and, lacking the natural resources such as oil, iron, and rubber it needed to become a dominant world power, it seemed bound to go looking for those resources elsewhere in the Pacific region, threatening U.S. territories including Guam and the Philippines. And so U.S. Navy ships began intercepting Japanese messages and building more Pacific intercept stations. In New York City, in 1923, naval intelligence officers surreptitiously raided the office of the Japanese consul general, where they rifled a steamer trunk, found a 1918 naval codebook, stole it, photographed each page, put the book back, and sent the pages to Washington. All of which ended up in the competent hands of Agnes Meyer Driscoll.

By then, a "research desk" had been set up in Room 1645 of Navy headquarters, located in a large, low wooden building along what is now Constitution Avenue. Cryptanalytic offices always have vague names, to disguise their true purpose. The desk was staffed by a small number of people, civilians as well as some naval officers, but the problem for the officers was that paging through codebooks, however helpful to the larger mission, was bad for their careers. This was even truer in the Navy than in other branches of the military. Cryptanalysis is shore duty, an office job. In the U.S. Navy, if you were a career naval officer with any ambition, shore duty was not what you wanted. What you wanted was sea duty and a position as "line officer" and commander. Being a specialist of any sort was not a viable career path. And so officers came and went from the research desk, studying for a couple of years and then shipping out to sea, as their careers required.

It was Agnes Meyer Driscoll who sat perpetually at the research desk, a civilian and a woman and therefore sentenced to permanent shore duty, unlocking the secrets of the naval fleet code that Japan was honing during the 1920s and 1930s. A fleet code is the main system that bases, ships, and organizations use to talk with one another about strategy and tactics, logistics, intelligence, morale, ship movements, situation reports, even weather: anything and everything a commander deems important. Agnes Driscoll studied the stolen codebook day after day, year after year. One of her trainees, Joe Rochefort, later remembered her turning pages with the tip of a pencil eraser, flipping back and forth. She cursed like a—well, like a sailor. She was seen as aloof and sensitive to any kind of treatment that smacked of being patronizing, a trait that Edwin Layton attributed to her discomfort at being a woman so outnumbered by military men. She liked to say that "any man-made code could be broken by a woman." She and her husband didn't socialize much, which is a handicap in the prestige-conscious Navy, where men's careers often are aided by wives who entertain admirals and throw dinner parties.

Even so, the men fetishized her genius. Studying the codebook, never having seen the Pacific Ocean or a single Japanese vessel, Agnes Driscoll became fluent in Japanese ship names and—importantly—cryptographic habits.

She also figured out how the Japanese disguised their fleet code, using a method called "superencipherment" that involves both a code and a cipher. For the main fleet code, the Japanese were using a large codebook containing thousands of three-character code groups that stood for Japanese words, syllables, phrases, and even punctuation marks. Once a clerk wrote out a coded message, he then enciphered each character, somehow, so that the code group would be sent as a different set of characters entirely. The "research desk" knew what the code groups stood for, thanks to the theft of the codebook, but that didn't do them any good. When they intercepted an actual message, the groups they were looking at had been enciphered. They couldn't make heads or tails of a message unless they could figure out how to get rid of the encipherment and restore each code group to its original form.

The tiny Navy team—Driscoll, one or two officers, a couple of clerk-typists, a translator—worked for years to achieve this. Joe Rochefort said that the process literally made him sick. It destroyed his appetite and caused him to lose twenty pounds, sitting every day in what he called the "staring process," smoking everything he could get his hands on: cigars, cigarettes, a pipe. After work he would have to lie down for several hours before he could eat. That the team succeeded he attributed to Miss Aggie—as Agnes Driscoll was called—who discerned that the encipherment was accomplished by transposing or switching the position of the characters.

Successful code breaking often comes down to diagnostics—the ability to see the whole rather than just the parts, to discern the underlying system the enemy has devised to disguise its communications. The Japanese, Agnes diagnosed, were encoding their messages and then using something called columnar transposition, which involves writing the code groups out horizontally but transmitting them vertically, aided by a grid with certain spaces blacked out, whose design changed often. "Mrs. Driscoll was responsible for the initial solution and for most of the solution of the new ciphers and 'transposition forms,'" said Rochefort, who, as a naval officer, technically headed the desk. The intelligence gleaned provided insight into Japanese fuel supplies, ship accidents, aviation advances, naval maneuvers, and—importantly—strategy for conducting combat operations against the United States. It also revealed that the Japanese were alarmingly well versed in America's own naval war plan.

Mastering the fleet code was a never-ending undertaking, for the Japanese and for Agnes Meyer Driscoll. As a security measure, the Japanese Navy periodically changed its codebooks—burning the old codebooks, printing new books, and distributing them to every ship, office, and island. When this happened, each word would be assigned a new code group, and the American code breakers would have to start from scratch. Sometimes, the changes were even bigger: In 1931, Thomas Dyer, then a trainee, was puzzling over a new intercept; Driscoll walked up behind him, took his work sheet, looked at it, and said, "The reason you're not getting anywhere is because this is a new code."

She was right. The Japanese had changed their system, coming up with code groups that were longer and organized in a tougher, more complex way. Puzzling out the new system took three years, and once again it was Driscoll who did the bulk of the work in cracking one of the most complicated systems ever seen. "Mrs. Driscoll got the first break as usual," said Rochefort. Her success "was the most difficult cryptanalytic task ever performed up to that date," Laurance Safford wrote later. When Edwin Layton rotated into the unit, he came to appreciate what he called the "magnitude" of Agnes's contribution and described it as "spectacular."

Her feat had enormous real-world impact. In 1936, Driscoll's efforts revealed that the Japanese had refitted a battleship that now could travel in excess of twenty-six knots. The United States didn't have a ship that fast, so the Navy upgraded a new class of battleships to exceed that speed. It was a major piece of intelligence and one that justified the entire expense of setting up a research desk. The naval code-breaking office was gradually enlarged, but in 1937, Agnes Driscoll was the only person who knew something about everything. "There is...only one fully trained individual among the permanent force who is capable of attacking any problem," an official history noted.

Then the code system changed again, and this change was even bigger. On June 1, 1939, the Japanese fleet began using a code that the Allies came to call JN-25. The Japanese—who had moved to using numbers rather than characters—now employed a massive codebook containing about thirty thousand five-digit groups. They also had a new way of enciphering. Before the code was sent, each code group was enciphered by using math to apply an "additive."

Here is how the additive method worked: When a Japanese cryptographer began encoding a single message, he would look in the codebook and find the five-digit group that stood for the word (or syllable or phrase or punctuation mark) he wanted. He would repeat that process until he got to the end of the message. Then he would get out a different book, called an additive book, turn to a page—selected at random—pick a five-digit number, and add that to the first code group. He would add the next additive to the second. And so on. The Japanese code makers used a peculiar kind of math called noncarrying or "false" addition. There was no carrying of digits, so 8 plus 7 would equal 5, rather than 15. If the code group for "maru" was, say, 13563, and the additive was 24968, the resulting group would be 37421 (1 + 2 =3; 3 + 4 = 7; 5 + 9 = 4; 6 + 6 = 2; 3 + 8 = 1). That was the group of digits that would be radioed. To crack a message, the Americans had to figure out the additive and subtract it to get the code group. Then they had to figure out what the code group stood for.

Once again, it was Agnes Driscoll who diagnosed the new system. Neither she nor anybody in the Navy operation had seen an additive cipher—everything up to then had been transposition, or switching—but she figured it out. It took her less than a year to make a dent. A March 1 status report for the unit "GYP-1" stated that for the "5-number system"—an early title for JN-25—"First break [was] made by Mrs. Driscoll. Solution progressing satisfactorily." She worked on it for several more months before being transferred in late 1940 to German systems—a promotion in the sense that the Atlantic was beginning to emerge as the hot spot. The research team continued working their way through JN-25, using her methods.

The process of stripping additives and discerning the meaning of code groups was laborious and excruciating. Years after World War II ended, American code breakers who worked in Hawaii and Australia were still arguing with their D.C. counterparts over what certain code groups stood for. Much like the women who trained the men who would get to do the wartime flying, much like Elizabeth Friedman over at the

Coast Guard, Agnes Driscoll taught the men in the field who did this. "In the Navy she was without peer as a cryptanalyst," wrote Edwin Layton, who headed naval intelligence for Admiral Nimitz, the chief naval commander in the Pacific during the war. In December 1940, both code and cipher were changed, to a system the Allies called JN-25B; the team stripped the additives and built a partial bank of code words. Then, in early December 1941—days before Pearl Harbor—the additive books were changed. The codebooks were not. The U.S. Navy was able to recover a certain amount of the new system—but not enough—before the attack on Pearl Harbor happened and all hell broke loose.

"If the Japanese Navy had changed the code-book along with the cipher keys on 1 December 1941, there is no telling how badly the war in the Pacific would have gone," said Laurance Safford.

As crushing as Pearl Harbor was, it was thanks in large part to Driscoll's decades-long detective work—and to the example Elizebeth Friedman set for other women—that America did not enter the Second World War quite as blind as it might have seemed.

CHAPTER THREE

The Most Difficult Problem

September 1940

Poland was occupied and had been for a year. Czechoslovakia had put up little or no resistance to being partitioned. The Nazi war machine had overrun Norway and Denmark, defeated Belgium and so many others, and proceeded to march into Paris, where Nazi officers were drinking café au lait and popping champagne corks in the finest restaurants on the Champs-Élysées. England was holding out, but barely, pummeled by German air raids and braced for an anticipated sea invasion launched from the shores of occupied France. Japan was shouldering its way through China and around the Pacific, chasing resources and seeking to establish a "new order" in which Asian nations would be rescued from Western domination and dominated, instead, by Japan. And here—in crowded U.S. Army offices in downtown Washington—a young civilian woman was patiently standing, waiting for a group of men to stop talking and notice her. She had something urgent she needed to tell them, but, shy and reluctant to interrupt, she waited for a pause.

The office was nothing to brag about—just a few rooms tucked away in a Washington eyesore known as the Munitions Building, erected in 1918 as a headquarters for the War Department. The Munitions Building and the U.S. Navy headquarters were side by side, as it happened, both constructed as "temporary" wartime structures during World War I and both still in service even now that the Great War was long over, together dominating the part of the capital city between Foggy Bottom and the National Mall. The twin buildings had concrete facades and a series of thin wings that stretched backward, perpendicular to the facade. Working in them was like working in a multistory warehouse.

On the first floor, the wings of the buildings had long narrow corridors and doors through which bicycle messengers often burst, so people had to be careful not to be knocked down by bag-wielding boys on bikes. On upper floors, wooden desks lined open workrooms, and large windows admitted the Washington air, for better or worse, depending on the season and humidity level. Technically the whole edifice was known as the Main Navy and Munitions Building. Washington's physical wartime expansion had not—yet—been effected; the Pentagon did not yet exist, nor had Arlington Hall been requisitioned, and in 1940 these two squat ugly buildings housed, in effect, the country's entire military brain trust.

On an upper floor, occupying the back of one wing of the Munitions Building, a handful of rooms had been allocated to the U.S. Army's code-breaking operations, which had grown considerably in the past year but were still modest enough to be contained in such a small space. One room held a bank of office machines used for tasks like sorting and collating; others held locked file cabinets with intercepted messages. In most of the others, people sat quietly at tables, heads down, sometimes smoking or drinking coffee, working with pencils on lined or crosshatched paper. Apart from the machines, the offices consisted of the usual government-issue assortment of scuffed desks, battered cabinets, and rotary phones. Somewhat by chance, the U.S. Navy's code-breaking force was located in an adjacent wing. While capable of cooperation, the rival code-breaking units were still marked by infighting, paranoia, and personality clashes. Unlike the Navy's code breakers, the Army team was mostly civilian, a hodgepodge of mathematicians, ex-schoolteachers, linguists, and clerical workers. While the Navy tackled Japanese naval ciphers, the Army code breakers were attempting to penetrate systems used by military and diplomatic officials in Italy, Germany, Japan, and Mexico.

It was clear enough that sooner or later the United States would formally enter what was shaping up to be a second world war, and on this Friday afternoon, in the heat of late summer, the atmosphere in the rooms was thick with urgency.

At the center of the Army's operation was William Friedman. Originally hired to ensure that the U.S. Army developed codes that were more sophisticated and secure than wigwag, Friedman had learned to break codes better than almost anybody in the world. He had hired most of the people in this office. The people laboring at the tables revered him. Friedman's Army superiors sometimes called him Bill, but the people who worked for him always called him Mr. Friedman. Sometimes in private they called him "Uncle Willie," but none would have dared to do so in his presence. He was sensitive, easily offended. Meticulous in his work habits, he was good about entrusting important jobs to capable others but did not give compliments easily. An excellent tennis player and ballroom dancer, he had a thin mustache and liked to wear a bow tie and two-tone shoes, and he was fanatic about precision; he hated it when people used "repeat" as a noun and insisted they say "repetition."

Friedman, now in his late forties, was a legend among the still-small global community of people involved in the making and breaking of codes and ciphers. After leaving Riverbank he had assembled one of the few known libraries on the topic. The literature included *Cours de Cryptographie* by General Marcel Givierge of France; *Manuale di Crittografia* by General Luigi Sacco of Italy; and *Elements of Cryptography* by a French captain, Roger Baudouin, that had been smuggled out of France just before Paris fell. Friedman had written many of the most important treatises himself—top secret monographs including *Elements of Cryptanalysis, The Principles of Indirect Symmetry of Position in Secondary Alphabets and Their Application in the Solution of Polyalphabetic Substitution Ciphers*, and his masterwork: *The Index of Coincidence and Its Applications in Cryptanalysis*. He also had written training manuals that were treated as kinds of bibles.

Over the course of a decade Friedman had assembled a tiny band of acolytes. In 1930, shortly after the abrupt closure of Herbert Yardley's office, his bosses had given him funds to hire three young mathematicians: Frank Rowlett, a southerner who was teaching in Rocky Mount, Virginia, best known for its production of moonshine; and Abraham Sinkov and Solomon Kullback, friends who had attended high school and City College of New York together. Friedman wanted his staff to be young, because he knew it would take them years to master their discipline. Together with John Hurt, a Virginian who knew Japanese and could translate deciphered Japanese messages into English, the men had spent nearly a decade studying Friedman's methods of "attacking" codes and ciphers. Funding was never abundant, and in the worst of the Depression, they sometimes had to supply their own penny pencils and bring in scrap paper from home.

As the staff expanded, Friedman had done something else: He had begun hiring women. There were several reasons for his willingness to do so. One was plain availability. In the 1930s—well before the war started—Roosevelt's New Deal had begun drawing women workers to Washington, where the expanding federal government proved more of an equal opportunity employer than the private sector. Discrimination existed in government hiring, to be sure, but for a woman, the advantage of applying for a federal job was that it entailed taking a standard civil service examination. Women took the same exam men did. Federal agencies were given access to test scores and made job offers accordingly. The 1920 census found that nearly 40 percent of employed people in Washington were female.

Equally important was that William Friedman was the sort of man who liked working with intelligent women, as evidenced by his own marriage to Elizebeth, now employed at the Coast Guard as both a code breaker and what might today be called a communications security consultant. At the moment, the Coast Guard's mission was enforcing "neutrality," which meant that Elizebeth's unit was deluged with message traffic from all sorts of ships plying the Atlantic. A valuable utility player for other agencies as well, Elizebeth also designed the code-making unit for the Coordinator of Information—the nation's new spy service—soon to be renamed the Office of Strategic Services.

Elizebeth's example was even more valuable to her husband than William Friedman knew. In October 1939, following the outbreak of war in Europe, the Army had given him the funds to further enlarge his cryptanalytic staff. One of these early hires was a woman named Wilma Berryman, who had been attracted to the field thanks to Elizebeth's renown. Berryman hailed from Beech Bottom, West Virginia, and graduated with a math degree from Bethany College. Though she had been trained to teach high school math, during the Depression the only job she could find was teaching first grade to a classroom of forty-five children. When her husband took a job in Washington, Berryman went to work in the payroll office of the department store Woodward & Lothrop—near the pocketbook shelves—then at the Census Bureau, and at a succession of other agencies. But when she read in the Washington *Evening Star* about Elizebeth Friedman's exploits (the article also mentioned William, but it was the wife's example that struck her) it awakened something in her. She began to envision another future.

Asking around, Wilma Berryman found out that the U.S. Navy had developed a correspondence course, to train its own officers but also to permit hobbyists and other potential civilian applicants to homeschool themselves in breaking codes. The purpose was not so much to teach the subject as to locate talent and winnow out those who had none. Wilma Berryman spent several years tackling the course, sending away for lessons, completing the exercises on her own time, and sending them back. The two code-breaking operations were porous enough that her answers found their way to William Friedman, who was always scrutinizing civil service rolls and whatever other sources he could gain access to, evaluating test scores and looking for the right abilities. Upon being hired, Wilma Berryman was put on the Italian desk—which is to say, she was given a beginner's textbook on the Italian language and plunged willy-nilly into the secret communications of the fascist government of Italy. Every morning, one of Berryman's colleagues liked to come over to her table and ask her, "How's Benito doing this morning?"

That was how the Friedman operation worked: It was a teach-yourself kind of place. Newcomers would spend the morning studying training manuals, puzzling out the answers to questions like "What four things were thought by Captain Hitt to be essential to cryptanalytic success?" (perseverance, careful methods of analysis, intuition, and luck) and "What two places in every message lend themselves more readily to successful attack by the assumption of words than do any other places?" (the beginning and the end). They spent afternoons attacking actual codes.

Wilma Berryman loved it.

So did Delia Ann Taylor, a tall, brainy midwesterner who graduated from Sweet Briar College in Virginia and had a master's degree from Smith College. Working near her was Mary Louise Prather, the daughter of a genteel family that had fallen on hard times. It was Prather's job to work the office machines—sorter, reproducer, tabulator, keypunch—that Friedman, a master of winkling things out of a closefisted government bureaucracy, had persuaded his bosses to buy him. While Prather's might seem like a menial job—running office machines was considered women's work—these were not conventional machines but had been modified to assist with sorting enemy messages.

Prather also filed the intercepts, which themselves were a kind of contraband. Since the United States was not, strictly speaking, at war, collecting radio and cable transmissions of foreign diplomats was not, strictly speaking, legal. The Communications Act of 1934 imposed "severe penalties for interception of diplomatic traffic," as one memo noted, but the code breakers had decided to overlook that. Friedman's Army superior, the now major general Joseph Mauborgne, felt the law could be ignored. Even so, intercepts were hard to come by: The Army did not yet have many clandestine radio intercept stations of its own, so they got some messages from the Navy and some from friendly cable companies who handed them over under the table. Prather kept a careful log of every last one.

And there was twenty-seven-year-old Genevieve Marie Grotjan, hired as a "junior cryptanalyst" in October 1939 for a salary of $2,000 per year. It was Grotjan who was standing waiting for the men in the Munitions Building to notice her. A native of Buffalo, New York, Grotjan had been a brilliant all-around student at Buffalo's Bennett High School, where she delivered the salutatorian's address in the customary Latin. She received a Regents scholarship to attend the University of Buffalo, where she majored in math and belonged to the International Relations Club. Graduating summa cum laude in 1938, she won a math prize, received a teaching assistantship to do graduate work, and aspired to teach college math. Like so many women of her day, however, Grotjan was unable to find a university math department willing to hire her. So she came to Washington and was hired as a statistical clerk at an obscure agency called the Railroad Retirement Board, where it was her happy task—she enjoyed it—to calculate pensions. When she took a math exam to secure a routine pay raise, her score attracted Friedman's attention. She received a call from the Signal Intelligence Service and was asked if she would like a job in the "code section." Grotjan didn't know what any of that meant, but she said yes.

Many of the code breakers were social with one another, but Grotjan was not one of these. Shy and introverted, she favored rimless eyeglasses, high-collared blouses, and a pragmatic hairstyle that consisted of tight blond pin curls crimped in a halo around her forehead. She rented a room in a boardinghouse at 1439 Euclid Street, a modest section of northwest Washington.

After less than a year on the job, however, Grotjan was shaping up to be one of the team's most promising code breakers. She was known for her thoroughness, powers of observation, and attention to detail. Humble and reticent, she possessed the pure soul of one who lives for numbers, oblivious to office politics and rivalries, which did exist. Based on the aptitude she demonstrated, she was assigned to the most pressing problem Friedman's office had undertaken: the cipher system used by Japanese diplomats around the world. It was a completely different system from that of their military counterparts: While the Imperial Japanese Navy often used laborious pen-and-paper systems, which involved a lot of adding and figuring, Japanese diplomats favored the newer machine-generated ciphers. The small team that Grotjan belonged to was trying to do something that almost certainly had never been done: reconstruct an unknown machine without having seen it or even a piece of it—not so much as a blueprint or drawing. They were attempting to penetrate the machine's inner workings by scrutinizing its pilfered output, sitting at tables looking at strings of random-seeming letters.

There were many challenges to this task, chief among them that the Japanese machine had been produced in an environment to which they did not have access. This was an era when governments and businesses alike were turning to cipher machines to keep their messages secure from intermediary parties (anybody from a Morse operator to an actual spy) and inventors were always designing new machines to enable them to do that. Friedman's office kept its own "nut file," recording the outlandish systems that hobbyists tried to sell them. Usually, the inventor wanted either a job or a million dollars and threatened to sell to the Russians or Germany if the U.S. government didn't bite. Friedman, like his Navy rival Agnes Driscoll, was a master at finding the weaknesses of these machines, and his acolytes often could break a nut's systems in a matter of hours.

But some machines on the Western market were first-rate. One of the best was called Enigma. Envisioned as a tool for bankers, the Enigma, invented by a German engineer and marketed by a German company in the 1920s, had been adapted for military use by the Nazis. In 1933, Hitler ordered it taken off the commercial market, however, so his military could have sole access. Many military machines, like Enigma, were small and light and sturdy, not much bigger than typewriters. Enigma in particular was a durable, portable, battery-powered device that could be lugged around and used during battle, or welded to the command center of a submarine, where it had one job and one job only: to change each letter of a message to a different letter.

This new Japanese machine had the same letter-changing mission, but nobody knew how it worked. No Westerner had laid eyes on it, or even a facsimile or prototype. The machine was not as mobile as the Enigma—not mobile at all. Unlike Enigma, it ran on electrical current and needed an outlet it could be plugged into. Only the most important Japanese embassies were given access to it—those in Washington, Berlin, London, Paris, Moscow, Rome, Geneva, Brussels, Peking, and a few other major cities.

A machine ciphering system worked well for diplomats. In the 1930s, phone calls were costly and vulnerable to being tapped. The foreign office in Tokyo often needed to send the same message to all its far-flung ambassadors, and rather than pick up a phone and make the same call, over and over, it was easier to craft a message and hand it to a clerk, who would write it out in Romaji, a phonetic version of Japanese that used roman letters to spell out the syllables, as in *ma-ru*, for merchant ship. The clerk would run "maru" through the machine, producing a new stream of letters—say, "biyo"—that could be cabled. The enciphering mechanism could be set in different positions, according to a key, or setting. The machine could be used in either enciphering or deciphering mode, so the diplomats could use their own machines to restore the message to its original meaning. They also could use it to write back to Tokyo.

The Japanese diplomats were, of course, discussing their country's war plans. They also were meeting with Hitler, Mussolini, and other key Axis leaders. If the Americans could uncover the machine's workings, they would have access to a priceless stream of insight, gossip, and strategy, involving not only Japanese intentions but those of every tyrant in Europe.

But cracking the Japanese machine was proving elusive. By the time that Genevieve Grotjan was assigned to the project, the Americans had been struggling for months. The first message in the new machine cipher had been intercepted in March 1939, emanating from the Japanese embassy in Warsaw. The code breakers had known it was coming thanks to the fact that they had broken a simpler machine cipher that the Japanese used for much of the 1930s. The Japanese referred to the first machine by the prosaic name Angooki Taipu A—Cipher Machine Type A—and so this new one was called Angooki Taipu B. The Americans called the first one Red and the second one Purple. Purple didn't work the way Red did. It was more complicated, which was why the code breakers were having such trouble with it.

The small number of Westerners who knew about the existence of the Purple cipher thought the Americans in William Friedman's tiny Japanese unit were wasting their time. The British had tried to break the Purple machine—as had the Germans—but both abandoned the job as undoable. The U.S. Navy, in the wing next door, worked on Purple for four months but decided to concentrate on JN-25. William Friedman's group of civilians were the only ones who refused to give up, or were temperamentally incapable of doing so.

The men Friedman had hired in 1930 had benefited from years of training. Now, with war raging in Europe and Asia, and America's involvement looming—ever since the fall of France, it had become inevitable—new hires like Genevieve Grotjan were flung directly into the work. Frank Rowlett, the southerner, was supervising the Purple effort. He was a big man, friendly, and tended to play up his rural-boy persona as a way of masking his strategic intelligence and competitive instincts, saying things like, "I'm just a country boy from the sticks, but..." Grotjan found him personable and easy to work for.

As he built his team, Friedman had scoured the civil service rosters. The kind of person he wanted was hard to define. He sought intelligence but also persistence. Though a penchant for crossword puzzles is sometimes seen as an indicator of code-breaking talent, Friedman scoffed at the idea that breaking codes is truly akin to solving newspaper crossword puzzles. Crossword puzzles are easy; once you get a clue, you feel spurred on, you feel encouraged. Small victories and incentives are built in. Crossword puzzles are designed to be solved, while codes and ciphers are designed to prevent solution. With codes, you have to be prepared to work for months—for years—and fail.

In September 1940, failing was exactly what they seemed to be doing. After more than a year of frustration, the only thing the code breakers knew for sure was this: One weakness of the Japanese Purple machine stemmed from the fact that Tokyo had been a little too eager to save money. In the 1930s, when Japanese cryptographers were designing the earlier Red machine, messages often were transmitted in groups of four or five letters. Groups that could be pronounced were cheaper to send. (Friedman attended the conferences where telegraph companies in different countries laid down rules of the road like these, coordinating things like costs and structure and allocation of frequencies.) To be pronounceable, a five-letter group had to contain at least two vowels. The Red machine therefore transformed vowels into vowels, and consonants into consonants, to ensure that "marus" ended up as something like "biyav" and not, say, "xbvwq." That way, the messages remained pronounceable.

Friedman's team had figured out that the old Red machine employed two mechanisms to achieve this, one of which transformed the six vowels, the other the twenty consonants. They referred to these mechanisms as the "sixes" and the "twenties." The Friedman team had managed to build a facsimile of the Red machine, using Western parts. Their facsimile worked so well that Friedman's code breakers often were able to decipher a Red message and deliver the contents to U.S. military intelligence before the Japanese code clerks had gotten the same message to their own bosses. When the Red machine began to go off-line, in 1939, American officials found it frustrating to be deprived of the fruits they had become accustomed to enjoying.

By the time the Purple machine came along, cable companies had relaxed the rule about groups needing to be pronounceable, so there was no need for sixes and twenties. Even so, new systems often contain elements of older ones: This is known as "cryptographic continuity." Banking on this, the code breakers hypothesized that the Purple machine also used two mechanisms, one that transformed six letters—any letters, not just vowels—and one that transformed twenty. Sure enough, when the

Purple intercepts started appearing, Friedman's code breakers were able to see that six letters appeared more often than others. But the twenties were the stumbling block. No matter what kind of system they conjectured, the Americans could not discern how the remaining twenty letters were enciphered.

Every code breaker had his or her method of coping with frustration. Frank Rowlett liked to go to bed early, then wake up in the middle of the night and see if inspiration struck him. William Friedman often thought of solutions while shaving; he was a big believer in the problem-solving power of the subconscious. Genevieve Grotjan was one of the most patient team members. She would sit for hours contemplating streams of letters, making notations, creating charts.

William Friedman had taught his students that if you scrutinize a cipher long enough, from as many angles as possible, a pattern must declare itself. The goal of any code maker is to come up with a system that is random and therefore unbreakable. But this is a hard thing to do. Most machines used switches or rotors—set in new orders each day or couple of days, according to the key or setting—to transform one letter into another, often several times, so that *A* might become *D*, and then *P*, and emerge as, say, *X*. The next time, the same letter would follow a whole new path. But wheels and rotors will eventually work through an entire cycle; at a certain point, they will come back to the beginning and encipher the same letter the same way. *A* will again become *D*, and then *P*, and then *X*. The more elaborate the mechanism—the more wheels involved, the more complex the settings—the longer the interval before the repetition occurs. But at some point, something, somewhere, will repeat.

What Friedman also understood, and managed to teach his team, was that there are mathematical ways of detecting the underlying behavior not only of language but also of individual letters. In English, *E* is the most frequent letter. If you are making a cipher and turn every *E* in a message into a *Z*, then *Z* will behave exactly as *E* does: It will become the most frequent letter. One of the first things a cryptanalyst does is take a "frequency count" of all the letters in an enciphered message. If Z appears most frequently, this likely means Z stands for E. Ciphers quickly get much, much trickier, but statistical methods always help. It's remarkable what can be done with math.

What Friedman had also taught them during their training is that you can break a foreign cipher without understanding the language, as long as you know how the letters in that language behave. Certain letters, like S, often travel alongside certain other letters, like T, and he taught his staff to count how many times certain pairs—digraphs—appeared together, as well as trigraphs like *ing* or *ent* or *ive* or tetragraphs like *tion*. He knew on average how many vowels—between thirty-three and forty-seven—typically appear in one hundred letters of plain English. He knew which letters rarely appear side by side. He had even figured out how many blanks—or letters not occurring—tend to appear in one hundred letters. He had identified which consonants (D, T, N, R, S) are most frequent in ordinary English and which are least frequent (J, K, Q, X, Z). He studied how French letters behaved (common digraphs: *es, le, de, re, en, on, nt*) and how English behaved when sent over the telegraph. Since "the" is often omitted from a telegraphed message, the statistical behavior of E changes slightly in a telegram. These are the kinds of nuances—random variations, standard deviations—that statisticians live and breathe for.

Over a span of months, the code breakers had come up with every attack on Purple that they could think of. They had mastered the behavior of romanized Japanese, in which pairs of vowels often occurred, such as *oo, uu, ai, ei*, and they knew that Y almost always was followed by O or U, often doubled, as in *ryoo, ryuu, kyoo*, and *kyuu*. They reviewed the workings of known machines on the Western market, in case the Japanese had borrowed ideas from them. Among these was the Kryha, a noisy thing with a gear-like mechanism that resembled clockwork; something they called the "Damm machine," an easily penetrated contraption named after its inventor, a Swedish engineer named Arvid Damm; and those invented by the horse thief Hebern. All used devices that could take a letter and turn it into another letter. Some advanced step by step. Some would skip forward several letters, or skip once and then not skip the next time. When the Purple machine was being installed in Japanese embassies, Friedman's team was able to follow the itinerary of the installer—a Japanese expert identified in memos only as Okamoto, who traveled city by city putting in the new machines—by reading updates he sent back to Tokyo over the old Red system. They kept hoping he would use Red to send home a report, some kind of clue as to what the Purple machine was and how it worked. Alas, Okamoto did not.

Friedman's team was under enormous pressure. When Purple first came online, they thought they could break the machine in a matter of months. As 1940 progressed and Jews in Europe were rounded up, more concentration camps opened, the blitzkrieg advanced, Roosevelt was anxious to know whether Japan would join in a formal alliance with the Axis powers of Germany and Italy, and if so, what the terms might look like. Emissaries from military intelligence visited Friedman every day, nudging him, filling him with anxiety, asking whether he was doing everything he could. The code breakers talked to radio intercept operators in the field, urging them to ensure that the circuits carrying Purple messages were fully covered. They set up more IBM machines—tabulators that could count and sort very fast—that had been modified to sort the Purple messages they were getting. And still: nothing.

Friedman liked his team to do their own pen work—copying out each letter—so as to have a palpable, physical connection with the cipher. One technique was to write out the text of an enciphered message and print above or below it something they called a "crib." Cribbing is an essential component of code breaking—perhaps *the* essential component. Cribs are educated guesses about what the message says, or even what just a word or phrase probably consists of. Some minor Japanese ministries and embassies were still using the old Red machines, and sometimes Tokyo would send a message to all embassies—known as a circular—using both Red and Purple. Circulars were a great source of cribs. The code breakers could decipher the Red version and set it against the Purple cipher, hunting for correlations.

They also knew that Japanese diplomats, like diplomats everywhere, relied on formal beginnings—"I have the honor to inform Your Excellency" or some such. Sometimes they would jot something like that below their Purple cipher and fiddle around to see if it worked. The fact that they had broken the sixes meant they had a few skeletal letters to work with to confirm the position of the crib, as in a game of hangman. It also helped that the U.S. State Department was negotiating with Japan over a commercial treaty, so messages sometimes came through that contained quotes in English. At Friedman's urging, the State Department would quietly slip the code breakers the originals, to use as cribs.

The code breakers formed a hypothesis about Purple, without quite being able to say why. They theorized that the Purple machine was using some kind of switching device (rather than wheels) to transform the letters. They thought these devices likely resembled the kind of "stepping switch" employed in ordinary telephone circuitry, which routed calls by passing electrical pulses from one switch to the next, using something called a wiper. The design they hypothesized "envisaged a set of four twenty-five point, six-level stepping switches, operating in tandem," Rowlett later wrote. They thought there might be more than one set of four switches, using a cascading rhythm to suppress repetitions. They had a hypothesis that, buried in a stream of message text, it might be possible to spot coinciding letters that would show this; juxtaposing a cipher against a crib, a code breaker might detect a pattern showing the work of the switching devices. If this was true, there would be many letters between each cyclical repetition. But the repetition would exist. Somewhere. You needed a long message to find this; you needed more than one long message, really, and the messages had to have been sent on the same day, so as to have been enciphered by the same key.

Their progress thus far had consisted of conjectures like these, followed by feverish attempts at confirmation, followed by disappointment. Their hopes renewed by this latest theory, Frank Rowlett and his Purple team eagerly looked for three long messages sent on the same day and, after ransacking their file cabinets, managed to find them. Now they needed a crib. Mary Louise Prather—keeping her meticulous files—happened to recall a message transmitted on the same day in a lesser Japanese system they had broken. It was a marvelous feat of memory and gave them the crib they needed.

Frank Rowlett had work sheets made up with the same messages and cribs. He assigned the same sheets to different people, to see if anybody could find anything. They were sitting at tables in a room of about thirty by fifteen feet, scanning and studying. "We were looking for this phenomenon," he would later say, "without actually being aware of precisely what we were seeking."

It was September 20, 1940, at around two o'clock in the afternoon. Rowlett, who was one of the more mechanically minded team members—he was a tinkerer and a hoarder and tended to scrounge spare telephone parts, which he kept in his basement behind a woodpile—was talking with some of the other men. Sitting there engrossed in what Rowlett later rather sheepishly called a "gabfest," they looked up and saw that Genevieve Grotjan, the would-be math teacher and former railway annuity statistician, had materialized beside them. As Rowlett later recalled, she was holding her work sheets clutched to her chest. "Excuse me," she told them shyly. "I have something to show you."

They looked at her with interest and a measure of hope. She was "obviously excited," Rowlett saw. "We could see from her attitude that she must have discovered something extraordinary."

Laying the work sheets on the table, Grotjan took her pencil and circled a place where two letters came together, one from the coded message, one from the crib, one above, one below. Then she went to a second work sheet and circled another coincidence, of two letters whose occurrence confirmed the very pattern they were looking for. Then, at the end of a long stream of letters, she circled a third. And a fourth. And she stood back. There it was. She had found the repetitions. She had uncovered the cycles and confirmed the hypothesis. She had broken the twenties.

Grotjan was a junior mathematician armed with a college degree, an uncompleted master's thesis, and less than a year of on-the-job training.

Many of the men she was working for had far more experience—years, decades. They had written the textbooks she had studied from. Nobody quite understood how she'd done it, then or ever. Grotjan had a powerful ability to concentrate and, in that state of concentration, to see in a different way. In code breaking, counting and making charts and graphs and tables are part of the process. But when you have exhausted that, sometimes, in a deep moment of concentration, pure insight happens, and you just, simply, see the thing you are looking for. And you apprehend that it is right.

The men knew instantly what they were looking at. Grotjan had given them their entering wedge. While she stood quietly, they erupted in cheers. Frank Rowlett began yelling, "That's it! That's it! Gene has found what we were looking for!"

Others crowded around to see. William Friedman came in to see what the noise was about. Grotjan, "obviously thrilled," as Rowlett described her, removed her eyeglasses and was unable to speak. Rowlett started talking to Friedman, narrating what had occurred and pointing out the cycles. It took a while to convince their boss they had succeeded. Friedman slumped, placing his arms on the table and collapsing against it, as if all the stuffing had gone out of him. He congratulated Grotjan, with whom he had barely spoken before. "I was just doing what Mr. Rowlett told me to do," she replied. She was already thinking about the next steps they would need to take, such as figuring out how to break the daily keys. But everybody knew this was the victory they needed. Celebratory Cokes were poured. Friedman went into his office to collect himself, and the others gathered around the table while Grotjan recounted her discovery and explained how she arrived at it.

The team could now construct a machine to decipher the messages. "When Gene... brought in those worksheets and pointed out these particular things," Frank Rowlett would later say, "we knew that we were into the Purple machine and that it would be solved."

Here is the thing about a machine cipher: It's hell to break, but once you break it, you're in. In the aftermath of the Purple breakthrough,

William Friedman—harrowed—spent the first three months of 1941 in Walter Reed General Hospital, recovering from exhaustion. It was a nervous breakdown. During the long ordeal he had not been able to say anything to anybody outside the office, not even his wife. Elizebeth would find him in the kitchen in the middle of the night, making a sandwich and unable to sleep. Even on the day of Grotjan's breakthrough, he went home for dinner and said nothing. He couldn't. "My husband never opened his mouth about anything," she said. The bottled-up stress broke him. He was never the same.

Three years later, Friedman wrote a top secret memo praising Genevieve Grotjan, Mary Louise Prather, and other members of the team in the highest possible terms. He described the Purple cipher as "by far the most difficult cryptanalytic problem successfully handled and solved by any signal intelligence organization in the world."

Never before, he pointed out, had a team of cryptanalysts managed to reconstruct a machine that nobody, apart from the enemy, had laid eyes on.

And here is the other thing: The Purple cipher didn't just give the Allies insight into Japanese thinking. As Friedman pointed out, the ability to read messages produced by the Purple machine provided "the most important source of strategically valuable, long-term intelligence" available to the Allies as World War II unfolded, including the thinking of fascist and collaborationist governments around all of Europe.

The team's breakthrough was held in the strictest secrecy. They would receive no public recognition. Only a handful of people could know the Purple cipher had been broken, because if the Japanese learned what had been accomplished—or even got an inkling—they would stop using the machine.

The code breakers took a week to test their discovery. Friedman then shared their success with the small number of officials in military intelligence and Roosevelt's inner circle who were entitled to know about it. His private announcement was made on September 27, 1940, the day that

Japan signed the Tripartite Pact, signaling that the world's belligerent nations would "stand by and cooperate" in pursuit of their "new order."

Within two weeks the code breakers built a facsimile of the Purple machine. Streams of messages were pouring in from Japanese diplomats in Berlin, Rome, Warsaw—all the key rumor capitals of Europe. Often they were reporting back to Tokyo on conversations with Axis leaders. The messages were lively, opinionated, and informative. They were full of detail and often went on for pages.

For most of the war, it would be the Japanese Purple machine that gave Allied nations their best information about what was being thought and said—and purchased and developed and manufactured—in Europe, especially Germany. This was largely thanks to General Baron Hiroshi Oshima, who served as Japan's ambassador to the Greater German Reich. Oshima was a former military man and confidant of Adolf Hitler who enjoyed wide-ranging talks with the Führer. The Japanese ambassador admired the Nazis, toured German military facilities, and wrote reports back to Tokyo that were long, erudite, and precise. Oshima's painstaking description of German fortifications along the French coast would be invaluable when Allied commanders were planning the D-Day invasion.

All of the dispatches were frank, written by men who had their ears to the ground all over Europe. Going forward until the end of the war, the Japanese diplomats used the Purple machine to convey what Hitler was saying to his French collaborators; what people on the streets of Europe were feeling; what newspapers were writing; what Albert Speer, Nazi minister of armaments and war production, was reporting about munitions; what transpired when a team of German officers tried to assassinate Hitler. ("What was really mysterious was the fact that the Chancellor, who was nearest to the bomb when it exploded, was unhurt with the exception that his clothes were torn to pieces by the blast and he sustained a few burns," reported one Oshima message.)

Early in 1941, several members of Friedman's team quietly boarded

Britain's newest battleship, the HMS *King George V*, which had stopped in Annapolis to drop off the new British ambassador. They stowed one of their precious homemade Purple deciphering machines, hidden in a crate, on board, and—at great peril—took it across the sea, passing through the rattlesnake nest of lurking U-boats and presenting it to their astonished British colleagues.

Read today, the language in the diplomatic messages feels fresh and intimate and vivid. To take an almost random sample, consider a series of messages sent in 1943 between Japanese diplomats in Europe, writing to one another and to their home office in Tokyo, using the Purple machine and a few other diplomatic ciphers.

"England and America are jingling money in their pockets," wrote Tokyo headquarters to the Japanese ministry in Madrid, at a time when Spain's putative neutrality was in question and the Allies were trying to prevent Spain from entering on the side of the Axis. "We have got to make Spain change her mind right away if we can."

"This London report twisted the facts, presumably to give the impression that there is a serious rift between Finland and the Axis," wrote Helsinki to Tokyo.

"HITLER said, 'When this war is over, we Germans are going to start founding a new Europe,'" reported the Japanese ambassador in Vichy France. Pierre Laval, a top French official in the Vichy government, then "retorted coldly: 'Why not found a new Europe first?'"

"The enemy's air bombardment of all Italy is being carried out with extreme violence," Rome wrote Tokyo.

"I would say that if this time Germany does not win, the song will be out and the jig up," warned Oshima from Berlin, as Germany girded itself for an effort to capture Leningrad and push its offensive in Russia.

"Passing over French territory night after night, British and American planes wreak havoc on Italy without any signs of a let-up," reported the ambassador in Vichy. "The French people have always fervently hoped for a victory for the British and Americans, and now they revel in the conviction that such a victory is coming sure as death.... Also, this business of sending Frenchmen to work in Germany is heinous to the French."

"Now that the Axis forces have been cleared out of Africa the question of an Anglo-Saxon invasion of Europe has become very real," Vienna warned Tokyo.

Each day, messages like these were deciphered and a summary was typed on special paper with TOP SECRET printed at the top and bottom. The intelligence from the Purple machine came to be known as "Magic," likely because Friedman's Army bosses referred to the team as their magicians. The Magic summaries were put in a briefcase and taken by a messenger to the few people with the clearance to see them. When intelligence was attributed to a "highly reliable and trusted source," this usually meant it came from Magic. In a 1944 memo, the Army noted that the Purple messages were "the most important and reliable source of information out of Europe." The sheer quantity was overwhelming to the translators who worked closely with the code breakers, converting messages from Romaji to English. "Through their almost naïve confidence in the security of their cryptographic systems, the garrulous Japanese unwittingly admitted us into many of their most solemn conclaves," noted one internal history of the translating unit. Between 1943 and 1945, more than ten thousand Purple messages were delivered to American military intelligence.

For a time, the Purple break would exacerbate the rivalry between the U.S. Army and Navy. After the machine was broken by Friedman's team, the Navy figured out how the Japanese were varying the key settings, and thus how to predict them. So eager were the two services for credit that an absurd compromise was reached in which the Navy took responsibility for breaking Purple ciphers on odd days, and the Army on even, so that neither would have an "unfair advantage." This led to wrangling about whether "even" meant the day a transmission was sent or the day it was received. For important transcripts, both services would crack them and race to be the first to deliver them.

* * *

In April 1941, seven months after her historic break, Genevieve Grotjan received a raise of $300 and a promotion to "principal cryptographic clerk." Friedman's team rapidly began to expand. The Purple machine could not predict the attack on Pearl Harbor, for the simple reason that Japanese diplomats were not clued in by their military as to what was about to happen; a fourteen-part message containing a precisely worded communication (in English) ending negotiations was deciphered by the Americans, it's true, but it did not give concrete warning of a naval attack. At the time, there were 181 code breakers working downtown for the Army. More began to pour in. The Munitions Building outlived its purpose—for everybody in it. The War Department prepared to move into the Pentagon, now under rapid construction. Friedman's operation also needed to relocate. The boom and expansion of the country's military administration had begun.

By then, a clandestine Army intercept station was being established in a converted barn at a place called Vint Hill Farms in the Blue Ridge Mountain foothills. Driving back from a tour of the proposed installation, a group of Army officials noticed the spacious grounds and elegant buildings of a place called Arlington Hall Junior College, where, as it happened, the code breakers Delia Taylor and Wilma Berryman had rented rooms one summer, when the school was offering itself as a "resort hotel" in a vain effort to stay solvent. The two-year finishing school—offering lessons in music, typing, homemaking, posture, and other subjects to girls of "good character" and deportment—was not yet twenty years old and had never been financially sound, nor well regarded in terms of academics. It had gone bankrupt during the Depression, and now war had decimated attendance.

The school lay nestled in a central part of the small county, on former swamp- and pastureland situated near the villages of Ballston and Clarendon, along an old streetcar line that connected Washington with Falls Church in an era when residents fled the city in the dank and awful summers for the barely more hospitable suburbs. The streetcar line had been replaced by a thoroughfare for cars. The one-hundred-acre grounds included a hunt course, riding rings, a hockey field, a golf course, cottages, and a teahouse. The location was convenient to Washington but far enough away to escape enemy bombing and the notice of secret agents. There was talk of putting the Signal Intelligence Service in the Pentagon, a few miles away, but space there was a concern, and the code breakers felt it would be better not to have the military breathing down their necks.

And so the thing was done. The War Department filed a Declaration to Take and paid $650,000, which was less than the Arlington Hall trustees wanted. The faculty and 202 students were evicted.

On June 14, 1942, a small guard detail including one Army second lieutenant armed with a .45 pistol and fourteen enlisted men shouldering sawed-off broomsticks took possession of Arlington Hall. Rifles were in short supply. The move was so hasty that schoolgirls were still clearing out their rooms. Convoys of vans secretly departed the Munitions Building, transporting machines and file cabinets stuffed full of intercepts. The intent was that the very existence of Arlington Hall not be mentioned outside its gates, but in time-honored government fashion, a press release was accidentally issued. The sunny ground floor of the Colonial-style main schoolhouse had formal drawing rooms and parlors, along with a chapel, a library, and an auditorium with a concert grand piano and pipe organ. There were dorm suites on the second and third floors, and classrooms on the fourth. Code breakers set up operations in the suites, storing intercepts in bathtubs. Some dorms still had beds and dressing tables that had to be moved. Oriental carpets were rolled up, curtains dismantled, and "the finishing school atmosphere was shattered by a regime of brisk efficiency," as one memo put it. Fences were erected, guard stations built.

The Purple machine was installed on the second floor but had to be draped whenever somebody who was not working on Purple used the nearby bathroom. People were allowed to do so once every hour. When the tabulating machines proved too heavy for an upper floor, they had to be moved to the basement—and then to new buildings. The indoor riding hall was paved with concrete and used for storage.

At Arlington Hall, Genevieve Grotjan would stay abreast of modifications and changes to Purple, with a colleague, Mary Jo Dunning. The two women became familiar with the intricacies of the Japanese diplomatic cipher; it became like an old friend. As the war wore on, a number of other ciphers were attacked, and Grotjan would be dispatched to solve challenges they presented.

Arlington Hall soon found itself working the codes of some twenty-five nations, enemy and neutral: Finnish, Portuguese, Argentinian, Turkish, Vichy French, Free French, Chinese, Thai, Belgian, Haitian, Irish, Hungarian, Liberian, Mexican, Chilean, Brazilian, and those of many Middle Eastern countries. Some were codes; some were ciphers; some were both. There was a French code they called Jellyfish, a Chinese enciphered code they called Jabberwocky, another they called Gryphon. Some were important; some were merely interesting. Each week, top secret reports detailed breakthroughs, and it's striking how often they were made by women. "The outstanding solution of the week was that of the SAUDI cipher, accomplished as a result of the 'golden guess' of Mrs. Flobeth Ehninger," read one report in September 1943. "The system appeared to be a two digit substitution with multiple variants. Mrs. Ehninger guessed that a certain repeat might mean 'the Arabian land.' This assumption proved correct, and within two hours all but 4 letters of the Arabic alphabet had been determined."

The tenor of the operation was changing. Top men like Sinkov, Rowlett, and Kullback received Army commissions and went into uniform. Military men took charge of some units, usually with a civilian "assistant," inevitably female. William Friedman was gently pushed aside: When he returned from convalescence he was given an office at Arlington Hall, working in an advisory capacity, but no longer ran the place. "He never really came back," as Wilma Berryman put it. But the majority of incoming workers remained civilians, following what had always been the hiring strategy in the Army's code-breaking unit. The hardy band of brothers and sisters from the old Munitions Building would retain their informal camaraderie, but they—including women like Wilma Berryman and Delia Taylor—would quickly find themselves in positions of enormous authority.

Sometimes, when she was riding the bus between her boardinghouse and the new Arlington offices, Genevieve Grotjan would look back on her moment of insight and remember it with "satisfaction and pleasure." Not often, though. She was too modest about her own contribution, and too busy.

CHAPTER FOUR

"So Many Girls in One Place"

Soon after Dot's arrival

Dot Braden hated living at Arlington Farms. The dormitories, thrown together so quickly by a federal government eager to house its overwhelming influx of women workers, were flimsy and shoddy. The walls were so thin that they shook when a person walked down the hall, and the spiritual effect of living in makeshift quarters constructed out of a substance called "cemesto" was depressing, WPA paintings or no paintings. The women in the dorms were always having to stand in line for something—the mailboxes, the showers, the cafeteria food, the phone, the bus. The county of Arlington had been transformed by the numbers of government girls. As Arlington Hall was getting up and running, top code-breaking officials had gone from door to door begging local residents to offer a basement, a bedroom, a cubbyhole, an attic, anything, to house a hardworking g-girl, despite the fact that she would be doing top secret work at all hours. Residents opened their homes; Arlington Farms was constructed; even so, it wasn't enough.

Sensing an opportunity, developers began building garden apartment buildings around Arlington Hall and advertising them in local newspapers. One day Dot's friend Liz, from Durham, pointed out an ad for a new complex called Fillmore Gardens, built nearby along Walter Reed Drive. Liz proposed that they move in together and set up house.

They approached a colleague, Ruth Weston, to see if she wanted to go in with them. Ruth had been hired one week before Dot—October 4, to be exact—and went through the same orientation classes, attending the same big Christmas party at which William Friedman and other code-breaking bigwigs made appearances, to spread cheer and celebrate the newly recruited girls upon their arrival. The two young women chatted on the bus back and forth to work and began to socialize in their free time. Ruth also lived in Idaho Hall. Parts of Arlington Farms were still unfinished, and the women would creep through an interior shortcut, a kind of utility tunnel, to meet in Dot's room.

Ruth Weston agreed that it would be nice to get away from the crowded dorm conditions. The Fillmore Gardens apartment was on the second floor of the building, a walk-up tucked away in a discreet corner just off of a dark stairwell, and boasted a single bedroom, a single bathroom, a kitchen, and a living room. As of yet the apartment complex, which was not finished, consisted of little more than a building in the middle of a field, but the place seemed palatial—and well constructed—compared to Arlington Farms. The women could cook and eat when and where they wanted to and would have to share their bathroom only with one another. Their new place was situated a mile and a half from Arlington Hall, so they could walk to work rather than waiting for the bus. "We'll pool our money," said Liz. They filled out an application, and they were accepted.

Finding furniture was not easy. Materials were scarce, as were funds. Dot's mother sent a bedframe on the train from Lynchburg. The women decided that Liz would sleep on a cot and Dot and Ruth would sleep in the bed together. The problem was that the bed lacked a mattress. So Dot and Ruth looked in the paper and found a department store that sold mattresses. They called up and confirmed that they were available—not a guarantee in wartime—and took the bus and streetcar downtown after work. They paid for the mattress and were told it would be waiting at the back door. When they went around back to receive it, however, they realized their predicament: The store didn't deliver. There was no way the two women could hand-carry a mattress on public transportation and get it all the way back to Arlington, some five miles away. By now it was getting late. There they were, two lost souls, struggling to hold a mattress between them. So Dot went inside, leaving Ruth, who was just five feet tall, to hold the thing upright, and found a salesman who looked as though he might be closing up for the day.

"We've got this mattress and we don't have any way to get it home," Dot told the man. "We live in Arlington. You all didn't deliver it."

"We weren't supposed to deliver it," the salesman told her. Then he relented. "Well, I'll tell you what—I live in Arlington and I'm going home. I can put the mattress on the top of my car. If you, where you live, have got some eggs, I've got a pound of butter with me. You all can cook me some eggs and use my pound of butter and I'll take your mattress for you."

So that's what the women did. They bartered a plate of scrambled eggs for the delivery of a bed mattress, and the salesman strapped the mattress to his car and drove them home. The women weren't worried about having a strange salesman sitting in their small apartment eating eggs at their table; with three women in the place they figured they could take care of themselves.

The mattress escapade—their first experience fending for themselves in the big city—was the beginning of Dot Braden's great friendship with Ruth Weston.

* * *

Ruth Weston had an even more pronounced southern accent than Dot did. This was one of the many reasons Dot liked her. Ruth hailed from the delta region of Mississippi, and when they caught the bus together, Ruth would ask for a "transfuh," and Dot would laugh and laugh. Dot herself had a strong and distinct manner of speech; coming from Virginia's Southside, Dot said "tomahto" and "auhnt" for "tomato" and "aunt," and pronounced "mouse" and "house" with a long *o*, as if they rhymed with "gross," a remnant of colonial settlement in the region. But

Dot still loved the languid way that Ruth said "transfuh," and she took to repeating it, to Ruth's mild annoyance, though in general Ruth reacted well to teasing. Dot also liked Ruth Weston because—quiet and reserved though she might be—she was always willing to go along with the escapades Dot suggested, be it letting a strange man deliver a mattress to their home, or traveling all the way to the beach and back during their one day off, after working seven days of eight-hour shifts. You might not be able to tell it from looking at her, but Ruth Weston was game.

Ruth was short and dark-eyed and olive-complected and somewhat round-faced. She had grown up in the crossroads of Bourbon, Mississippi, where she was one of seven children. Boy, girl, boy, girl, boy, girl, boy, the Weston sibling series went. Each girl had an older and a younger brother. Their mother was open about the fact that she didn't want girls and was trying to have as many boys as possible; the girls in the Weston family were basically just interstitial accidents. "Boys were more important to her than girls," recalled her youngest daughter, Kitty. "She didn't say, 'I didn't want you.' But it was there." Their mother did sometimes allow that, fortunately, the Good Lord had known better than she did.

Bourbon was little more than a postal address, in truth, surrounded by cotton plantations. Ruth's father served as postmaster and owned a general merchandise store. He also farmed, but not well. The Depression was hard on the Weston family, as it was on most people in Mississippi. The children worked in the store, and if they sold $5 worth of items between opening and closing, that was a big day. In June 1931 their father suffered a cerebral hemorrhage, so the younger Weston children never knew him except as an invalid.

Despite Ruth's mother's bias against daughters, education in the Weston family was important, even for the girls. Ruth's maternal grandparents came from Germany and saw higher education as a way for the family to become assimilated quickly. Ruth's mother had attended some college, as had her sister—the children's aunt—who lived with them and ruled the roost. Ruth had an older sister, Louise, and a younger sister, Kitty, and all three Weston girls went to college. They attended

Mississippi State College for Women, a public women's college founded after the Civil War—as the Industrial Institute and College for the Education of White Girls—when the South needed educated women to help rebuild the economy. Many female students majored in home economics or "secretarial science," but the Weston girls majored in math.

Despite her gifts, Ruth had struggled to find a teaching job. That was just before the war, when male teachers enjoyed hiring preference. When she did find one, the conditions were tough and low paying. Ruth's younger brother, Clyde, drove her in the family car to her first teaching job, about sixty miles from their home, in a place called Pleasant Grove. It lay in the northern part of the state and was the most rural spot he had ever seen—more rural, even, than Bourbon. Ruth boarded with another teacher and taught at a school with no heat, electricity, or running water, making $71 a month. She lasted one year, then found a job in Webb, Mississippi, which was also low paying, if slightly less primitive.

Once the war started, the Weston family's German roots became problematic. People would ask Ruth's brothers how their mother felt about the war and whose side she was on. Hearing that "hurt my mother terribly," Ruth's sister Kitty later said. Their mother felt proud of her German heritage, but the whole family was staunchly American and intensely patriotic. Ruth's father flew an American flag every day and instilled his strong sense of civic responsibility in all of his children. People in the area looked up to him as a leader even after he had his stroke. Ruth was close to him; she resembled his mother—her paternal grandmother—and he had been delighted to see the resemblance when she, their fourth child, was born. He adored her, and she him. His stroke had been very hard on Ruth, who shared his patriotism and commitment to public service. For all these reasons, the Arlington Hall job was perfect for her.

In addition to her math acumen, Ruth Weston had a musical streak. She was a talented piano player, and unflappable. Once, during a recital, she was wearing a pink evening gown and a June bug started crawling up her bare back. During a pause in the score, she stopped playing for an instant, reached back, and flicked it away without missing a note. Her little sister Kitty was watching admiringly from the audience and never forgot Ruth's casual, self-possessed flick.

But what Ruth Weston was most famous for was this: You could tell her anything and she'd keep it to herself. She was the most reserved, most closemouthed person imaginable. Her shyness made teaching hard for her. Ruth had many objections to teaching, chief among them that it paid so badly—she called it a "respectable way of starving to death"—but also because she disliked being the center of attention.

The family was never quite sure how Ruth had found out about the job at Arlington Hall. In the chaos of many brothers shipping out to military service, Ruth managed to slip away without attracting as much notice as the boys did, making the two-day train ride from nearby Leland, Mississippi, to Washington. Her mother had not finished altering some clothes before she left, and sent them along behind her. Her brother Clyde was surprised when investigators called to inquire after his sister's background. The agents called her mother to find out if she'd had childhood diseases like mumps and measles, and her mother reported that her children had contracted—and fought off—every childhood disease you could imagine.

Apart from her skill at keeping secrets, the other good thing about Ruth Weston was that you could tell her anything and she would not judge. For Dot, this was a relief from the snobbery that prevailed among old-line Virginians, and even among branches of her own family, who tended to keep close track of which members were doing better than others. Ruth knew that Dot's parents were separated. She knew that Dot's father was a good person, as was her mother. After her own father's stroke, family struggles were familiar to Ruth. The Depression had been hard on people. Hard on families. The two young women confided to each other as they lay in their shared bed. Yet certain things they did not share. Close as they were, Dot and Ruth confided nothing to each other about their work, even though they sometimes had lunch together in the

Arlington Hall cafeteria and were, in fact, working on different areas of the same Japanese Army code-breaking effort. They went around terrified that they would let something slip. "We were scared to death," as Dot put it.

* * *

Ruth's full name was Carolyn Ruth Weston. Sometimes she went by Ruth, sometimes by Carolyn. One day the milkman delivered the bill to the new Arlington apartment and addressed it to "Crolyn," and Dot thought the misspelling was the funniest thing she had ever seen. "Get the bill, Cro-lyn!" she implored, laughing. She started calling Ruth "Cro-lyn," and over time the nickname shortened into simply "Crow." The name stuck: Crow. At least, that's what Dot called her. Nobody else did. It was somewhat like a secret code name.

The apartment became more crowded. Crow's older sister, Louise, wrote them six months after they moved in, saying she too wanted to come to Washington to look for government work. Crow was not happy to hear this. She had been glad to get out from under the thumb of Louise, known in the family simply as Sister. But what could she do? Crow and Dot went to fetch Sister from Union Station, a trip that involved several, as Crow would put it, trans-fuhs.

They arrived amid a downpour to find Sister, who was tall and fair and redheaded, standing bedraggled outside the arched facade of the train station, near the plaza fountain with its statue of Christopher Columbus, having been caught in the rainstorm. She was wearing a lacy hat and looked very much like somebody from Bourbon, Mississippi. Her dress, which was linen, had shrunk, and her slip was showing. "Look at that," muttered Crow. They were city girls now, and Sister looked unspeakably country. They showed her how to navigate the bus and the streetcar and took her back to Fillmore Gardens, where she would sleep in the living room on a daybed. (Liz had a cot in the bedroom.) But there were advantages to the Mississippi influx. The women cooked for one another, and

Sister would make red beans and rice. Dot had never eaten Cajun food, and it was a treat on a cold day after finishing an eight-hour shift and walking a mile and a half home from work in the rain.

The women paid their own bills and cooked their own food. At any given time there might be five or even six women in the one-bedroom apartment. Sister stayed; Crow's little sister, Kitty Weston, came up for the summer. Dot's mother often took the bus up for visits, as did friends and family members, including her own little sister, "Mary B." Crow's brother Clyde was stationed with the Navy in New York and came down to see them. He would flirt with Dot, who he thought was "a real cute girl," "outgoing," and a good friend to Crow. Having all those people in one small place did not seem hard. None of them was used to being coddled—or "sitting on a cushion and sewing a fine seam," as Dot put it. They ate Sister's red beans and rice, and frozen peaches, which was the kind of thing you had for dessert when you couldn't get sugar. They would laugh about how their mouths froze and puckered from the peaches, and they would sit around the apartment talking with frozen lips.

Liz's mother visited from North Carolina and made green beans using fatback, which was a method Dot had not heard of. Liz's mother then fished the fatback out of the beans and wrapped it up to use again. "I'm saving it," she explained. That too seemed very country to Dot and Crow, and "I'm saving it" became a mutual joke. "I'm saving it!" they would say, and fall over laughing.

Dot and Crow got along perfectly in every way. They became part of each other's routine. Crow was slow in the mornings, so Dot would fix her breakfast, usually just coffee and toast, or cereal. Crow would wait for her to make it, and then they would go on together to work.

* * *

At Arlington Hall, the presence of so many code-breaking girls from small towns presented a challenge to their bosses. The women came from remote places where their own experience of the war effort had consisted mostly of enduring rationing, listening to the radio, and worrying about the fate of boyfriends and brothers. Local newspapers were rich in war news, but even so, the top brass felt the women needed educating and motivating. And so emissaries were brought to Arlington Hall to school the young women in the geography of the world and impress upon them the realities of the fighting and their own contribution to it. The goal of the lecture series, This Is Our War, was to expose young ex-schoolteachers from Durham, North Carolina; Lynchburg, Virginia; and Bourbon, Mississippi, to the full contest they were now part of; to sketch out just how large the theater of World War II was and the important role that coded messages played in it.

It was heady stuff. One of the first dignitaries to address the young women was none other than Rear Admiral Joseph R. Redman, director of Naval Communications (his brother John was also a naval communications officer and headed up the code-breaking unit; the two were often confused), who graciously agreed to cross the Potomac to address the code breakers of the other service. In his lecture, "The Navy Attacks," which he gave on September 7, 1943—a bit before Crow and Dot arrived—Redman tried to evoke the sheer enormousness of the Pacific, which was bigger than anything the women had seen or could imagine. Newspaper reports, he told them, gave the impression that ships from opposing navies were always just somehow finding each other and commencing to shoot. "I don't believe you can visualize how much water there is out there," Redman told them. "You take a fast ship and it takes about three weeks to get from San Francisco over to Japan. And that's a lot of water and you don't see many people on it." The point was that the task of locating an enemy ship was a bit like finding a needle in a haystack.

"You think, from what you see in the newspapers, you bump into a submarine or see a ship—you have an engagement out there—that's easy," he said, but that was only because the newspapers didn't understand the secret work being done by people like them to ensure the engagements happened.

Ensuring that two enemy forces met in a maritime battle was hard.

And it only occurred, Redman said, thanks to the work of code breakers who could help pinpoint enemy movements. "The work you are doing—as dull as it may seem to you—is very exciting to someone else, and the information you are able to fire [to] the operating agencies in the field contributes a great deal to that success," he told the women. "And some days when it is hot, and you feel tired and sleepy, remember that the delay of a few hours in picking up some important information may have a vital bearing upon the operations that are going on out in the ocean."

The speakers were all high-level. Major General Wilhelm D. Styer, chief of staff of the Army Service Forces, delivered a lecture entitled "Fighting to Win" during which he also endeavored to make clear to the young women "just what a great show we are in." Though they were by definition fighting on land, he pointed out that Army troops had to cross water to get to their battle engagements in Europe and Asia, and they had to transport supplies across both the Pacific and Atlantic Oceans. During the prior war, he told the young women, the U.S. Army had one sea route, going from America to Europe. Now it had 106 overseas ports, 122 sea routes, and supply lines in the "Asiatic Theater" that were as much as twelve thousand miles long. He talked about trips he had taken to Marrakesh, Casablanca, and Algiers, and about his first sight of the Chinese soldiers who were U.S. allies. "They are considerably smaller in stature than our own troops, but they are very stocky. And as these are a handpicked lot, they are a very fine-looking bunch of troops."

He also shared his view that many of the world's people "do not look like a happy people. They are sullen—look like they lack ambition. When you get back you are just proud of the fact that you are an American, and you are willing to do anything you can to preserve the standards."

Perhaps most thrilling was a talk revealing the inner workings of the Federal Bureau of Investigation, whose assistant director, Hugh H. Clegg, came to deliver a lecture called "The Enemy in Our Midst." Clegg talked about the FBI's own war—against criminals, informers, kidnappers, spies, and fifth columnists. His remarks contained a half joke that shows just how ferociously the agencies were competing for the women sitting before him. "Immediately upon arrival, I was threatened," he remarked. "Threatened that if I undertook in any way to try to recruit any of the lovely young ladies of this audience to come down and work in our fingerprint identification bureau, that there would be another casualty of war hobbling back in the general direction of Washington."

But even as the Arlington women were being educated and courted, they were receiving a subtle message that their very involvement in the war effort—these apartments they were renting together, the furniture they were buying, the meals they were cooking, their newfound independence—was creating troubling social changes. A talk by Charles Taft, titled "America at War," centered on this idea. Taft, son of the late president William Howard Taft, now served as director of the Office of Community War Services, a new agency created to cope with the disruptions the Second World War was tearing in America's social fabric.

"I was a little startled as I came around the curtain to see so many girls in one place," Taft began when he took the podium to talk. He then launched into a speech in which the word "problem" appeared over and over. These problems, he explained, had been developing since the first new industrial plants were built in the United States in response to the bombing of England in 1939 and 1940. As Taft sketched it out, America now was one roiling mass of chaos. Factories and construction projects were drawing workers to communities that were hardly able to hold them. Minority groups were moving to find higher-paying work in places that had never seen them. Black churches from the South were moving entire congregations to California. New industries were setting up overnight, Taft said, without the infrastructure they required: schools, housing, playgrounds, hospitals. He talked about other changes in America's quotidian rhythms. People were working around the clock, not just men but women too, and it was impossible to do chores in normal hours. "I send my washing home every week and I get it back every week," confided Taft, whose family lived in Cincinnati. But even he had to contend with grocery shopping. "If you don't provide some sort of extra night when you can shop out of hours, then you are going to find yourself with a serious problem."

But the main problem seemed to be—well, them. The women. He pointed out that many people working in factories were female—he mentioned a bag plant "which was to employ almost entirely women"— and 20 percent of those had young children. "And that," he said, "gets you into some very troubling kinds of problems." To deal with what he called the "child-care problem," he said, the government had created a range of federally funded child-care options: "nursery school projects" for small children, aftercare for school-aged kids, and even home care, set up in boardinghouses, for infants. But American mothers were suspicious of child care, he lamented, because it was a "new idea" and nobody had ever offered it to them before. As a result, Taft said, children were running amok, and the government was sending social workers around to try to convince working mothers that putting their children in child care was better for everybody. "Of course, you have among mothers what is perhaps a very natural feeling, whatever trouble may be caused by little Willie, who is the neighbor's child, that little Johnny, who is our child, doesn't cause anybody any trouble."

The other problem was sex. Taft launched into a vivid description of Cincinnati's railroad terminal, which had become a national crossroads, the center of rail traffic passing along the north-south route between Chicago and Florida or the Gulf Coast, and east-west traffic passing between Washington, D.C., and St. Louis. The station was jammed by travelers, many of them men. And you know who travels to find traveling men: traveling women.

In Cincinnati's Union Terminal, he told them, predatory women lurked. It's possible to imagine the audience at this point uneasily shifting in their chairs, as Taft blamed their gender for much of the nation's moral ills and social dislocations. "Starting with some of the professionals, and then extending down among large numbers of amateurs, girls go down there and wander around in the station and find themselves a soldier and go out and sit in the park. And the park is a very large one with trees and bushes and everything else. And it's gotten to be a kind of a bad situation." Taft did not elaborate on what he meant by "professional" and "amateur," but he did return often to the topic of prostitution, talking about "camp followers" who traveled to construction sites and military encampments, spreading "prostitution and promiscuity" as well as venereal disease.

At the end of this soliloquy on laundry, grocery shopping, troublesome children, pox, prostitution, and other developments caused by wartime changes in their own behavior, the female code breakers of Arlington Hall were invited to rise and sing the first verse of "The Star-Spangled Banner."

* * *

It was true that newfound freedoms were changing the women's lives. Men, now, were the ones avid to get married. Men were the ones who wanted to have someone back home to write to; someone to produce an heir by; someone waiting when they came back from war, wounded or whole. Women—often—were the ones holding out for a bit more time to think. Dot Braden herself was in a bit of a pickle in that department.

At the time, of course, the only real way for men and women to stay in touch was to write letters. Phone calls were feasible, but only rarely and not for long distances, or not often: Long-distance calls were expensive and soldiers often didn't have access to a phone, nor did women living in dorms or boardinghouses, where, at most, there might be a bank of pay phone booths or a lone public phone used by all residents. At Fillmore Gardens, there was a sole telephone in the basement. But everybody could write letters: weary mothers writing to faraway sons—and, now, faraway daughters—late at night after chores were done; young women scribbling on paper held to their knees while riding crowded city buses; soldiers waiting in camps and aboard ships. Everybody had stationery and pen and pencil, and everybody, everywhere, was writing letters. The censorship department read the letters to make sure secret locations were not revealed; loose lips sink ships, as everyone knew, but that didn't stop the letters from traveling back and forth across thousands of miles of land and ocean.

Dot, for her part, had no fewer than five men she was writing to. Two were her brothers. The third was her putative fiancé, George Rush, a tall young man with a prematurely receding hairline whom she had dated while she was at Randolph-Macon. George had been a good college boyfriend: He was an avid dancer and liked going to mixers. It was true that he'd once given her a pink corsage to go with a stop-sign-red dress, and it was true, Dot had to admit, that she had thrown the corsage against the wall—everybody knew pink and red did not go together—but overall he had been a good companion for that period in her life.

Dot and George had not seen each other much in the past year, however. He entered the Army in April 1942, four months after the attack on Pearl Harbor, and kept getting moved farther west. He was stationed now in California and while away had sent Dot a small package, which she had opened to see—to her dismay—a diamond engagement ring. Doubtless he expected her to be pleased, but the ring's arrival was not welcome. Dot liked George but never envisioned spending her life with him. She was inclined to send it back, but young women were told not to do anything to upset soldiers who were away from home, so she kept it. She had never worn it teaching; it would have created too much of a stir. And she never truly considered herself engaged. At one point, perhaps sensing this, George had made the cross-country train ride back to Lynchburg to secure her affections. He showed up on her doorstep and announced his plan: He and Dot would elope. They would drive to Roanoke, an hour away, and get married. Then Dot would move to California to be near him.

Dot had no intention of doing any of this. She disliked being pressured, and refused. George insisted. As a stalling tactic, Dot said she had to talk to her mother. It was a ploy to get her mother involved, and it worked; Virginia Braden sat on Dot's suitcase to prevent her from going with George. The gesture was unnecessary. Dot turned down the marriage proposal and George went back to his base but continued to write. Once her own job offer came from the War Department, Dot felt even more eager to avoid being tied down. She did not want to move to California. She wanted to move to Washington, and she wanted to serve the war effort. Her patriotism had been aroused along with her sense of adventure.

Meanwhile Dot had begun exchanging letters with Jim Bruce, an Army meteorologist whose family owned a dairy farm in Rice, Virginia. Jim was tall and laid-back and four years older than she was. She had met him during a casual dinner date with mutual friends and had known him now for several years. Even when she was dating George Rush, Jim had always been there in the background, keeping quiet track of where Dot was, dropping by, coaxing her to go out. For quite some time, Jim had been after her to take off George Rush's ring. There was something about Jim that was steady and reassuring. He was a college graduate and before the war had been working at a DuPont chemical plant. The problem was, he ran with a rowdy crowd of drinkers. But her mother liked him. The Army put him in meteorological training at the University of Michigan, but from time to time he would come to Lynchburg and look Dot up. She would walk with him, always a little worried that friends of George Rush would see them.

At one point Jim Bruce had persuaded Dot to ride to Richmond to see his sisters. She accepted in part because she liked his blue-and-white Chevrolet. They'd eaten at a hamburger joint and gotten up to dance. "You know, if people see you with that diamond on, they'll think I'm engaged to you," he murmured as they were dancing.

In Richmond they drove into Byrd Park, where he parked the car. Dot didn't protest because she knew he was honorable and would never try anything. "I bet you're going to marry me" was what he said. She put him off. But she always had Jim Bruce in the back of her mind. It was hard to say why. He did look spiffy in his uniform.

There was another soldier she was writing to. There was nothing serious about it. He had the somewhat improbable name of Curtis Paris and she'd met him at a Washington dance. Curtis was a fellow southerner and had asked if he could write to her. This was a common request—men were always asking to write—and Dot didn't see any reason to say no.

Early in December 1943, about two months after she arrived in Washington, Dot got a letter from Curtis, who was stationed in San Francisco. "Dearest Dottie, Greetings and salutations and stuff and junk," it began, and he thanked her for a letter she had sent him that "had a soothing effect plus the faint aromma of apple blossom (face powder??) which is rare and outstanding as compared to the usual g.i. chow hall etc." Curtis chatted about what the soldiers at the Presidio were doing: hiking ten miles two or three times a week and playing a lot of football. "I'm getting some good muscles," he informed her, commiserating a bit about the senselessness of the U.S. Army. They had gone to play "our usual football game" in a local field, only to find out that the field had been plowed. The men had placed stakes with football helmets for the boundaries "so now I'm sure I know what Flanders Field looked like. No wonder I could not run with the ball." He said that the men in his unit teased him for saying "y'all," and he was sure she, as a southerner, would understand what that felt like.

In her last letter, Dot had sent him a photo. Curtis was very taken with it.

The picture!! Wow! I was never more pleased in my life. I don't see how it could be better. The fellows all looked it over and seemed astounded at the Virginia talent. Of course they ask all the particulars and where you were from and well to my regret there just isn't much of a story for me to tell—at least, not as much as I wish there had been. I have a pass today so I'll get the portrait framed in good form so as to lend moral support to the military personnel around my bunk.

He asked her to keep writing him. "I like your letters and the phrases you coin."

Dot didn't take it seriously. All the girls were writing letters, often to lots of soldiers, and many women received three or four or five letters a day. One was writing to twelve different men. It was fun and it was something to do and it felt like they were helping the war by keeping up morale. The women sent snapshots: small black-and-white pictures showing them in front of the U.S. Capitol, or sunbathing, with a handwritten inscription on the back. To be sure, not all male-female encounters were epistolary: Soldiers and officers streamed in and out of Washington, and it was feasible for a woman to have a different date for every meal of the day. The magazines and papers put it about that women in Washington were lonely—"they can get a wonderful job, but they can't get a wonderful man," one article said about g-girls—but nothing could have been further from the truth. There was a master sergeant from Massachusetts Dot went out with from time to time; he would take her to dances and bring her back on the streetcar, and while the whole thing was chaste and lighthearted, she did get attached to the corsages he would bestow. She wore George Rush's engagement ring, or didn't wear it, depending on the situation. The ring was good for fending off men or for keeping dates light and tentative.

But—as she soon learned—it wasn't just women who were dating more than one person. Around the same time she got that letter from Curtis Paris, Dot received a phone call from Jim Bruce, who was being sent overseas and wanted to know if Dot would come down to see him off. "I'll have to let you know," she told him, in some confusion, then called her mother. "Jim Bruce wants me to see him off. What do you think I should do?"

"Well, now, Dorothy, you know I like Jim and I don't like George," came Virginia Braden's voice on the line. "But you still have that ring. You'll just have to make up your own mind." So Dorothy called Jim back, placing a long-distance call. When she got through, the operator said:

"I have another person calling Lieutenant Bruce. I let her go first. If you don't mind waiting I'll put you through afterward."

"You certainly are popular," Dot said when she was put through.

"That was my sister," Jim Bruce told her unconvincingly.

Dorothy wasn't fooled. There was another girl in the picture. "Well, I'm sorry, I can't go," she told him tartly.

This occurred when she was living at Arlington Farms. The next morning there was a message tacked to the door to her room. It said, "Lieutenant Bruce is coming to Washington this afternoon." She read the note and went on to work thinking she'd see him at the end of her shift, but while she was working, an administrator came to say she had a phone call. Security was so tight at Arlington Hall that it was hard for an outsider to get a call through, so Dot knew it must be urgent. The call was from one of Jim Bruce's sisters.

"Have you seen Jim?" his sister asked breathlessly. Dot said no, but that she expected to see him that evening. "He's wanted back here!" said Jim's sister. He was being shipped out sooner than he thought. This was not uncommon. Orders were often dispatched and changed at the last minute, to keep the enemy from anticipating troop movements. "I have talked to every girl he knows, and you were last on the list."

Jim showed up at the end of Dot's shift, unaware that his unit was trying to reach him. It emerged that he did have another girlfriend, who had been willing to see him off. But after he'd put that young woman on the train, he'd rented a car and driven up to Arlington in a spontaneous burst of passion to see Dot, the one girl he really wanted to see before he left for war. "You've got to be back at the boat," Dot told him. They said a hasty good-bye, and Jim, having returned the rental car, scrambled to get a seat on a train back to Richmond. And that was the last time Dot Braden saw Jim Bruce for almost two years.

Life was like that. Men came and went. At night, air raid sirens would sometimes go off, signaling residents to turn off their lights and pull down their blackout shades. This happened when Dot and Crow were having dinner with an old friend of Dot, Bill Randolph, who had taught at the military academy in Chatham and now held a diplomatic post. Bill's mother lived in Alexandria, nearby. The sirens started wailing, so Bill pulled out his guitar and they all sat on the porch in the darkness, Dot and Crow and Bill and his mother, and sang into the night. Life was strange now, and often oddly pleasant, even as Dot kept big framed photographs of her brothers, Teedy and Bubba, and George Rush and Jim Bruce and worried about all of the men in her life, all the time.

PART II

"Over All This Vast Expanse of Waters Japan Was Supreme"

CHAPTER FIVE

"It Was Heart-Rending"

June 1942

The two diplomats were hatching plans to sow discord among the American people. More than five thousand miles away from each other, working on different continents, the men strategized by long distance, pinpointing vulnerabilities in the enemy's social fabric that could be inflamed through propaganda. Japanese foreign minister Shigenori Togo, writing from Tokyo in June 1942, asked Baron Oshima, the ambassador to Berlin, to share "material and background" suggesting ideas for propaganda that could be directed at the United States. Togo cited such possibilities as inflation and the "treatment of negroes." Writing back on July 12, Oshima ventured that Japan should target isolationists and make every effort to further unsettle U.S. citizens chafing at the hardship and deprivation of war. The unity America displayed after Pearl Harbor had now dissipated, Oshima opined from his perch in Berlin, and the national mood was vulnerable and sour.

"The successive retreats of the American and Allied forces since the thunderbolt at Pearl Harbor have completely upset the moral equilibrium of the American people who had been taught to believe that Japan could be completely whipped in six months and that the American Navy had no peer," Oshima ventured smugly. The missive was enciphered and sent on. The men had no idea that their high-level musings were being read by the enemy, much less by young enemy women.

It was true that the first six months of 1942 were a dark and disheartened time for the United States, especially for the U.S. Navy. In retrospect it is easy to underestimate how fragile that period felt. The Japanese were a formidable naval foe. At the outset of the war, the Japanese Navy controlled one-quarter of the Pacific Ocean and had not lost a naval battle in more than fifty years. Japan had a brilliant top commander in Admiral Isoroku Yamamoto, who masterminded the Pearl Harbor attack. But Yamamoto—who had initially opposed going to war—was not foolhardy. He had spent time in the United States, studying economics at Harvard, and well knew America's industrial might. He understood that Japan had a limited window before U.S. factories could unleash their full power, producing ships and planes in overwhelming numbers. Yamamoto warned his superiors that he could "run wild" in the Pacific for six months or even a year, but needed a knockout blow—what the Japanese called a "decisive sea battle"—so that the war would be over before the Americans could augment their fleet.

The point of Pearl Harbor had been to deliver that blow. The attack didn't succeed on that level—U.S. aircraft carriers were safely out of the harbor, and some stricken battleships could be recovered and repaired—but attacks elsewhere in the Pacific followed so quickly as to feel simultaneous.

Just hours after Pearl Harbor, the Japanese launched air attacks on the Philippines. They captured Guam, in the Marianas, two days later, and took Wake Island before Christmas. Meanwhile the Japanese Army was stabbing westward and southward, capturing British and Dutch colonial holdings. Hong Kong fell on Christmas Day 1941, Singapore two months later, Burma in May. The Japanese cut a merciless swath through the Dutch East Indies (whose defending troops could not expect reinforcements from their Nazi-occupied home country), taking island after island along the Malay Peninsula—Java, Borneo, Celebes, Sumatra—which stretched through the lower Pacific down toward Australia.

February 1942 brought the worst blow to the American forces. After months of fighting, Roosevelt ordered General Douglas MacArthur to leave the Philippines. The U.S. Navy had a small cryptanalytic team holed up in the Philippine island fortress of Corregidor. The men were smuggled out by submarine and taken to Australia before Corregidor too fell, marking a breathtaking series of victories for Japan. The juggernaut of attacks also devastated the British, who lost their warships *Repulse* and *Prince of Wales*. "As I turned over and twisted in bed the full horror of the news sank in upon me," British prime minister Winston Churchill wrote in his memoirs. "Over all this vast expanse of waters Japan was supreme, and we everywhere were weak and naked."

In the Atlantic, things were going equally badly, if not worse. The late spring of 1942 marked the low point for Allied powers in both of the world's great oceans. The United States now was an active partner in the Battle of the Atlantic, the deadly six-year contest between German U-boats and Allied convoys. The Battle of the Atlantic began on the first day of the war and did not end until the last. Winston Churchill described it as the thing that worried him most. Whoever won the Battle of the Atlantic, it was believed, would win the European war. Beginning in the spring of 1941, President Roosevelt authorized that supplies could be sent to England, and U.S. Navy ships could escort merchant vessels in the North Atlantic. This meant American sailors were put in harm's way even before America was a formal combatant.

In 1942 the United States had begun to feel the U-boat peril in a much more violent and intimate way. After Pearl Harbor brought America into the war in earnest, German Admiral Karl Dönitz saw a ripe opportunity: the vast and unprotected Atlantic coast of the United States, from Maine to Florida. The U-boat commander dispatched his submarines to cruise the East Coast, where they roamed startlingly near shore—the Germans called this the "Happy Time"—sinking freighters, tankers, trawlers, and barges. The goal was to destroy supplies being produced to feed the Allied war effort. The U.S. Navy was slow to organize an escort system for coastal shipping, and ships were sunk in full sight of horrified

American citizens, who could stand on beaches and watch freighters burning. The Outer Banks of North Carolina became known as "torpedo junction" because of the number of ships destroyed there. The men on the U-boat crews clambered on top of their subs and took keepsake photos of the wreckage.

U-boats were like terrorist cells in their ability to sow fear. They were invisible, ubiquitous, noiseless. To pluck off ships making the Atlantic crossing, the U-boats would place themselves across the convoy lanes and lie in wait. When a U-boat spotted a vessel, it would radio central command, which would alert other subs to close in. Some U-boat commanders were so daring that they would submerge and surface in the middle of the convoys, shooting at the Allies from the inside out.

The Battle of the Atlantic was a war of lives and of commerce. England needed food. The Allies needed troops and war materiel to press their campaigns in Italy and North Africa. American shipyards were churning out Liberty ships—low-cost cargo ships that were being massproduced in unheard-of numbers—but the U-boats in 1942 were able to sink ships faster than America could make them. Making things worse was the fact that the Germans were reading the cipher the Allies used to direct their convoys, something the Americans suspected but the British were slow to admit.

To be sure, the Allies for some of this time were reading the German cipher, so the Battle of the Atlantic also was a battle of code-breaking prowess. U-boat messages were enciphered using Enigma machines, which the Germans believed could not be broken. To send a message, an Enigma operator inserted three rotors and positioned them in a certain order. When a single letter was typed on the keyboard, the rotors—which were facing one another, like hockey pucks stacked sideways—would turn, transforming the letter over and over. A light on the top side, an ordinary flashlight bulb, would illuminate the letter as it emerged in its enciphered form; that letter would be radioed. Each Enigma had more than three rotors to choose from, and each rotor could be set in twenty-six positions. The rotors were surrounded by movable outer rings, and there were plugs, called steckers, attached to a board. The upshot of all this gadgetry was that there were millions of ways a letter could travel through the encipherment process.

One major strength of Enigma was the setting order for the rotors and other movable parts, which was known as the key and changed each day. The Germans knew that commercial Enigmas had been circulating around Europe in the 1920s and early 1930s and that the Allies might have some idea how they worked. But they believed their enemies could not break the key, which was somewhat like guessing a computer password. Even if it was theoretically possible, the Germans figured, breaking the key would require a building full of machines to run through all the potential combinations, and they did not believe the Allies could produce a building full of machines.

Before the war, a team of Polish cryptanalysts had in fact figured out the workings of the Enigma. Small, vulnerable nations surrounded by big potential enemies—Poland is bordered by Russia and Germany—tend to be hypervigilant about their neighbors, and the Polish Cipher Bureau was remarkably good. The Poles broke the Enigma during the 1930s, in part thanks to a German who passed schematics and decrypted messages to French intelligence, who passed it to them, and to a commercial model they obtained. The Polish mathematician Marian Rejewski solved the wiring, and in 1938 they built six "bomby" machines that could detect possible daily settings.

In July 1939, before the Nazis overran their country, the Poles shared their discovery with the British and French. At Bletchley Park, Alan Turing and others refined the design, developing a method in which cryptanalysts could use a crib and write it beneath the cipher, then figure out, mathematically, what combination of rotor settings, wheel settings, and steckers might produce the cipher. This "menu" permitted the machine to check for possible settings, and was in some sense an early form of a computer program. The British built sixty "bombe" machines, which, beginning in 1941, were run by some two thousand members of the Women's Royal Naval Service, or Wrens. The bombes would test a menu to see if it could be a viable key setting. If the bombe got a "hit," then a smaller machine—an Enigma facsimile—was programmed with the setting and a message fed into it. If coherent German emerged, the code breakers knew they had the correct key setting for the day.

The British at first kept their bombe project secret, even from their allies, for fear the enemy would find out and change the codes. Churchill called his Bletchley code breakers the "geese that laid the golden eggs, and never cackled." In February 1942, however, the hypercautious German Navy added a fourth rotor to the naval U-boat Enigma machines, increasing the possible combinations by a factor of twenty-six. The Allies called this new four-rotor cipher "Shark," and initially it proved impenetrable. The Allies lost the ability to read U-boats. The whole system went dark. This crushing turn of events occurred just months after the United States entered the war, and it began an eight-month period of death and destruction and helplessness, a time when ship after ship went down and it felt very much as though the war could swing the wrong way.

* * *

This, then, was the demoralized atmosphere that the young women from the Seven Sisters colleges were entering when they traded their May Queen festivals, their end-of-year hoop rolls, their amateur theatrics, and other hallowed traditions for service in the hot downtown offices of the U.S. Navy. The Navy code-breaking program, as ever, operated separately and apart from that of the Army: While Arlington Hall across the Potomac River pursued its attack on diplomatic ciphers, the Navy, still in its downtown headquarters, wrestled with the task of breaking enemy naval messages in the two major oceans, with lead responsibility for code breaking in the Pacific. It felt, at the outset, like an undoable task. The women in the summer of 1942 were signing on with a Navy still reeling from Pearl Harbor and the swift Japanese victories that followed. Officers were being reassigned and lines of command reshuffled. America was losing the war on all sides—or so it felt—and the atmosphere was chaotic. In January 1941, naval code breaking consisted of just sixty people occupying ten office rooms in the sixth wing of the Navy building. By mid-1942 the number had increased to 720, with more arriving every day. The rooms were starting to overflow.

Up to that point, it's worth noting, the Navy code-breaking office had employed some civilian women apart from Agnes Driscoll but treated them differently from men in terms of pay. According to a November 1941 proposed salary memo, female clerks, typists, and stenographers were paid $1,440 per year, while men doing the same job made $1,620. Women college graduates who had taken an elementary course in cryptanalysis made $1,800; men with those qualifications made $2,000. Women with master's degrees made $2,000, compared to $2,600 for men. Women PhDs made $2,300; men with doctorates made $3,200. The early women came from a variety of backgrounds: Some officers' wives and daughters liked to dabble in cryptanalysis, and there was a civilian, Eunice Willson Rice, who came from a staunchly naval family and worked on the codes of the Italian Navy. When she became pregnant, the men liked to call her Puffed Rice.

Despite these historic inequities, the young women's arrival could not come soon enough for the Navy. In May 1942, none other than Commander John Redman—head of the code-breaking operation, known as OP-20-G—wrote each female student, begging her to get herself to the Navy building as fast as she could.

"Could you start within a week or two after the close of College?" Redman asked Ann White and Bea Norton and the other Wellesley seniors. He sent the same letter to each woman at Goucher and the other cooperating schools. "There is important work here waiting to be done," Redman told them, adding that "it is a good opportunity for you, particularly since you are getting into this work in its early stages." He gave each woman the address of the office and begged her to "keep me posted as to the approximate date of your arrival."

By now the women's ranks had winnowed. The students from the Seven Sisters and Goucher had been selected on the basis of ability, willingness, and loyalty, but tenacity was something they had to prove during the months-long correspondence course. Some had become discouraged and dropped out; others married and relocated to follow husbands; others did not answer enough problems correctly; others were rejected by the Civil Service Commission based on some aspect of their background. Back in the patriotically fervent winter of 1941, Barnard had enrolled twenty women, of whom seven stuck it out and showed up at Main Navy, the Constitution Avenue headquarters. Bryn Mawr started with twenty-seven and ended with twelve. Goucher's ranks fell from sixteen to eight; Mount Holyoke's, from seventeen to seven. Radcliffe had a bounty of fifty-nine women at the start and just eight at the finish. Smith's first class fell from thirty to twelve; Wellesley's, from twenty-eight to twenty.

In all, 197 young women had received a secret invitation. A hardy band of seventy-four survivors found their way to D.C., where they were employed as SP-4s, assistant cryptanalytic aides making $1,620 a year. Goucher graduates Constance McCready and Joan Richter were among the first to arrive, showing up at the front desk on June 8, 1942. Viola Moore and Margaret Gilman, from Bryn Mawr, walked through the doors of Main Navy on the fifteenth. The rest trickled in toward the end of June and beginning of July.

The Navy didn't want to lose a single one. Fearful the women might quit if they couldn't find housing, the Navy wrote each college president, seeking help in locating alumnae for the women to stay with. Some took rooms at the Meridian Hill Hotel for Women, a Washington residential hotel constructed to house g-girls. Others were scattered throughout northwest Washington, staying at homes on Klingle Road and Euclid Street and elsewhere. Vi Moore and Margaret Gilman lodged at 1611 Connecticut Avenue NW. Anne Barus, from Smith, found herself living at 1751 New Hampshire Avenue NW, with Bea Norton and Elizabeth "Bets" Colby and other Wellesley women. Their addresses changed often. Many would live in six or seven different lodgings during their wartime tenure. Navy memos show that clerks were constantly typing updated addresses as the women scrambled for rooms in basements, boardinghouses, and—in one case—the back half of the Francis Scott Key Book Shop in Georgetown, where a group of women were allowed to borrow books and use the telephone, in return for letting the bookstore staff use their lone toilet.

As quickly as they arrived, the women from the Seven Sisters found themselves put to work. The Navy operation was already on a twenty-four-hour, around-the-clock basis, and the women were divvied up between the three shifts, known in the Navy as "watches." Fran Steen from Goucher and Ann White from Wellesley were among those who drew the midnight watch, from midnight to eight, while luckier souls drew the day watch, from eight to four, or evening watch, from four to midnight.

The summer of the Navy women's arrival was punishingly hot. The women would start each day in high heels and clean cotton dresses and take the bus downtown. By the time they arrived—or after working for half an hour—they would be dripping with sweat and the thin fabric of their neat dresses would be plastered to their skin. They would lift their forearms from the table and find the paper beneath was soaking wet. Salt tablets were kept in dispensers—they were a fad of the time; it was mistakenly thought the tablets prevented perspiration—which made many of them sick. The old Navy headquarters was not just crowded but unclean. Vi Moore, a French major from Bryn Mawr, was assigned the task of reporting how many cockroaches were crawling around in the women's bathroom.

Back at their colleges, the women's training had been rigorous and they had taken it seriously. The naval course included exercises in which they had to memorize the most common English letters—*E, T, O, N, A, I, R,* and *S*—and take frequency counts. They had been introduced to old-fashioned methods like a grille, which is a template that can be put over an ordinary letter, with little holes that make certain words pop out to reveal a hidden message. They were instructed that "the motto of the cryptanalyst should be: 'Let's suppose,'" and that "the most important aid in cipher solution is a good eraser." Weekly problems tested their mastery of "numerical cipher alphabets," "polyalphabetic substitution," and "diagonal digraphic substitution." Each packet contained one problem that could not be solved, to show that sometimes, a jumble of letters or numbers doesn't stand for anything; sometimes, a code breaker fails. The fine institutions the women attended had never encouraged them to fail, and they found the idea disconcerting.

It was a good course, and they had worked hard at mastering it, but the problems often didn't dovetail with the actual work they found themselves doing; lots of the tasks they were facing had not been covered. And all of this was no longer an academic exercise. They were responsible for men's lives, and the responsibility felt awful and real. Most of the women started out on the Japanese desk, but those few who knew German soon found themselves helping fight the Battle of the Atlantic. The British still had lead responsibility, but the Americans—who also had a stake in the outcome—were doing what they could to help crack Shark, the four-rotor Enigma cipher. Without the aid of bombe machines, the women used hand solutions to try to guess the day's key setting. They were in constant communication with the British, trading notes and cribs.

Margaret Gilman, who had majored in biochemistry at Bryn Mawr but studied German in high school, was given a proficiency exam in German and put to work in a small sealed room attacking Shark. Her unit consisted of all women, with a lone male officer supervising. In a room guarded by Marines, Gilman labored over Nazi messages transmitted in the Bay of Biscay, the body of water off the coast of occupied France, where huge U-boat bases now were located. The U-boats had to cross the Bay of Biscay to reach the Atlantic convoy lanes. Before the subs left base, the Nazis would send out weather vessels to report back on conditions, using Enigma machines. There are a limited number of weather-related words—wind, rain, clouds—so it was sometimes feasible to come up with cribs. "BISKAYAWETTER" was a crib the women often would try as they made charts and graphs of common cribs and the places in messages where Germans were most likely to nestle certain words.

The urgency of the work was harrowing. American men, they knew, were trying to cross the ocean where the U-boats waited. "German submarines were literally controlling the Atlantic Ocean," Margaret Gilman recalled later. "Can you imagine sending out American troop ships loaded with soldiers through an Atlantic ocean riddled with submarines? It was heart-rending, oh my God." In the unit's workroom, a detailed wall map displayed the Atlantic Ocean, with pins for every U-boat whose position they could locate. Margaret couldn't stand to look at the map and would position her head to keep it out of her peripheral vision. The morale of the whole country would suffer when a troop ship was lost, and the women felt the burden of responsibility. "If we had any doubts about whether what we were doing was important," she recalled, just let a few days go by with no progress, "and the brass were down there yelling at us—what are we doing, neglecting our duty."

Ann White also was assigned to the Enigma unit, having majored in German at Wellesley. The work brought her into uncomfortable contact with the humanity of the enemy. The British were sending over items to help develop cribs based on things like lengths of transmissions, the locations of the boats they were addressed to, and instructions for returning to port. From time to time, the British would send documents found in a sinking or captured U-boat. These included personal effects, such as family photos, belonging to German sailors who now were drowned or captive. Once, Ann's team broke a message from a Nazi commander announcing the birth of his son. One code breaker composed a snatch of doggerel as a translation: "From here to Capetown / be it known / A little Leuth / has now been bo'n," which of course rhymed only if you had a southern accent.

But mostly, the work was frustrating, and it imbued the women with sadness and a sense of failure. Ann White's job was to translate German messages into English, so she knew what the contents said. During the winter of 1942–1943, her unit partially cracked a message from Dönitz, alerting a wolf pack to a convoy of Allied ships passing the southern tip of Greenland. The code breakers, American and British, desperately tried to determine the location of the U-boats that lay in wait but could not. Later, they learned that most of the ships had been lost. "We worked on the Enigma desperately," Ann White would later say. "Blindly." It was a relief to be doing something: "Everyone we knew and loved was in this war. It was a Godsend for a woman to be so busy she couldn't worry," she reflected. But "we knew men were dying."

The mood in the Japanese code rooms was equally grim, as women struggled to learn their work and to master the newest iteration of JN-25, the Japanese fleet code. It was such a daunting effort that, in the unit Vi Moore was assigned to, more than one commanding officer expressed the view that America might lose. Another liked to say that even if the Allies did prevail, "every war was a preparation for the next one." Fran Steen lost a fiancé, shot down in the Pacific, early on. Many American men who had been stationed on Pacific islands before the attack on Pearl Harbor were now prisoners of war. Erma Hughes, a psychology major recruited out of the University of Maryland—her father, a bricklayer, sold land to finance her tuition—was sending care packages to classmates in POW camps. The ROTC students in her class had mostly become para-troopers. For graduates of the class of 1942, the attrition of friends and classmates was stark and acute. Erma could never be certain her care packages arrived and did not even know whether the intended recipients were alive, but she kept sending them just in case.

* * *

At this point, naval code breaking was still laboring under a very dark cloud. The disaster of Pearl Harbor had called into question the value of cryptanalysis, and many top naval officials felt that even when it did work, code breaking took too long to be of use in the heat of an ongoing battle. There were other disadvantages beyond their cramped and cockroach-ridden conditions. As the women streamed into the D.C. headquarters, male officers were sent to the Pacific to work on smaller teams set up near intercept stations. The Pacific field units could begin tackling messages as soon as they were plucked out of the air. These field teams sometimes deciphered more quickly, but Washington—with more machines and a bigger staff—would eventually produce more solutions. Often, though, Washington had to wait a long time for intercepts to arrive. The Navy headquarters had some teletype lines, but not enough. Some messages were sent by air, stowed aboard the luxurious Pacific Clippers operated by Pan-American Airways, but many more were sent by boat and took weeks, sometimes even a month, to make the journey. The teams—D.C. headquarters and field units—were cooperative but also competitive, solving JN-25 by agonizing bits and pieces, sharing recoveries of code groups, additives, or pieces of intelligence.

Despite the Americans' frustrations in 1942, the Japanese were more vulnerable than it might have seemed. It was one thing to capture so many islands and bases, and another thing to supply and defend them. The U.S. Congress had passed the Two-Ocean Navy Act, funding a massive fleet-construction program. And it was important that America's aircraft carriers had been safely out of harm's way during the attack on Pearl Harbor. World War II was the first war whose naval outcome would turn on aircraft carriers and the planes flying off their decks.

Things had begun looking up for the code breakers just before the women's arrival in June. The first indication of the vital role code breaking would play in the outcome of many epic Pacific engagements had begun to emerge in early May, when decoded JN-25 messages tipped off Admiral Nimitz that a Japanese fleet was steaming back from an assault on the British in Ceylon and now aimed to capture Port Moresby in New Guinea. The Japanese were astonished when two carrier task forces of the American Navy materialized to meet them. The Battle of the Coral Sea, from May 4 to May 8, 1942, was the first naval battle in which the opposing ships never saw each other—the fighting was all done by aircraft—and it was the first Pacific contest where code breaking played a key role in the outcome. The result was a tactical draw—the Americans lost the *Lexington*, and the *Yorktown* was badly damaged, but the Japanese losses, including many of its best-trained pilots, were bad. It also checked Japanese expansion toward Australia.

By mid-May, the U.S. Navy got wind of an even bigger Japanese operation. Thousands of messages began flashing back and forth in JN-25, suggesting Japan was sending a massive flotilla somewhere. "Vague indications of an operation to be launched," noted one early message. Joe Rochefort, one of many officers working in the Pacific who had been trained by Agnes Driscoll, supervised the Pearl Harbor code-breaking team. A lot of information was obtained about the planned Japanese operation, but there was one key puzzle piece that stumped everybody: In mid-May the Americans intercepted a message saying that the Japanese, who often used two-letter geographic designators, were headed to "AF." The code breakers had recovered some other designators but could not be certain where AF was. Rochefort and his team felt sure AF stood for Midway, a tiny atoll where the United States still maintained a base, crucial for defending Hawaii and the West Coast and indeed for maintaining any U.S. presence in the Pacific. Others thought the target might be Hawaii or the Aleutians.

So Rochefort and Edwin Layton, Nimitz's chief intelligence officer—also trained by Agnes Driscoll—hatched a plan. They instructed the men at the Midway base to radio a message—not coded, just plain English—saying their distillation plant had broken down and that Midway was short on water. The idea was that the Japanese would intercept the bulletin and pass it on. Just as they hoped, a local Japanese unit picked it up and sent its own message saying AF was short of water, and the message got passed along to the fleet. The Americans intercepted it. The trick succeeded. The Americans had confirmed that AF stood for Midway.

Admiral Yamamoto's aim was to achieve the knockout blow that had eluded him at Pearl Harbor. Mustering a mighty flotilla of more than two hundred warships, transports, and auxiliaries, Yamamoto intended to split his forces and send one contingent—a smaller one—to the Aleutian Islands, off the mainland of Alaska, to mount a subsidiary attack that would serve as a decoy. Nimitz, he reckoned, would hasten to counter that attack. By the time Nimitz got back, the Japanese would be at Midway in great numbers, prepared to ambush him and finish him off.

But Nimitz didn't take the bait. He let the Japanese head to the

Aleutians and set about reinforcing Midway, in order to ambush the ambushers. Thanks to cryptanalysts reading JN-25, Nimitz knew more about the planned attack than most Japanese officers did. "He knew the targets; the dates; the debarkation points of the Japanese forces and their rendezvous points at sea; he had a good idea of the composition of the Japanese forces; he knew of the plan to station a submarine cordon between Hawaii and Midway," noted an internal history.

The Japanese duly showed up on June 4. Their carrier task force launched an air strike on the island, but this was no Pearl Harbor: American fighters scrambled to meet the incoming planes in midair, taking heavy fire but pushing the enemy back while four waves of U.S. bombers took off toward the Japanese carriers. The Japanese had expected to be attacked by planes from Midway, but didn't realize the Americans also had aircraft carriers nearby. They soon would: From the decks of the *Hornet*, the *Yorktown*, and the *Enterprise*, torpedo and dive bombers took off. The American planes caught the Japanese carriers as they were preparing for another attack on Midway; the Japanese decks were crowded with bombs and fuel hoses, setting off massive fires. The Japanese had expected a quick victory; by the end of the first day, with its attack force stunningly crippled, it was clear this would be a different story. By the time the Japanese withdrew, calling off the operation, the U.S. Pacific Fleet had lost 2 ships, 145 aircraft, and 307 men; the Japanese endured devastating losses including 4 carriers, almost 300 aircraft, and more than 2,500 men.

The four-day Battle of Midway was an unparalleled American victory. One of the most storied naval battles in world history, it marked the end of Japan's expansion in the Pacific and a major turning point in the war. Slowly but surely, for the United States, the Pacific War would turn from a defensive to an offensive one. The fact that an outnumbered American fleet had scored a resounding win over an armada of enemy attackers did much to lift naval morale—not to mention the spirits of the country. Commanders are often loath to share credit, but Nimitz allowed that code breaking had provided a "priceless advantage" at Midway. To be sure, it was mostly a victory for the code breakers in the Pacific unit;

Washington was still bogged down in acrimony between code breakers and top brass. Even so, "the Battle of Midway gave the Navy confidence in its cryptanalytic units," an internal history noted, and it gave the code breakers confidence in themselves. More staff would be funneled into Washington, more funds freed up, more teletype lines established. The Americans had delivered payback for Pearl Harbor.

"Midway," the history noted, "was a vindication and an incentive."

The Midway victory also set in motion one of history's great bureaucratic backstabbings. Joseph Wenger and John Redman, two of the top intelligence officers in Washington, had believed the attack would happen a week later than it did. Joe Rochefort and the team in Pearl Harbor had gotten the date right. To cover their mistake, the Washington bigwigs (who feared Rochefort was building a unit to compete with theirs) let it be known that they were the ones who had pinpointed the correct date, and Pearl Harbor had gotten it wrong. This shocking lie found its way all the way up to Ernest King, commander in chief of the U.S. Fleet. Redman and Wenger continued to conspire against Rochefort, who eventually was relieved of his command and put in charge of a dry dock.

* * *

The great Agnes Driscoll by now had also been sidelined. In 1937 she had suffered a car crash that broke her leg badly, as well as both jaws. It took her a year to recover, and in some ways she never did. Many people felt her personality changed following her ordeal. In 1940, before Pearl Harbor, Miss Aggie had been taken off JN-25 and put in charge of an independent U.S. solution of the Enigma. It was not a good match. She seems to have suffered from an excess of pride; at one point the British made early overtures of cooperation but she rebuffed them, apparently wanting to succeed on her own. After more than twenty years as a cryptanalyst, Agnes Driscoll now occupied a strange position, revered by many in the Navy yet also marginalized. The Navy did not seem to know what to do with her. Military men willing to brave enemy fire were wary of crossing her; she seems to have been treated with a poisonous combination of deference and dismissal.

"I never felt that I should go tell her that the world had fallen, times had changed," said Prescott Currier, a young officer assigned to work with her. Driscoll had recruited her sister to help her and "had two cronies, Mrs. Talley and Mrs. Clark," both "mediocre office-type clerks," as Currier described them, taking her frequency counts.

Code breakers, like poets and mathematicians, often do their best work when they are young. Even the great ones, like William Friedman, at some point go off the boil. In later interviews, the men described Agnes Driscoll as resorting to extreme measures to retain her authority, enforcing a rule of silence in her office, hoarding intercepts so no one could track her progress. It's possible that Agnes Driscoll by now was past her mental prime. But it's also possible that secreting intercepts and surrounding herself with loyal henchwomen was her way of preserving authority as the world around her was becoming bigger, more competitive, and more male. With a war on, her civilian status was more of a liability than ever. "She became fearful that she wouldn't be able to do things," Howard Campaigne, another newcomer, later reflected. Even the most junior officer was her superior. "As an officer, I fitted into the organization better." This much certainly can be said: She was not as gently treated as William Friedman.

The order was changing. The Navy was bringing in male graduate students and college professors, reservists who provided fresh expertise and thinking. Many were mathematicians, like Driscoll, but unlike her they had enjoyed the benefit of attending institutions like Yale, Princeton, and MIT, which would never have admitted her. They were big men—literally—and several took one look at her and thought: "witch." Their oral histories, taken years later, obsessively use this one word to describe her. Before her accident, "she was a very strikingly beautiful woman in her early forties," one of them, Frank Raven, said. "When she came out she looked like a witch in her seventies who could only walk with a cane and with her sister holding her arm."

It was Raven who did Agnes Driscoll in. Frank Raven was a smug Yale graduate who seems to have arrived spoiling for a fight. He was a brilliant cryptanalyst; it was Raven who figured out how to predict the daily key settings for the Purple machine. But he was also a malcontent and an instigator. Raven felt the old Navy admirals were in thrall to Driscoll and decided to do something about it. "You can't visualize the climate around Aggie," he told historians. "There wasn't a regular Navy officer except Safford who had the guts to say boo to that gal." In the early months of 1942, Raven was directing a unit of twenty men working in a room next door to hers, assigned to "German Navy miscellaneous," a term for any German messages that weren't U-boat. But he was eager to get his hands on some real Enigma intercepts. So Raven decided to pillage Agnes Driscoll's safe, which he had the keys to. During an overnight watch he rifled her papers and saw some from England; this convinced him the British had their own Enigma solution. Hers seemed to be a laborious paper model, which to succeed would have required what he called "a trial of exhaustion."

Subsequent historians have discerned that Miss Aggie's Enigma solution might have worked, but only with the kind of supercomputers that came along much later. Raven figured it would take a whole war to get one message. He claimed later that she set the U.S. Enigma project back by three or four months, because the mathematicians who were being brought in to work in earnest on the four-rotor Enigma had to tiptoe around her. Driscoll was the "curse of the Enigma effort well up into 1944," he said. "The old Navy considered Aggie as sort of a god, some sort of a goddess."

Raven didn't just despise Agnes Driscoll; he despised many of the people he was working with. It was a tense atmosphere in the downtown naval code-breaking offices, full of politics and subcurrents. There was a strong caste system: Career naval men distrusted the new, educated reservists; reservists thought they were smarter than careerists; everybody looked down on civilians. If you were a woman, you had three strikes against you. One of the officers had a hair-raising collection of graphic pornography, which he kept in a drawer and which the security guards liked to come look at.

Raven managed to engineer Miss Aggie's downfall. As the war went on, Agnes tended to be given projects that seemed hopeless— busywork—and at one point was assigned to work on a machine-generated Japanese naval attaché cipher. Naval attachés are military men whose time-honored duty is to serve as spies under a kind of flimsy cover, hanging out affably in foreign embassies, reporting back on the weapons of other countries. During a lull in his own work, Raven returned to his old habits. He opened Agnes's safe, "bootlegged copies" of her naval attaché intercepts, and solved them. He had a machine built to decipher them, and he tapped Agnes's intercept line so he could get incoming traffic. He tried to hide the machine from her, but Driscoll eventually saw it, whereupon, Raven said, she "demanded that I be court-martialed." But he had solved the machine, and she had not, and that did Agnes in. She was not fired, but she was put out to pasture.

After the war, Raven claimed that he saw her downfall as one of the tragedies of the war and believed she should never have been readmitted after the car accident. "In retrospect I am convinced that Aggie Driscoll is one of the world's greatest cryptanalysts," he added. "I am convinced that the same accident that moved her from a beautiful woman to a hag affected her mind and that when she came back she couldn't solve a monoalphabetic substitution."

Nobody knows how Agnes Driscoll felt. Nobody bothered to take an oral history from one of the greatest cryptanalysts in the world.

* * *

This was the dynamic at work in the careerist Navy facility: enmities, lies, thefts, insecurities, power plays. In so many ways, the young women who joined it had no idea what they were getting into. But they were delighted to be there. The number of workers was so small—and the task so large—that the women from the Seven Sisters instantly took on real responsibility. As the women dove in, cryptanalysis was being done by a group called OP-20-GY, which soon was divided into GY-P, for Pacific, and GY-A, for Atlantic, and then subdivided. Room assignments were made and remade as GY overflowed into three wings of the old Navy building, an arrangement that was "neither convenient nor economical," as one memo put it. The size of a shift might be determined by how many chairs were available. There was nowhere to stow workbooks, no secure telephones between rooms; papers had to be shoved aside to make a little training area; and with so many people stomping in and out of the main naval headquarters, secrecy became a constant headache. The group working on JN-25 soon occupied two wings. Vi Moore found herself doing additive recovery in Room 1515, while Anne Barus, Louise Wilde, Ann White, Fran Steen, and others were doing the same thing in Room 3636, an awkward distance away.

Most of the women were assigned to JN-25, at a time when the complex supercipher had become more complex. The Americans weren't the only ones galvanized by Midway. In the wake of Japan's shocking naval defeat, the Japanese decided to divide the fleet code into five "channels," so that certain regions or kinds of communications—Singapore, the Philippines, operational, administrative—had their own code and additive books. The volume of messages grew and grew. The naval code breakers received 18,000 JN-25 intercepts per month in the first six months of 1942, and more than double that, 37,000, in the second half of the year. By the fourth quarter of 1943 they would be getting 126,000 messages per month.

The women rose to the challenge. Anne Barus, the Smith history major, was assigned to recover additives, a task her collegiate training course had not covered, and one that involved ceaseless mental math performed day after day, week after week, for more than three years. The women in her unit were given big sheets of paper, about a yard long and two feet wide. Each sheet was filled with rows of five-digit numbers—14579 35981 56921 78632 90214, say—that also lined up in vertical columns. The messages were placed so that each code group was directly above a code group enciphered by the same additive, a symmetry determined by women trained to evaluate a key group at the beginning. It was Anne's job to figure out what the additives were, so that the Japanese additive book might be reconstructed.

To do this, Anne had to master the same "false math" the Japanese used—only in reverse. She and her colleagues had to start with the enciphered numbers and work backward to find the underlying code group. And they had to do it fast. Looking down a vertical column, Anne was tasked with finding the lone additive used to encipher all the code groups in that column. Aside from her own wits, she had one thing to help her: a quirky feature designed to cope with radio garble. Garbling was a huge problem in radio transmissions, so the Japanese developed clever "garble checks" so the person at the receiving end could do a bit of math to be sure the message had transmitted correctly. It was a sensible enough tactic, except that it also was an insecure one: Many of these checks—and the ghostly patterns they left—helped with breaking the messages.

One such JN-25 garble check was the rule that a valid code group was always divisible by three. Looking at her work sheet, Anne would conjecture a possible additive, then go down the vertical column, quickly, in her head, stripping the hypothesized additive out of each group she saw, and looking to see if the remainder was divisible by three. If she conjectured an additive, stripped it, looked at the row of code groups, and saw that all were divisible by three—17436, say, or 23823—then she knew she had gotten down to valid code groups and had therefore conjectured a viable additive. It took all this work to get one single additive, which would be recorded in the book they were building. Whenever the Japanese changed the JN-25 cipher books, the unit would start all over again. It was truly like sweeping the sand from a beach.

Anne, like the other women in her room, learned to look for common enemy mistakes. More than seventy years later she would remember them clearly. In such a massive fleet system, it would sometimes happen that an oblivious radioman would send a message in the clear—plain Japanese—that others were sending in code. The women could use the plain Japanese as a crib. The Japanese, like the Germans, also tended to send out fleet messages that were formulaic and patterned. Japanese merchant ship captains often sent a *shoo-goichi* message, stating what their exact position would be at noon. Anne learned the code groups for "noon position," and she learned where the phrase was likely to appear. When she saw an enciphered group in that place, she could subtract the code group and obtain the additive.

Many things the Japanese did with the intention of making the code harder to break made it easier. Sometimes, enemy cryptographers liked to begin a message in the middle. When they did this, they would include a code group that stood for "begin message here" to show where the message started. The women learned the code groups for "begin message here"—there were several—and gained another point of entry. There also was something called "tailing." Japanese encoders were told not to end a message at a certain point in an additive book and then begin the next message at the next additive number—they were supposed to choose another random starting point—but often, they were lazy or harried, and they did. The women mastered these ins and outs and quirks. Whenever they saw a mistake, they pounced. The *shoo-goichi* messages did more than help recover additives; the noon position would be swiftly radioed to an American submarine captain, who would be waiting for the Japanese ship when it appeared on the horizon.

It was boring, tedious work, except when it wasn't. Elizabeth Bigelow, an aspiring architect recruited from Vassar, also began working on JN-25 when she came in with a later university-trained cohort. She at one point was given an urgent but badly garbled cipher and asked to decipher it, which she did, within a matter of hours. It told of a convoy sailing later that day. When she was told that her work had helped to sink the convoy, she said later, "I felt terribly pleased."

The operation developed the swiftness and efficiency of an assembly line. To the extent that space provided, the women were assigned to groups, or "rooms." The write-up room would prepare work sheets. The "key" room placed the messages; the classification room salvaged garbled intercepts; a priority room with "expert additive workers" attacked "hot or priority messages." Hotlines were set up to convey additives to translators. Some messages were tagged "routine," others "urgent." There was a special category marked "frantic." As the number of messages increased, the ranks of people solving them grew steadily more female. By the fourth quarter of 1943, 183 men and 473 women were working on JN-25 in Washington—more than twice as many women as men. One memo noted that it was impossible to keep the women in the dark as to what the messages said. The memo added that the most important secret was the fact that JN-25 was being worked at all, and this secret was at the "mercy of the humblest worker who ever glanced at a work book."

The women kept that secret and became integral to the operation. And they shared the outrage when the truth about Midway's success made its way into the press. On June 7, 1942—while the battle was still going on—the *Chicago Tribune* published a blockbuster story in its Sunday edition, headed: JAP FLEET SMASHED BY U.S. 2 CARRIERS SUNK AT MIDWAY: NAVY HAD WORD OF JAP PLAN TO STRIKE AT SEA; KNEW DUTCH HARBOR WAS A FEINT. The article noted that the makeup of the Japanese forces "was well known in American naval circles several days before the battle began." The article appeared in several other *Tribune*-connected papers but was hushed up by the office of censorship out of fear that the Japanese would take note.

Then the syndicated gossip columnist Walter Winchell compounded the risk, saying in a July 5 radio broadcast that "twice the fate of the civilized world was changed by intercepted and decoded messages"—meaning Coral Sea and Midway. Two days later he wrote in his "On Broadway" column, which ran in the *New York Daily Mirror*, that Washington was abuzz over the *Tribune*'s item, which, he argued, "tossed safety out the windows—and allegedly printed the lowdown on why we won at Midway." Of course, he had done the same thing.

The Navy was so apoplectic that it ended up making things worse. When it emerged that a reporter aboard the USS *Lexington*, Stanley Johnston, was the source of the Midway story, the Navy decided to go after Johnston. The hearings resulted in more publicity, and the code breakers worried that there was no way the Japanese could ignore it. The Japanese made another major overhaul of JN-25 not long after, and many code breakers were convinced this new changeover was the result of the John-ston investigation. "Our crucial battles in the Solomons were conducted without the aid of enemy information that had been available up to this moment," wrote Laurance Safford bitterly.

Whether the Japanese did take notice is a matter of dispute. The Jap-anese periodically changed JN-25 books anyway, and their mid-August change most likely had been planned for some time. Regardless, it occurred just as the U.S. Navy began to press its new advantage in the Pacific, taking the offensive and launching an invasion to retake the Solo-mon Islands. This brave push began with the Battle of Guadalcanal, an amphibious invasion that went well, at first, then bogged down into a bloody months-long quagmire. During the battle, U.S. Marines came upon a cache of codebooks buried six feet in the ground; gallingly, they were in a version of JN-25 that was no longer in use.

Fortunately—given how often JN-25 changed over—the U.S. Navy had made the wise decision to set up a smaller unit to tackle what were called "minor ciphers." The minor ciphers were lesser systems, but they were anything but unimportant. In the vast Pacific Ocean, not every message could travel in the main fleet code. The Japanese used scores of auxiliary systems, some brief and temporary, to communicate between captured islands, or between weather lookouts and rice ships, or even just to broadcast water levels and fishing conditions. They also devised temporary "contact codes" for use in battle. The ability to read some of these systems had been useful during Midway; the Japanese changed their JN-25 cipher on May 28, just before the battle, so all through that actual engagement, the Americans were able to follow the conversations of Japanese combatants only thanks to contact codes and other minor systems.

As luck would have it, the minor-cipher unit was under the charge of Frank Raven, the very code breaker who destroyed what remained of the career of Agnes Driscoll. His "German Navy miscellaneous" team was now working "Japanese miscellaneous," plucking intercepts out of random piles in the Navy building that were accumulating in a junk box marked "W." The crew was a good one—according to Raven, they broke at least one system per week beginning in March 1942—and now the men on that crew were replaced by women. In May 1942 Raven had twenty-three male sailors working under him, and by June, just one month later, "approximately ten civil service girls" came on to form the nucleus of the new team.

Among these were Bea Norton and Bets Colby, both members of the first Wellesley cohort. Fortunately, Raven was not as unpleasant toward them as he had been toward Agnes Driscoll; he later described his new crew as "damn good gals," though he did also see fit to point out that they were "damn pretty gals." The unit's main ongoing task was deciphering something the women called the "inter-island cipher," which was known in most official documents as JN-20, and, like many minor systems, was far more important than it ever got credit for. "The Navy never mentions the inter-island cipher," wrote Bea Norton many years later, saying that it was in fact the inter-island cipher that had carried the no-water-at-Midway message. "The plain text of the water shortage was picked up by a Japanese island operator, wired in the inter-island cipher to fleet headquarters and thence to the main Japanese fleet," she asserted, saying that the reason Raven's team never got credit was because the Navy was "traditionally of the view that all meaningful accomplishment was strictly by Regular Naval personnel, and along with this, distrusted any civilian, even Naval Reserve, results."

Her assertion is plausible, and at any rate, there is no doubt about this: During the many times when the big fleet code went dark—meaning the JN-25 books changed and the code breakers could not read it—the island cipher proved a rich alternative source of intelligence. "Whenever the main code was not being read, a feeling of frustration and exasperation permeated the radio intelligence organization and spurred them on to each new success," noted one internal history. "Even during these periods the darkness was not complete. Minor ciphers were usually being read. Frequently the information gained from the minor ciphers rivaled in importance that gained from the main naval code."

For the women working in Raven's unit, the inter-island ciphers gave vivid glimpses of the warfare unfolding on volcanic beaches and in thick island jungles thousands of miles away, as the Navy commenced its post-Midway Pacific pushback. When U.S. Marines hit the beaches of Guadal-canal in August 1942, Raven's crew began to work a cipher set up by the Japanese as an emergency form of communication between the island occupiers and the fleet at sea. As the U.S. Marines pursued them, a small band of Japanese retreated into the jungle, sending twenty or thirty mes-sages a day in the tiny makeshift cipher. It gave the women a plaintive image of what it felt like to face certain death. "I have not seen the sea for two weeks," said one message. "I have not seen the sky for three weeks. It is time for me to die for the Emperor." This band of Japanese resis-ters eventually shifted to the inter-island cipher; their numbers dwindled until, as Raven put it, the "three or four men who were left got into a motor-boat; we followed them daily in JN-20 as they described the bad conditions, etc. We sank the boat."

When she started working on Raven's team, Bea Norton was assigned the tedious job of taking frequency counts of individual letters. The mes-sages arrived on Western Union tapes. Armed Marine guards stood out-side her door, and she was forbidden to keep photos or anything personal on her desk. Her college training course did come in handy, as this cipher was a "substitution transposition" cipher, which involved changing let-ters to new ones by using a table, then switching some of the new ones to further scramble the cipher. Once Raven's team constructed the table—that is, once they figured out how the alphabets were stacked—the table didn't change. The only thing that did change was the monthly key tell-ing how some of the letters were mixed up.

The changing of the key imbued the minor-cipher unit with a curi-ous work rhythm. The women would race to break a new key and got so expert ("JN-20 ciphers were broken with increasing speed and exploited with increasing efficiency," noted one memo) that they could then enjoy lulls of inactivity. They took advantage of the downtime: Bets Colby, a math major from Wellesley, was a favorite of Raven, who described her as a "real brilliant gal" and fondly remembered that she liked to throw epic parties, which stopped just short of being orgies. "One of the standing orders on the boards was that she had to get approval from me before she had a party because she'd take the crew out of action for ten days. She'd come and say she wanted to throw a party, what dates were available?" Raven later remembered.

There was a wall calendar in Raven's office, pinpointing when the island cipher would change its key. He would select a date ten days before the key change. "You can have your party in there," he would tell Bets Colby. That way, the women would have ten days to recover from their hangovers before they had to apply their minds to a new key.

Despite the bureaucratic wars raging all around them, the women loved their work. "I felt so lucky to be in this small interesting unit," said Bea Norton later, "and to feel my work had some value." Many felt they had been preparing for this all their lives. "Never in my life since have I felt as challenged as during that period," reflected Ann White. "Like Hegel's idea about when the needs of society and the needs of an individual come together, we were fulfilled."

The only hitch was the heat. From time to time the minor-cipher unit would get instructions to cover their desks because workmen were coming to install air-conditioning. Their hopes would be dashed when Frank Raven would tell them the air-conditioning was for the top officers' private offices, not for them.

They were doing such valuable work that Donald Menzel—the Harvard astronomy professor who helped recruit them—wrote Ada Comstock about the good things he was hearing from their bosses. "The women are arriving in great numbers and...they are proving very successful. Those who have written me are delighted with the work and find it interesting and exciting beyond all expectation." Preparations were made for the next cohort of female trainees, from the same schools plus Vassar and Wheaton. There was less attrition as instructors began to master the material and everybody settled in. Of the 247 seniors in the class of 1943 who took the course, 222 would finish. The Radcliffe women of '43 had been meeting on Friday afternoons on the fourth floor in a Harvard building, and toward the end, a dramatic show was made of burning their training materials, to impress the women. Even given this new influx, the Navy began to perceive that ever more girls would be needed to get the job done.

CHAPTER SIX

"Q for Communications"

July 1942

Women were proving so useful to the war effort that a new field opened to them: military service. By 1942, allies such as England and Canada had admitted women into their military, and key U.S. women's groups began pushing for America to do the same. There were only a few women in the U.S. Congress, but one—Massachusetts Republican Edith Nourse Rogers—made women in the military her passion project. As early as 1941, Nourse Rogers met with General George Marshall, informing him that she was introducing a bill creating a Women's Army Auxiliary Corps. Nobody took her seriously, at least not until Pearl Harbor.

But even once planners grasped the difficulty of fighting a two-ocean war using only men, the idea of putting women in uniform remained controversial. "Who will then do the cooking, the washing, the mending, the humble homey tasks to which every woman has devoted herself; who will nurture the children?" thundered one congressman. People worried that military service would imperil women's femininity and render them unmarriageable. Many believed servicewomen would be, in effect, fully embedded "camp followers," a euphemism for prostitutes and hangers-on who followed soldiers from post to post.

But General Marshall did see the advantage of having women doing clerical and encoding work. Like so many, Marshall believed women were well suited to telecommunications, being dexterous and willing to do work that was boring and routine, and he felt they would make fewer mistakes than men did. President Roosevelt signed the WAAC bill into law in May 1942, a mixed victory in that women were allowed into the Army on an "auxiliary," or inferior, basis. WAACs were paid less than men and did not hold the same ranks or receive the benefits. Some of this disparity would be rectified when "auxiliary" was dropped in 1943 and the WAACs became WACs, but women were by no means equal. The WAACs, coming first, bore the brunt of negative publicity, enduring gibes about their chastity and criticism of their morals and motivation for joining.

Even so, they fell over themselves to enlist. 10,000 WOMEN IN U.S. RUSH TO JOIN NEW ARMY CORPS, wrote the *New York Times* on May 28, 1942, noting that at a single recruiting office in New York, fourteen hundred women put in requests for applications in person by the end of the first day, and another twelve hundred by mail. "Mild brute strength was used to combat the feminine forces," the reporter noted with a flourish of purple prose. "A guard's broad shoulders held back the tidal wave of patriotic pulchritude." The Army women were barred from serving as combatants but did fill important ancillary posts. They served as drivers, accountants, draftsmen, cooks, occupational therapists, encoders. They dispelled stereotypes. Despite fears that women would become hysterical in emergencies or that female voices were too soft to be heard, WACs worked in airplane control towers and did well.

Emboldened, Congresswoman Nourse Rogers also began working on the U.S. Navy, which was a tougher nut to crack. As early as December 1941 she had called on Admiral Nimitz (who at the time was chief of the bureau of navigation, which handled personnel) to urge the Navy to establish its own women's unit. He was not enthused, nor was the rest of the old guard. When Nimitz polled the naval bureaus—the branches of the Navy—only two were receptive. These were the code-breaking operation, which of course already had civilian women, and the Bureau of Aeronautics. The open-mindedness of the aviators was thanks in part to the efforts of Joy Bright Hancock, a former yeomanette who worked to persuade the Navy to allow women to train as mechanics for airplane engine repair and maintenance. She was also a trained pilot.

In May 1942, President Franklin Roosevelt urged the Navy to get a move on. So did First Lady Eleanor Roosevelt, joined by advocacy groups such as the American Association of University Women. Barnard's Virginia Gildersleeve and other college leaders also pushed. It was an uphill battle. "If the Navy could possibly have used dogs or ducks or monkeys, certain of the older admirals would probably have greatly preferred them to women," Gildersleeve acidly remarked later.

The chair of the Senate's Committee on Naval Affairs argued that "admitting women into the Navy would break up homes and amount to a step backward in civilization," as Gildersleeve put it. Elizabeth Reynard, a tiny but flinty English professor from Barnard, was appointed special assistant to the chief of naval personnel, Admiral Randall Jacobs. It was her job to try to make this work. But even she was shocked when she received from Jacobs what would become a famous telegram: "Women off the port and starboard bows. Visibility zero. Come at once."

In July 1942, Roosevelt signed the law creating a women's naval reserve, which aimed "to expedite the war effort by releasing officers and men for duty at sea." The women were politic in victory. It was Barnard's Elizabeth Reynard who came up with the acronym WAVES—Women Accepted for Volunteer Emergency Service—and every word was chosen with care. "Volunteer" assured the public that women were not being drafted, and "emergency," as Gildersleeve characterized the strategy, "will comfort the older admirals, because it implies that we're only a temporary crisis and won't be around for keeps."

The Navy women were not an "auxiliary," a term that overtly confers lesser status, but a naval reserve like the men's. They were "in" the Navy, not just "with" the Navy, a key win. But there remained many inequities. Women reservists were entitled to the same pay as men, but not to retirement benefits. At the outset they could not hold top ranks. Mildred McAfee, the charismatic president of Wellesley, accepted appointment as director of the WAVES. She was a lieutenant commander at first, promoted to captain in 1943. She was often cut out of decisions, however, and deprived of the support needed to navigate a crafty and hidebound naval bureaucracy. Good-humored and beloved by the women who served under her, she joked that the attitude of her male colleagues was like that of the eighty-eighth psalm: "Thy wrath lieth hard upon me, and thou hast afflicted me with all thy waves."

The architects of the effort bumped up against lingering subterranean resistance during lengthy meetings over things like uniforms. Virginia Gildersleeve would later recall that at one, a "handsome young lieutenant" ventured that the women's uniform should not be navy blue like the men's. "I saw the faces of the women around the long conference table light up with faint, repressed smiles at this somewhat revealing opening gambit," wrote Gildersleeve. Even after it was allowed that Navy women could wear navy blue, the service balked at letting the women have gold braid. It was proposed that the women wear a uniform piped with red, white, and blue. McAfee, appalled, thought this so gaudy that it looked like a "comic opera costume."

Josephine Ogden Forrestal, whose husband, James Forrestal, would soon become secretary of the Navy, took things in hand, approaching the fashion house Mainbocher. There were discussions about minutiae like pockets, which Virginia Gildersleeve felt were essential for any working woman. But the designers felt pockets would spoil the lines of the suit. "Utility was sacrificed to looks," Gildersleeve noted with some disgust in her memoir. "They certainly looked very attractive and no doubt won many recruits for the Navy; but I regretted those pockets. (A later model, I am glad to say, contained a good inside one!)"

But the end result was spectacular. The WAVES uniform consisted of a fitted navy blue wool jacket, with built-up shoulders and a slightly rounded collar; a flattering six-gored skirt; a white short-sleeved shirt; a tie; elegant dusty blue braid; a dashing little cloche hat with a detachable top that could be blue or white; and a square black pocketbook that strapped diagonally over the shoulders and had a white sleeve that could be fitted on it to go with dress whites. The women were issued raincoats and roomy hoods called havelocks. They were expected to wear gloves that were white or black. No pins, earrings, or jewelry could be worn; slips must not show; the hat must be worn without tilting it to one side; umbrellas were a nonmilitary item and could not be carried. They were expected to be in uniform at all times, except when wearing athletic clothing or being court-martialed. There also were dress whites, work smocks, and a summer seersucker shirtwaist dress.

All of this may sound silly and frivolous, but it wasn't. The uniform conveyed to a skeptical public that the Navy, however stubborn and reluctant to have them, cared about its women and how they were perceived. A number of code breakers admitted that the Mainbocher uniform was one reason they enlisted; some felt it was the most flattering piece of clothing they ever owned. Others chose the Navy over the Army because they preferred the classic Navy blue over the drab khaki that was the fate of the WACs. The Army women even had to wear khaki bras and girdles, which the WAVES thought was hilarious. In true competitive-service fashion, Navy women felt superior in being able to wear their own underwear.

* * *

What the creation of the WAVES meant, for the women working as civilian code breakers in Washington, was that they now would be commissioned as officers in the U.S. Naval Reserve. They were given a choice—a few would remain civilians—but the majority accepted commissions. This meant that in the fall and winter of 1942, just as they were truly settling in, the women had to depart the D.C. headquarters for officer training camp, to absorb the rudiments of what it meant to be a naval officer, instruction that contained nothing of use to code breaking. The men at Main Navy were reluctant to let them go. JN-25 had gone dark again. The U-boat Enigma cipher was likewise unreadable. It was a grim time, Midway or no Midway. "The work the women are now doing is too important to the war effort to risk a period of absence and disorganization," protested John Redman, head of OP-20-G. To appease the commander, the Navy agreed to stagger the women's departure. Six were sent to officer training in October, another handful in November, and so on. And off they went—in many cases, to the same colleges they had graduated from.

A WAVES officer training school was established at Smith College in Massachusetts, and another at Mount Holyoke. In late 1942, Bea Norton, Fran Steen, Ann White, Margaret Gilman, Vi Moore, and the rest were sent north. The number of women on the Smith campus doubled overnight, prompting the president to conclude a national radio broadcast with a special address to his own students. Noting that a thousand male officers had reported for naval training at Dartmouth, Smith president Herbert Davis ventured that his students would prefer the units be reversed: women training at Dartmouth, men at Smith. "Hide your disappointment," he urged his students, apparently thinking that in a time of global crisis, the only thing young women cared about was boyfriends. "And be as generous as you can to your rivals in the women's reserve."

The women didn't have uniforms yet, and so they drilled in civilian clothes, wearing black Oxford shoes with one-and-a-half-inch heels. The heels presented a problem. When they were marching and had to back up, the women sometimes fell backward on their rears, particularly if conditions were wet or icy. During classroom time, they received standard naval instruction. They studied *The Bluejackets' Manual*, memorizing the nuances of personnel and ranking; the fine print of Navy protocol; the names and acronyms (BUAER, BUSHIPS, BUPERS) of the many bureaus. They studied American history from the point of view of the U.S. Navy, absorbing what Admiral This or That had said about sea power. They learned the difference between stripes and chevrons and all the intricacies of lingo. A work shift was a "watch." You were "welcomed aboard" when you joined your unit. The thing you walked on was a "deck," not a floor, even if it was in a building. Personal possessions were "gear." To assemble was to "muster." A meal was a "mess." If you were out sick you were "on the binnacle list." The bathroom was, of course, "the head."

Absorbing the material was doable, if not always relevant. The women learned the lines of a battleship, the functions of a destroyer, how many guns were on a cruiser. They were taught to recognize the silhouettes of enemy ships and airplanes, a skill they would never need. WAVES were not permitted overseas (except a few who were sent to Hawaii), though many joined with the hope of going abroad. Even so, they had to get the same vaccinations men did: shots for diphtheria, smallpox, typhoid, and tetanus, administered by means of a "daisy chain," a system in which the women walked forward as they were jabbed with needles from both sides. The shots were frequent and potent. If you were about to faint, you were told to go outside.

The women were subject to naval discipline and committed to serve for the duration of the war plus six months. Wild parties and ten-day benders were a thing of the past, at least while in training. Reveille was at five thirty a.m. and lights were out by ten p.m. Drinking alcohol during the training period was forbidden. The women had to make up their beds smart, shipshape and seamanlike. That meant square corners and the blanket folded in half, then in thirds, then in half again, placed at the bottom of the bed. The cover had to be so tight a quarter would bounce on it. The woman on the top bunk was required to sleep with her head at the opposite end from the woman on the bottom. Shoes had to be lined up in the closet with toes facing out. At college, many of the women had maids to make up their beds and clean their rooms. College life now seemed like a distant memory.

Women's medical issues did not exist in the view of the Navy. The women were told not to complain of menstrual cramps. The need for Kotex boxes was acknowledged, and they got a demerit if their box was not square in the drawer. They had to do the same calisthenics men did. If a woman couldn't shinny up a rope, others helped. One unit included a woman named Lib who was working on a top secret joint project between the Navy and private industry. Lib couldn't shinny to save her life, and all the women, knowing how important Lib's brain was to the Allied war effort, tried their best to hoist her upward.

They no longer existed as individuals. Everyone's hair had to be above her collar. If your hair was too long, your roommate cut it. The women bunked four to a room. They had to identify themselves as "seamen" when addressing an officer leading a class. They learned how to salute—not as easy as it looked. The first WAVES officers at Smith were reviewed by Eleanor Roosevelt. Ordered to salute the first lady, they put their hands up but permitted their thumbs to drift and ended up thumbing their noses at Mrs. Roosevelt. Subsequent classes were warned not to make the same mistake. They were told, when saluting, to tuck their thumbs firmly against their forefingers.

The women were warned that everything they did would reflect on the WAVES, for good or for ill. "As you no doubt have discovered, where a WAVE goes, all eyes go," they were instructed in a newsletter, which admonished them that a self-respecting WAVE does not "slouch over desks and counters when she talks to others; she maintains a neat, clean, well-pressed appearance at all times; she wears her hat straight...and does not wear flowers at any time on any part of her uniform."

And they marched. Everywhere. Erma Hughes, the bricklayer's daughter, came to Smith in the cold of February 1943. When marching, the women were arranged by height. Being short, Erma was in the last row and always having to adjust to the gait of the women in front of her, sometimes doing a little gallop or shuffle to get back in step. The women would go out in the gray dawn and make formation. At Smith, some were assigned to dorms, while others bunked in town at the Hotel Northampton. All took meals at Wiggins Tavern, which meant the women living on campus would march right into town three times a day. Sometimes Army men stationed nearby would come to laugh at them. The women retaliated by singing, with spirit: "Nothing can stop the Army Air Corps...except the Navy!"

They marched on the street, on the campus, on the playing fields. If a woman fell or fainted—woman overboard—they were told to step around her and leave her lying where she fell. While they marched, they sang. They sang sea chants, and they sang songs that had been written or adapted for the WAVES, often based on popular tunes, by some of the more creative and musical new ensigns. The songs included proto-feminist lyrics such as:

I don't need a man to give me sympathy
Why I needed it before is a mystery

And:

Honor a glorious past
Strive for a future bright
For, like our men at sea
We, too, will fight

And (written by code breaker Louise Allen):

If there e'er was a seaman that was struck by the moon
Oh, it's Ginny, the Ninny of the First Platoon
Oh, she flaunts femininity with curl and with frills
But her mates want to choke 'er when she drills
"Forward march, forward march," and she skids to the rear
"Column right, column right" and she stalls changing gear
But she's deaf to our curses, unaware it's a crime
That she drills, and always out of time.

The women loved every part of it. They loved eating at Wiggins Tavern, which had delicious breakfasts and legendary blueberry muffins. They loved the marching and parades. They loved having a purpose in the war. The Navy used extra drills as a punishment for infractions, but that didn't work with the women. They liked the marching too much to feel they were being punished. They sang and sang. They felt abstract love for the men they were replacing. They marched to chapel on Sundays, where they were joined by men doing officer training. The WAVES had a song written as a counterpoint to "Anchors Aweigh":

WAVES of the Navy,
There's a ship sailing down the bay.
And she won't slip into port again
Until that Victory Day.
Carry on for that gallant ship
And for every hero brave
Who will find ashore,
his man-sized chore
Was done by a Navy WAVE.

During services the men would sing the original and the women would sing the descant, and the harmony was so moving and powerful that Frances Lynd, from Bryn Mawr's class of 1943, always said it made the hair stand up on the back of her neck.

The women weren't the only ones who appreciated the pomp. People came from everywhere to take pictures of a WAVES officer graduation. In the winter, cameras froze. In the summer, the tar on the roads would melt and the women's feet would go *squish squish squish*, and whenever afterward they thought of the Navy songs, they felt their squishing shoes should be part of the music. When the women graduated, they received their uniforms, with bars to show they were ensigns. Their beloved Mildred McAfee might feel she was struggling against the Navy brass—which she was—but the women felt that they belonged to the U.S. Navy.

The women code breakers were not allowed to stay in officer training long. They were missed so badly that most were snatched after only four weeks—thirty-day wonders, they were called—and brought back to Main Navy. At Smith, Mildred McAfee came to watch the graduation of that first group. She recognized some of the Wellesley women—Blanche DePuy, Bea Norton—and hailed them by name, which overwhelmed them with a sense of their importance. The night before departing, they lined up to receive their uniforms. Goucher's Fran Steen, training at Mount Holyoke, was asked her size and said that she was a four. She was given a fourteen—the only size left—and would have it tailored later. Her skirt hung down to her instep and her new warm thick navy wool coat flapped around her ankles.

On the train ride back, some children thought the women were nuns. In Washington, nobody had seen a woman in military uniform, at least not since the days of the yeomanettes. The women stopped traffic. Cars would have fender benders. "Some of them seem to be nice girls!" one onlooker breathed, in astonishment, to her husband, when she saw her first uniformed woman. It took a while for the women to master the regulations of appearing in public. Edith Reynolds, rushing to catch a train in Grand Central Terminal, crossed in front of a male officer. Remembering she had been taught never to walk in front of a superior officer without saying "by your leave," she blurted it out and saw from the astonished way that he looked at her that he thought she was trying to pick him up.

Back in Washington, the first group got a mixed reception. Their bosses were glad to see them, but Bea Norton felt the Marines guarding each room took "pernicious pleasure" in making the women salute over and over. Blanche DePuy sensed veiled resentment that women were in uniform alongside men, an attitude that felt "obnoxious." Her father was a colonel in the Army, so she was used to military nonsense. The resentment, she felt, was not harmful, but neither was it well concealed. Nancy Dobson from Wellesley was asked by a male officer to sew on a button—a task every Navy man knows how to do for himself—and rebuked when she sewed it upside down.

But other women were gratified by their new status. In her office in the Japanese unit, Fran Steen found that even as an ensign—the lowest officer rank—she was the top officer. There was nobody who outranked her. Her hair was slightly longer than it should have been, but there was nobody to order her to cut it. The men were being shipped out that fast, and the women code breakers were coming in.

* * *

Jaenn Coz, a bored librarian, was working in California when she happened to see a mail truck emblazoned with a poster saying that Uncle Sam wanted her to join the U.S. Navy. The idea appealed to her, and she went to her local enlisting station. Now that the WAVES had been created, any and all women who met recruiting qualifications could sign up. There was no need for a shoulder tap or a secret letter. The U.S. Navy's basic requirements for female officers were a college degree or two years of college plus two years of work. Regular enlisted women, who made up the bulk of WAVES recruits, could get by with a high school degree. This opened up opportunity to women who had not had the advantage of college. Once again, more women than expected answered the call: While naval officials had anticipated there might be ten thousand WAVES in total, by the time all was said and done, more than one hundred thousand women would serve.

Women joined up for all sorts of reasons—because they didn't have any brothers and wanted to represent their family in the war effort; because they did have brothers and wanted to bring them home. At the outset there was a cap on the number of officers, so women were often overqualified for their ranks. Many college women enlisted as ordinary seamen, just to get in.

Depending on their qualifications—and the status of the quota—the WAVES were funneled into officer or enlisted training. Green recruits underwent physical exams along with aptitude and intelligence tests— math, vocabulary, even essay writing—as well as interviews and vocational exams, and were sent on for specialized training. A WAVES enlistee might end up rigging parachutes, training carrier pigeons, working as a "weather girl," operating a radio receiver, or learning the standard yeoman's duties of clerking and bookkeeping. But more than three thousand enlisted women who tested high for intelligence and loyalty as well as typing and secretarial skills would be quietly informed that they were headed to communications training and then on to Washington,

D.C., to perform work of an unspecified nature. They too had been selected as code breakers.

Troop trains now carried women across the country. WAVES member Ethel Wilson enlisted in Columbia, South Carolina, and soon found herself on a train to Stillwater, Oklahoma, where the campus of Oklahoma A&M had been converted to the site of a basic training school for enlisted women. The train traveled from Columbia to Washington, D.C., from Washington to Chicago, and from Chicago to Shawnee, where the women were put on a bus. She would never forget the sight of the train running alongside the Mississippi River. The banks had flooded and the land was deluged. From her window it looked as though the train was running on top of the river, rooster-tail flying, like they were going through pure water.

The country unfolded before them. Women from rural areas were astonished to ride through cities where people pulled clotheslines back and forth between apartment buildings. Georgia O'Connor joined the WAVES out of curiosity, attracted by the smart uniforms, the hope of adventure, and the desire to see whether she could pass the tests. She thought the WAVES were good-looking women and felt proud to be among them. Ava Caudle joined because she had grown up on a North Carolina farm so remote that the most exciting event of the month, she later recollected, was the arrival of the bookmobile. As a girl, she had never seen a movie. Out of high school the only job available to her was as a cosmetologist. All of these women were selected to work in the code-breaking operation. Many would find themselves doing the same work the college-educated women were.

Myrtle Otto enlisted even before her own brothers did. "I had such a yearning to do something," she said. En route to Cedar Falls, Iowa, where another basic training camp was established at Iowa State Teachers College, she got on a train that left Boston's South Station and traveled north to Canada; down through Kalamazoo, Michigan; to Chicago; across the Mississippi; and on to Iowa. It was the first time she'd ridden in a sleeper car. At Cedar Falls, the women were issued uniforms. Enlisted Navy women had to wear heavy lisle stockings that made their legs look as thick as logs. The cotton lisle was so stiff that—when the women kneeled and stood up—the shape of the knee remained in the stocking. As they stood for inspection in the freezing Iowa cold, their noses ran and they were not allowed to wipe them. They took showers by numbers, one girl at the first bell, another one at the next, with three minutes to dress after you showered. It was hard to tug a girdle up a wet body in three minutes, so sometimes the women wouldn't wear anything at all underneath their uniforms as they marched from the showers back to the barracks. It got so cold in Cedar Falls that when the women would put bottles of Coke on the windowsill to chill, the liquid would freeze overnight and pop the bottle top off, and the Coke would expand upward and freeze like a fountain.

A number of the WAVES selected as code breakers met resistance from their families when they enlisted. Ida Mae Olson was born in Colorado and attended a tiny country school where she was the only student in the fifth grade. During the Dirty Thirties her family lived on a farm in eastern Colorado, north of Bethune, where dust storms blew endless thistles that caught in the fence. Her father would stack the thistles and feed the cows with them, because there was no other food to be had. She was working as a nurse's aide in Denver when her roommate enlisted. She couldn't afford the rent on her own, so she enlisted too. Her mother objected that "only bad women join the service. You know, wild women." Ida Mae joined up anyway, and her mother came around. "When I would come home on leave, she'd take pictures of me in uniform. She'd be so proud."

Still, many Americans persisted in the view that military women were just prostitutes in uniform, admitted into the military to service the men. It was an old slander that had been used against the yeomanettes in World War I. There were other resentments. Sometimes WAVES were confronted by mothers unhappy that their sons were being sent into combat, thanks to the women coming in to work in the desk jobs.

But others thought the WAVES were wonderful and would invite them home for milk and cookies or holiday meals.

The Navy was sensitive to what was being said about its women. The fine print said an enlisted member of the WAVES had to be at least five feet tall and weigh ninety-five pounds. It also stated that she had to be a woman of "good conduct." Many got the sense that the Navy had some unwritten criteria regarding looks. Millie Weatherly was a telephone operator in North Carolina and went to the recruiting office with a friend. They took Millie but wouldn't take her friend, saying that "she wasn't pretty enough."

Basic training was a learning experience in more ways than one. On the train to Cedar Falls, Betty Hyatt, who up to then had never left rural South Carolina, wondered aloud "what a Jewish girl looks like" and learned to her mortification that the girl beside her, angry and offended, was Jewish. She hastily apologized. Up to that point she had never met anybody who was Jewish or Catholic. At Cedar Falls, the tables were turned and she took grief for being southern. Since southerners were considered slow, her teacher remarked upon how odd it was that Betty was the fastest typist in the class. She flunked the swim test, but—having taken the IQ test—was told she was being given special dispensation and sent to Washington. When she asked if she had passed the intelligence test, the commander replied, in effect: "Did you ever."

* * *

The women officers continued to train at Smith and Mount Holyoke, but by February 1943, boot camps for enlisted WAVES were consolidated on the campus of Hunter College in the Bronx, which could hold five thousand women at a time. Some ninety thousand women went through six weeks of basic training at the USS *Hunter*. Residents of nearby apartments were evicted to house them. The Navy now saw how valuable these women were. Many bureaus were clamoring for as many WAVES as could be made available. The women began working as gunnery instructors, storekeepers, pharmacists' mates, and instructors showing male pilots how to use the flight simulators known as "Link trainers." A single bureau might take an entire graduating class. The code-breaking unit had to compete to get them.

For the women, coming to New York was a revelation. Many southerners had never seen a northern city. Women from small towns were afraid the subway would swallow them up. Even women from Minnesota found the East Coast cold—the wetness, the way it went through your bones—to be shocking.

Jaenn Magdalene Coz, the librarian from California, traveled east on a five-day troop train and alighted into ankle-deep New York snow wearing only her thin civilian clothes, which were soon boxed up and sent back to her parents. She was left-handed but had been forced to use her right hand in school, so she had trouble discerning left from right, and this made marching difficult. On Christmas night, her unit was marching through wet slush and she felt so cold and homesick that she started to cry. The petty officer told her to shut up. As punishment for crying, she was made to mop the dirty snow from the hallways of the apartment building. During training, she asked to be sent back to California and stationed in San Francisco. She was sent to Washington, D.C., instead.

Other women were in for shocks of a different nature. Ronnie Mackey had grown up in a big family in Delaware; her father favored home remedies and she had rarely seen a doctor, so the physical exam came as a surprise. At the gynecological station, she put her feet in stirrups for the pelvic examination. The nurse said, "Put your head down and be quiet," but she tensed up so that the gruff Navy doctor ventured that with muscles like that, she must be a swimmer or a football player. It also came as a shock when they were sitting in a room waiting to go on "shore leave"—the naval term for a break—and an officer told them they could stop at the prophylactic station. She was convent educated and did not know what "prophylactic" meant.

The WAVES by mid-1943 were a big deal. New York mayor Fiorello

La Guardia loved to watch the Hunter College women on parade. He would call at the last minute and say he wanted to bring an ambassador or other foreign dignitary, and the women would drop what they were doing and muster. There were reviews every Saturday morning, with a Navy brass band, a WAVES drum and bugle corps, and a color guard proudly carrying the wave-blue flag of the USS *Hunter*. Barnard's Virginia Gildersleeve visited often. She found the parades to be useful sociological studies, involving "a remarkable cross section of the women of the United States of America, from all our economic and social classes, from all parts of the country, and from all our multitude of racial origins and religions." Glad of the chance to study what America's collective womanhood looked like, she found it looked different from what popular culture and the Hollywood movie industry led her to expect. Standing in the street watching the women march past, she was surprised to note there were not as many blondes as she thought there would be, and the women were shorter than she expected.

It was not only women from rural families who had their horizons expanded. Jane Case was the daughter of Theodore Case, a physicist whose own contributions to communications technology—he pioneered sound in movies and worked during World War I to develop the Navy's ship-to-shore communications—had made him a wealthy man. Jane grew up in a huge house in Auburn, New York, where she had been crushingly lonely. Her mother was insecure and belittled Jane, making her conscious of her imperfections, such as highlighting her nearsightedness by snatching the glasses off her face. Jane, to her relief, was sent away to the Chapin School on the Upper East Side of New York City. While she loved it there, she hated Manhattan high society. It was stuffy and boring, and elite boys at their boarding schools would lazily throw darts at women's dance invitations to determine which they would accept. The social scene was brutal. Pearl Harbor cut short her debutante season.

Jane had always had a visual mind and could see, in her mind's eye, the country from sea to shining sea, the United States and its citizenry rolling out before her from east to west: mountains, wheat fields, rivers,

Americans of all creeds and races. She found the vision of teeming diversity to be thrilling. As soon as the WAVES were created, Jane took the subway to Lower Manhattan to enlist. She memorized the eye chart and managed to make it through the eye examination without revealing that she wore glasses. Her strategy hit a snag, however, when the enlistees were told to strip.

In the Navy, men were subjected to naked group examinations, and now, so were the women. Jane was ordered to go into a little booth and remove her blouse and bra. After she emerged, a female officer took a red marker, drew the number 10 between her breasts, and told her to stand between numbers 9 and 11. Jane had never seen another woman naked. At the Chapin School, the word "breast" had not been uttered. Now, nearsighted without her glasses, she was obliged to go around peering closely at other women's bare bosoms.

Jane had expected to be made an officer, but the Navy did not consider her tenure at the Longy School of Music to be the equivalent of two years of college. Jane did not object. It suited her fine to be an ordinary seaman. Families with a son in the service put stars in their windows to show their sacrifice and contribution. Jane obtained a star, slapped it down in front of her mother, and said, "There. There's nothing you can do about it." She joined the singing platoon at Hunter and loved it. In Washington, she bunked with a mortician's daughter who was very proud of a music box her father had given her, in the shape of a casket.

"I would have to look at it every day, and say, 'Dottie, that's so beautiful!'"

* * *

By late 1942, the Navy's Washington, D.C., code-breaking operation—like that of the Army—had grown so large that it had to be relocated. In just under six months, the office had swelled from a few hundred to more than a thousand people. The Navy began to cast around for a bigger facility, and in true elite Navy fashion found a women's junior college in the most prestigious part of northwest Washington, a leafy neighborhood, Tenleytown, that boasted mansions as well as nearby landmarks such as American University, the Washington National Cathedral, and a number of foreign embassies. The school was called Mount Vernon Seminary. More rigorous than the junior college that formerly occupied Arlington Hall, and better connected, Mount Vernon educated daughters of diplomats, politicians, cabinet members, and other eminent Washingtonians, including Alexander Graham Bell. In a kind of circular irony, Ada Comstock, the Radcliffe president who helped launch the naval code-breaking program, had studied at Mount Vernon. The school occupied some thirty-eight acres on an elevated point from which it was possible to see the Pentagon in Virginia—even the Blue Ridge Mountains, beyond it—and Fort Meade in Maryland. The main building was a refectory of Georgian red brick, with cloisters closed in that "permit the girls freedom for exercise and are secluded from public view," as a school publication put it. Of course, a facility that shielded its girls from public view would be perfect for shielding code breakers.

Mount Vernon had been built early in the twentieth century and expanded. Students' rooms were located so that every bedroom at some point of the day enjoyed sunlight. There were dorm rooms with en suite bathrooms, music rooms, an art studio, a gym and an indoor swimming pool, and a great hall with a portrait of the founder, Elizabeth Somers. On walls were chiseled inscriptions of the school motto, *Vincit qui se vincit*, or "She who conquers self conquers all." One building had a "mathematical door" that was numerically perfect in its dimensions. There was a study hall with a cork floor to deaden noise; boxwood borders based on a colonial design; and a beautiful white-fronted Georgian chapel designed to reflect the look of a Methodist meetinghouse. Funding had been raised for all this from moneyed Washingtonians who believed it was worthwhile to educate women.

Mount Vernon's trustees had been watching with alarm the military's rush to take over schools in the area, including not only Arlington Hall but also National Park College in Maryland, now occupied by the Army Medical Corps. They hoped Mount Vernon would prove too small. They hoped in vain. On December 15, 1942, the Navy seized possession of its campus and nine buildings. There was talk of gutting the chapel and turning it into a two-story naval office building. This was widely seen as an offense against architectural history, and the American Institute of Architects filed a complaint. The chapel was spared. Before relinquishing it—students would attend classes in Garfinckel's department store while the school sought a new location—the Mount Vernon president, George Lloyd, snuck in after midnight and removed the altar materials, the founder's Bible, and the altar cross and candlesticks: "all the things we felt the Navy could do quite well without, and we knew we couldn't."

This was the same chapel where the WAVES would be sat down and told they'd be shot if they blabbed.

The Navy's new top secret code-breaking operation was located at 3801 Nebraska Avenue, a deceptively peaceful patch of land graced by trees and birdsong. Prior to the move, Navy leaders met to find a "harmless" name to call the facility. Cover names were proposed, such as "Naval Research Station" and "Naval Training School." In the end it was dubbed the Naval Communications Annex, but most people called it the Annex; the USS *Mount Vernon*; or WAVES Barracks D. It was located near where Massachusetts and Nebraska Avenues converge at Ward Circle—named after the Revolutionary War general Artemas Ward—which taxi drivers started calling WAVES Circle. In an unofficial Navy Annex newsletter, cartoons showed Ward's statue grinning and leering and trying to peep into the windows of the vast WAVES barracks, which in a matter of months was constructed across the street from the code-breaking compound: rows of glorified Quonset huts insinuated among lawns and mansions.

When Elizabeth Bigelow showed up for duty in 1944, the whole code-breaking compound "appeared to be a huge encampment of ugly temporary buildings surrounded by a high fence and secured by Marine guards."

Barracks D was the largest WAVES barracks in the world, and virtually all the women who lived there were code breakers. It soon had a beauty shop and a bowling alley. The women were assigned bunks and tall lockers. There was a mess hall catered by Hot Shoppes, the renowned local cafeteria chain known for its milk shakes and hot fudge sundaes. The women worked round-the-clock shifts, which made it hard to sleep. There were always people coming and going, always noise in the barracks. The women had never lived in proximity to so many other women. Jaenn Coz made sure to use the last toilet stall, fearful she would get "the clap" from a shipmate.

The Annex was perched on an incline slightly above the barracks. On rainy days, the women code breakers took off their shoes and walked barefoot to work, upward through streaming water. Women officers were given a weekly stipend and allowed to live off campus. One officer lived in a boardinghouse with a roommate who tried to stay up and listen to see if she talked in her sleep about what she was doing. There was a lot of curiosity in Washington about what went on at the Naval Communications Annex. Armed Marines inspected the women's badges and bags. Many of the Marines had seen traumatic duty at places like Guadalcanal and were given the guard job as a recuperative assignment. When she went in and out, Anne Barus made a point of looking the men in the eye as they saluted her. She had a brother in the Pacific and she knew the Marines had been through hell.

The enlisted women were given a naval rating, Specialist Q, which was inscribed on a patch. The Q did not stand for anything, but it did arouse a lot of curiosity. One day, Jane Case, the former debutante, was walking up Wisconsin Avenue when a car stopped and offered her a ride. The wartime rule was that cars should pick up members of the military. It was pouring rain and Jane accepted, gratefully, and clambered into the back. The driver was a man wearing a raincoat, and his wife was sitting beside him. During the ride up Wisconsin, he grilled Jane, asking her what went on in the communications annex and what she was doing for the Navy.

She replied with the answer she always had ready. "I fill inkwells and sharpen pencils and give people what they need," she told him.

"What does the Q in Specialist Q stand for?" he asked her.

Jane laughed it off. "It's Q for communications; you know, the Navy can't spell," she said flippantly.

When they got to the barracks and the driver reached across and opened the back door to let her out, the sleeve of his raincoat hiked up slightly. She saw one gold stripe, and then several more. She realized that the driver was a Navy admiral. He gave her a faint, knowing smile. He had been testing her. She had passed.

She wasn't the only one to whom this sort of encounter happened. After arriving in Washington, Ruth Rather and some other WAVES were processed and told they had a few days off before they began work. They were advised to see the sights. As they did so, they were struck by the number of male strangers who tried to pick them up, ply them with alcohol, even seduce them. When they showed up for their first day at the Annex, they were introduced to the same men—naval officers who had been testing their character and discretion.

At Mount Vernon the work remained the same but the surroundings were more capacious. Women were situated in old classrooms and in new temporary buildings. The gym was converted to a cafeteria. On the compound, an incinerator was constructed for burning papers, and a pistol range built. A path was cut in the rear of the compound, leading to an apartment building called McLean Gardens—built on the former estate of the wealthy McLean family, whose members included the publisher who stiffed William and Elizebeth Friedman when they were developing his private code—that housed officers and other war workers. As it happened, Elizebeth Friedman also moved into the Naval Annex, working in Coast Guard offices that were technically under the oversight of OP-20-G. In the run-up to war, the Coast Guard's assignment to monitor neutral shipping led to the receipt of a mass of intercepts from around the Atlantic; that, in turn, led to a full-blown mission to monitor spy communications between Germany and secret agents in the Western Hemisphere. Elizebeth belonged to a team endeavoring to break a version of Enigma used by a clandestine station in Argentina. Like Agnes Driscoll, despite her long years of public service, she was at a disadvantage due to her civilian status in wartime. A male officer replaced her as head of her unit. She often clashed with her military supervisor, whom she considered careerist and self-centered. The two of them, she later said, "frequently debated the proper mission of the unit," and she felt they could have been doing work that was more important. Her unit was highly secret, as it involved counterespionage, and did not intersect with the work of the Navy women, who remained unaware of it.

The Navy women now worked in every unit—breaking codes, drafting intelligence summaries. Some were put to work as a "collateral" desk, making trips to the Library of Congress and elsewhere, looking up names of ships and cities and public figures and whatever might provide cribs and shed light on the message content. The place had a pleasant, university feel; many male officers were reservists from academic peacetime occupations, including Fredson Bowers, a florid, hardworking Shakespeare bibliographer from the University of Virginia; Oswald Jacoby, the renowned bridge player; and Willard Van Orman Quine, a philosopher and mathematician from Harvard. There also was Richmond Lattimore, who taught classics at Bryn Mawr and later did a translation of the *Iliad*. The publishing world was represented by Charles Scribner Jr. and Pyke Johnson of Doubleday, and there was a woman named Elizabeth Sherman "Bibba" Arnold, a Vassar-educated mathematician who struck Elizabeth Bigelow as "brighter than anyone." The presence of so many learned men prompted the Navy to refer to one unit as the Office of College Professors. The women called it the Booby Hatch.

They meant it affectionately. Many of the professors treated the women like undergraduates. Suzanne Harpole, recruited from Wellesley, was assigned to work for Fredson Bowers. Bowers was married to Nancy Hale, a writer who published fiction in *The New Yorker*. They seemed to Suzanne a glamorous literary couple and she was excited to be in proximity to such luminaries. Bowers worked interminable hours, and one Sunday he came over to Suzanne's desk, perched on the edge, took a cigarette out of a cigarette holder, and thoughtfully tamped it on the desk. "Miss Harpole," he said, "I am going to give you the kiss of death."

"What have I done wrong?" she wondered anxiously.

"I am going to make you a watch officer," he said with dramatic relish. He was promoting her. A watch officer was in charge of a shift. Bowers was kind to her in other ways. When he learned of her love for opera and music, he would give her 78 rpm records of her favorite singers. He made her promise that after the war she would go to Glyndebourne, in England, to hear the famous opera there.

* * *

The number of women at the Annex steadily increased, and soon they outnumbered the men. In July 1943, there were 269 male officers, 641 enlisted men, 96 female officers, and 1,534 enlisted women. By the following February the number of women had nearly doubled and the number of men had shrunk: There were 374 male officers, 447 enlisted men, 406 female officers, and a whopping 2,407 enlisted women. Women's ranks would grow until there were 4,000 women breaking enemy naval codes at the USS *Mount Vernon Seminary*, comprising 80 percent of the total force there.

For the women, being privy to the message content was eye-opening, and not in a good way. By early 1943 many U.S. citizens were led to believe America was winning the war, or starting to. This was not wrong, necessarily, but the code breakers got a more sobering perspective in the sense that they understood the full cost. At the Annex, in addition to code-breaking units, certain code rooms received encrypted internal U.S. Navy messages coming from the theater of war. When Ensign Marjorie Faeder reported for duty, she found herself assigned to work the electric cipher machine, or ECM, a noisy, rugged piece of equipment used for transmitting and receiving American messages. The "incoming messages were telling us very clearly that we were losing the war in the Pacific," she later remembered. She had a vivid picture of what was going on. "Casualties were high, ships were going down, subs were lost." Faeder found the disparity between the public news and the private truth shocking. "When I would go off watch, the newspaper headlines were of how many Japs we had killed, how we were winning the war.... Growing up with the idea that our newspapers always told the truth, I quickly learned about propaganda."

Some of the male commanders did not see the women as proper sailors. Yeoman Ruth Schoen was put to work in a unit where one officer told her she had "bedroom eyes" and kept making passes at her. He ordered her to get him coffee, and Ruth nipped that idea in the bud by refusing. "I didn't want to start serving coffee to anybody," she said. He was taken aback. She was young but self-possessed and learned to steer clear of him.

But other women had commanders who cherished them as the precious resource they were. Jaenn Coz one day was whistling, and a new ensign—male—told her "no whistling aboard the ship." The officer told her he'd put her "on report," which meant he'd report her for disciplinary action. Her own commanding officer, whom she later recalled as a Texan named Hanson, went over the ensign's head and complained, and—this is how she heard it—the ensign got shipped out to sea. For security reasons, code breakers had to clean the rooms they worked in, a duty that included "swabbing the decks," or mopping the floors. Rather than make the women in his unit do it, Hanson would order his officers to mop them.

Another time, Jaenn Coz's family in California sent her a crate of oranges, and she went into the hall during a break to eat one. She was standing there when along came Admiral Ernest King and Secretary of the Navy James Forrestal, the two top men in the entire U.S. Navy. The hallway smelled of orange and there was juice running down her arms. She was mortified.

"Well, sailor, you look like you are having a good time," said Secretary Forrestal, stopping to observe her. Coz stood there wishing the floor would open up and swallow her. "Sir, I am a little California girl and my folks just sent me a box of oranges and let's face it, the oranges out here are lousy," she blurted. Forrestal laughed, asked her name, and began chatting with her. About then, Commander Hanson came along.

"Mr. Secretary, this is one of my girls," Hanson told the naval secretary.

"Commander, does she give you much trouble?" Forrestal asked.

"All the time!" said Commander Hanson, and they both laughed.

With that, Forrestal said, "Carry on, sailor," and Forrestal and King walked on. After that, Coz sometimes had to deliver secret dispatches to Forrestal, and whenever she did, the U.S. Navy secretary would greet her with the word "oranges."

One of the other commanders was Wyman Packard, who headed a unit of women working the mid-, or midnight, watch. Packard, fresh from active service in the Pacific, had been surprised when he learned he would be supervising women, but it didn't faze him. He liked to publish a little mimeographed broadsheet called *Midwatch Murmurs*, in which he kept the women apprised of softball games, cigarette shortages, the number of people in the hospital for sickness or stress, or name changes due to marriage. He asked them to send him news items and "spicy bits of gossip."

Commander Packard also reminded the women that it was his awkward duty to monitor their appearance. "When one is in charge of all men this task is not too difficult," he wrote, but "under present conditions, I think you'll all agree, I'm in a difficult spot" because he didn't know anything about female clothing or hairstyles. He begged them to "save me unnecessary embarrassment by abiding by the prescribed uniform regulations."

He also praised them, pointing out that communications had the reputation of "being the most strenuous but the most thankless job in the Navy" and that theirs was "the hardest working, least glamorous, most exacting, yet one of the most important organizations in the Navy Department." He wrote, "Your work has continually strengthened my admiration for your unquenchable spirit, your sincere devotion to duty and your unquestionable loyalty."

As they went about their assignments, women found competencies within themselves they had not known existed. Jane Case—told, growing up, that she was bad at math—turned out to have a ready facility with numbers. She sat at a desk in a large room where a conveyer belt brought messages to her; it was her job to use her math skills to evaluate a few enciphered numbers at the beginning and decide which JN-25 messages were important enough to pass on. She was very careful in making decisions, aware that an error in judgment—if, say, she failed to pass on a vital message—could be catastrophic. The work was repetitive, but it demanded hard, intense, constant focus. That was stressful. There were far too many messages to break every one. "God, the Japanese sent out messages. There was no such thing as a moment when you didn't have a stack," she recalled later. Working in landlocked northwest Washington, D.C., Jane could tell if a major Pacific Fleet action was under way, because the stack would grow even higher. "Because of the traffic you knew something big was happening."

She wasn't the only worker who could feel the accelerating rhythm of the war via the size and pace of her workload, which—as the Allies began to take the offensive in earnest in the Pacific—responded like a tuning fork to the action overseas. After a long slog through sand and jungle, the United States secured Guadalcanal in February 1943, and, over the course of many months, it was on to the Gilbert Islands, the Marshalls, and the Marianas: Saipan, Guam, Tinian.

As the men in the war theater did their brave and grisly work, the women at the Annex did their best to support them. In advance of every push, every landing, the tempo in the code-breaking rooms would accelerate, as if a string had been tugged, far away, and the women were feeling the pull. Furloughs would be canceled, memos sent. On May 25, 1943, the Naval Annex additive recovery room received a memo saying, "As all of you are doubtless aware, we are approaching what promises to be a period of intense activity." Anne Barus and the others were told to give no hint, outside the office, of the quickened pace within it—not even in other parts of the Annex. "Exactly when the period of maximum activity will hit, it is impossible to say," the memo noted. "Certainly within the next three or four days batches of traffic will be sent down which must be handled with the greatest possible rapidity."

The author, Lieutenant W. S. Weedon, noted that the unit had received a commendation for its work during Guadalcanal. "It would be very pleasant indeed if this sort of thing were to happen again." He also noted in a separate memo that their top boss, Commander Charles Ford, who was overseeing the JN-25 effort, was "anxious to push the 24-hour total once again over the 2500 mark." That is to say, Ford wanted more than 2,500 additives recovered in the course of a day.

"This is not a cracking of the whip," the memo stated, "but a request which, correctly interpreted, means that the workers on June 8 should choose the best traffic and give some extra effort to piling up the best score possible."

The women obliged, and then some. Writing them soon after, Weedon noted "the truly splendid effort made yesterday by all hands in the Additive Recovery Rooms." They had not only exceeded 2,500; they had broken their own record, with 2,563 additives recovered. Commander Ford wrote Weedon: "Please convey to all hands my congratulations for their performance.... The day by day individual and group records are truly amazing to me. New records have been established and then improved." This was more gratifying than some other memos the women received. "Pick up scraps!" Ford wrote in another. "Use ash trays! Dispose of empty bottles and cans! Keep your personal gear neat and clean."

The women quickly rose to trusted positions. Many female officers who started out as ensigns became "lieutenant junior grades," which was the next rank up, and then full lieutenants, and sometimes lieutenant commanders. When Betty Hyatt started out as a yeoman, she was assigned the tedious job of making frequency counts of five-digit numbers. Before long, she had clearance to go in almost any room. As American forces assaulted enemy-held islands in the Pacific, the U.S. men were sometimes able to capture codebooks left by retreating forces. Betty Hyatt was on duty in 1944 when a naval officer brought in a Japanese codebook, having traveled as a businessman to escape notice. It was up-to-date, intact but for some slightly burned pages. It identified code groups in the latest version of JN-25 and enabled the code breakers to read every message on file in that system. Betty volunteered to help, an exhausting job that took two days and two sleepless nights. The code, she later recalled, "gave them the location of Japanese ships, what was on each, who was on each, and the station."

During this effort Betty was assigned to take some code recoveries to a high-priority room, where she opened a door and was shocked to see a man waiting to receive it who was Japanese. She stood in the doorway, paralyzed and uncertain what to do. Every American citizen by then had been subjected to the most intense and cruel propaganda against Japanese people. "We had been taught that anything Oriental is your enemy and you cannot trust them," as she later put it. Clearly the worst had happened, Betty thought, and the enemy had seized the Naval Annex. The women had been warned that "this place could be invaded at any moment." Determined to hold out to the last, Betty refused to surrender the material she was holding. The man laughed graciously.

"I'm an American," he assured her. He was a Nisei, an American citizen of Japanese heritage, working as a translator.

"You don't look like one," she blurted in her exhaustion, and felt sorry about that remark for the rest of her life.

* * *

Many women ended up in units that were almost wholly female, to the disappointment of those looking forward to working among men. Fearing that so many women could not resist gossiping about their work, the Navy developed strategies—such as moving women from barracks to barracks—to prevent them from forming close friendships. The tactic failed. Completely. The enlisted women had undergone the ancient military ritual of being stripped of their old identities. "You were not your old self any more," as one put it. The women had been made anew by their wintry marching, their boot-camp training, their gynecological examinations, their vaccination daisy chains, their naked physicals, and now they were in it together.

Scores of enlisted women worked in the library unit—OP-20-G-L—where about a dozen formed a tight friendship. The library unit typed incoming messages, fresh from the teletype, on file cards, categorizing them and making careful note of coincidences and recurrences. Like the women in additive recovery, they responded to action in the war theater. When Betty Allen, a librarian from a small Illinois town, arrived in the unit, a group of women were busily indexing geographic place names from maps of Alaska, in anticipation of a possible Japanese invasion there. Betty Allen became a staunch member of the friendship group, as did Georgia O'Connor, a farm girl from Missouri; Lyn Ramsdell, who had worked in the office of a Boston wool merchant; and Ruth Schoen, a legal secretary and the yeoman who had put the "bedroom eyes" officer in his place.

Ruth Schoen was the only Jewish member of the group. Her grandparents on both sides had emigrated from Hungary. Like most Americans, she was not yet aware of the Nazi death camps, but she was patriotic and wanted to do her part so badly that she enlisted despite being underweight and underage. She had grown up on Long Island, in a religiously mixed community that was reasonably tolerant, though as a girl she did have one close friend whose father turned out to be a member of the Bund, a pro-Nazi group. An excellent student, she skipped a grade and graduated from high school at seventeen. She wanted to go to college, but her parents told her she needed to augment the family income. So Ruth went to work for a Manhattan lawyer and taught herself to execute depositions, summonses, complaints, and other legal forms. Part of her earnings went to pay her brother's college tuition. She herself enrolled in night classes at Brooklyn College. Enlisting in June 1943, she picked the Navy because her father had served in it in World War I. Being underage, she needed a parent's consent. Her father would not give it, but her mother did; she felt Ruth had lived a sheltered life, and getting out into the world would do her good. A pound too light, she ate as much as she could and managed to pass. The lawyer she worked for wept when she left.

Of all the women in the library group, Ruth was the one whose family lived the closest, and she would invite her new friends to travel home with her for visits. "My parents were so happy" to host the women, she said. They loved them all and made sure to find the churches they wanted to attend on Sunday. "They treated us like a nest of chicks," Lyn Ramsdell remembered.

In Washington, there was a synagogue that had weekly Friday night parties, and toward the end of 1943 Ruth met a soldier there named Dave Mirsky, who asked her out. They dated a bit, and at some point Dave's brother Harry came to visit him. When Dave was sent overseas with a unit of tank destroyers, Harry Mirsky stepped in and took his brother's place courting Ruth. All the women in the library unit loved Harry Mirsky, who was gregarious and funny. He had been injured when thrown from a jeep and so was still stateside, recovering. "Guess who's waiting for you downstairs!" they would tease Ruth, coming in as she was finishing her shift. On their dates he would sometimes ask her what she did, and she would change the subject. After two months Harry took Ruth to dinner and told her, abruptly, "I want you to be my wife."

"I hardly know you," she protested.

"I've made up my mind," he replied winningly. They met in December 1944 and married in May. Her friends threw her a shower in a French restaurant. Ruth applied for permission to wear a wedding gown, which was considered civilian clothing. She had six days of leave in which to find a dress, arrange the ceremony, marry, and take a Catskills honeymoon. All of her Specialist Q librarian friends managed to get to Queens for the wedding. Some were granted leave; some took a chance and went briefly AWOL.

Georgia O'Connor was next in the group to marry. Despite having grown up poor, she married a wealthy heir to a publishing fortune, a man whose wartime job was tracking Nazi spies in Chicago and whose family owned a villa near Cannes that was now occupied by Germans. When her high-society mother-in-law asked what her own father did for a living, Georgia O'Connor replied, honestly, "He slops pigs."

This kind of thing happened all the time. Amid the wartime upheaval, the wildest pairings became normal.

The women were living life in the moment, with little idea what the future held. If a woman went out on four dates with a soldier and he didn't ask her to marry him, she figured she had bombed. Romances, naturally, blossomed on site. Marge Boynton, from Wellesley, married Willard Van Orman Quine, from Harvard, a few years after the war. Their love was born in the Booby Hatch.

* * *

Together this group of Navy women broke and rebroke the fleet code that Agnes Driscoll had laid the groundwork for, and they broke the inter-island ciphers, and they helped keep track of the movements of the Imperial Japanese Navy. Men in the Pacific, such as Joe Rochefort, Thomas Dyer, and Edwin Layton, would get credit—well deserved— for famous victories like Midway, but most Pacific achievements were group ones. "The history of the Navy in the Pacific is the history of a system, and must be written up as a system," said Frank Raven later. "While there were a number of prominent individuals, you can't credit any individual with winning the Battle of Midway or of breaking any major cipher system. These were crew jobs."

And the women were the crews. They outnumbered men in virtually every unit of the Naval Annex. By 1945, Pacific decryption there consisted of 254 military men and 1,252 military women, plus 33 civilians. Though they did not receive public credit, some did become legendary in the small sealed rooms where they were working. One enlisted member of the WAVES "had such a knack for running additives across unplaced messages and recognizing valid hits that for over a year she was allowed to do almost nothing else," one internal memo admiringly noted.

The mostly female makeup of the Annex became a logistical problem—or rather, the fact that the women could not be sent overseas created a problem. Some men would have liked a break from overseas work, and many women dearly wanted to replace them. But neither could happen. Every year, Suzanne Harpole got a standard form asking if she wanted a transfer, and every year, she asked to be sent abroad. "Every year I'd get a response saying we received your request but we are only sending men overseas. That happened all the time. I kept thinking: If they can send nurses overseas, why can't they send us?"

Fran Steen, from Goucher, wanted to fly airplanes. She asked the Navy to send her to flight school, but her request was denied. So she took her ground exam at Washington National Airport in Virginia and learned to fly planes on her own.

As the women proved their capabilities, the few remaining male officers at the Annex were obliged to figure out, by improvising, what the women could and could not do. For example: Might women officers be taught to shoot? One male officer asked this at a staff meeting. Many code rooms at the Annex had pistols ready in the event of unwanted visitors, and pistols were worn by officers escorting "burn bags"—sacks in which all discarded papers were put—to the incinerator. When this question was raised, the men running the Annex realized that "there is no well-defined policy on teaching WAVES to shoot." Someone pointed out that some other bureau was letting its women shoot, so an on-the-spot decision was made: yes. Fran Steen learned to shoot on the pistol range, as did Suzanne Harpole and Ann White. Their commanding officers joked that they needed smaller pistols, since the big ones spoiled the lines of their uniforms. A September 1943 memo noted that "pistol practice was progressing satisfactorily."

The male officers also confronted the vexing question of how much authority the WAVES officers—many had risen to become watch officers—had over men working for the shifts they supervised. This was a controversial topic. Regulations were reviewed showing that the relationship of WAVES officers to enlisted men was like that of a civilian instructor. The WAVES officer had authority within her unit, as a teacher might, but did not have disciplinary power over the men. The minutes from the meeting did note that WAVES officers were entitled to salutes from enlisted men, "although it seems to be the exception rather than the rule when they are given."

* * *

It was hard, harrowing work—all of it—and the women took it seriously. Many got sick from the crowding and the pace. Jane Case contracted mononucleosis and spent a month at Bethesda Naval Hospital. The doctor treating her ventured that she "came into the service so you could get to a higher social level," an idea that struck Jane, given the debutante background she had fled, as hilarious.

Local institutions did what they could to honor the Navy women. There were ten thousand WAVES all told working in the Washington area, serving in varied capacities, including at the downtown Navy head-quarters. Jelleff's department store had a fashion show in which WAVES served as models, and Hecht's had a day to honor them. They enjoyed free entry almost everywhere. Despite wartime rationing, they could buy things like jewelry, cigarettes, and nylons at the ship's store. In their downtime they could visit the National Zoo and the Washington Monu-ment. The Capitol Theatre had piano playing and singalongs. The National Gallery had musicians playing in its rotunda. Vi Moore heard the Budapest String Quartet perform at the Library of Congress. Tickets were 25 cents and the women would line up at eight a.m. after working the night watch. The Washington Opera—a struggling outfit—gave free performances on a barge near what is now the Watergate. The women sat on the steps or rented canoes and paddled over, resting their oars in the mud while they listened. The Marine Band played behind the U.S. Capitol; the National Symphony played in Constitution Hall. The women visited a roadside joint called River Bend in Virginia, which was a daring place to go. They jitterbugged.

Jaenn Coz, whose mother had been a flapper, would sometimes put on civilian clothes, surreptitiously, and go to the "black part of town," as she put it, to hear Eartha Kitt. Frank Sinatra would sing at Club 400, where she went on Fridays. You could buy a pitcher of beer for a few cents. You could get good fish down at the wharf. She also learned to play bridge and to gamble.

There were splendid hotels in Washington: the Willard, the Carlton, the Statler, the Mayflower. Most held frequent dances in their ballrooms, often sponsored by individual U.S. states, with big bands and lots of swing dancing. The women would go to American University and watch the Navy men playing baseball. They would sail on Chesapeake Bay. Theaters were open all night. You could go to a movie anytime. You never knew whom you would see. One WAVES member went to dinner with a lieutenant and found herself chatting with the Eisenhowers. Another met President Roosevelt himself at a party for disabled veterans.

Many of the women had never had to fend for themselves, and now the ones living out of barracks had to figure out where to obtain pea coal or cook a big piece of meat that some Marines had brought to a party. A group of WAVES officers lived in a house where there were seven women and six beds. The women hot-bedded it, taking whatever bed was open. They would leave notes for one another on the pillows, sharing tips about hotels and restaurants that were giving away things free or had good discounts.

And they traveled. They could go anywhere on a train for a discounted fare. Trains around the country had romantic names like the Rocky Mountain Rocket and nicknames like the Grunt and Crawl. If they had a thirty-six-hour leave, the women would go to New York City. If they had seventy-two hours, they'd go farther. Ida Mae Olson invited her friend Mary Lou to visit her family in Colorado. Mary Lou had joined the WAVES because her parents had been killed in a car accident and her wealthy uncle had not known what to do with her. Mary Lou was terrified when she saw a group of Native Americans and asked Ida Mae if they would attack. When she saw the dryland farm, she asked if it was okay if she took off her shoes and ran in the soft fresh dirt, something she had not been able to do, growing up in a city.

The women hitched rides on military planes. When Jane Case learned that her father was dying, her commanding officer gave her emergency leave and got her on a plane to New York. Her father was too far gone to recognize her. Jane always thought that if she had just been able to tell him she worked in "communications," he—having done work for the Navy—would have known what that one word meant. It would have been a chance to make him proud of her, but she never got it.

Even just in the rooming houses, there was a lot of mixing and mingling. Suzanne Harpole settled into a boardinghouse run by a Mrs. Covington, from an eminent Washington family, and found herself eating alongside Canadians and White Russians. The conversation was endlessly stimulating. People of all backgrounds were thrown together. The war, for her, was "this period of very creative thinking and concern about what life was all about, and what societies were all about." At the Annex, a weekly current events presentation gave the latest war news. Suzanne's brother was a Marine who led a dog platoon—dogs were used to sniff out enemy ambushers—and when the presenters announced the landing on Bougainville of the first dog platoon, Suzanne felt like shouting, "That's my brother!"

The women's freedom brought other revelations, though, about the uglier side of their own country. Washington was in many ways a southern city, with a segregated public school system and black residents often consigned to the poorest and least well-served parts of the city. Nearby Virginia was worse. Northerners were shocked to encounter such strict racial divisions. When Marjorie Faeder boarded a train to Virginia Beach to take a quick honeymoon with her new husband, the couple sat down in a deserted car with plenty of seats. To their dismay they were told they were in the "colored" car—they had not known such a thing existed—and shooed into the whites-only car. Nancy Dobson was horrified every time she took a bus from Washington into Virginia. When the bus arrived in the middle of the bridge, all African American passengers had to get up and move to the back, and there was "just this deadly hush." Frances Lynd, from Bryn Mawr, needed to buy some furniture, so she engaged two African American men with a pickup truck to take her downtown to get it. When she jumped in the cab with them—being from Philadelphia, she thought nothing of it—the men were appalled and fearful to be seen with a white woman sitting next to them.

There was periodic unrest in the city, and in the country, as civil rights advocates pushed to advance social changes that were taking place as a result of the war. In 1944, Eleanor Roosevelt and WAVES director Mildred McAfee managed to gain entry for African American women into the WAVES. But there would be no racial experimenting at the stodgy and sometimes paranoid Naval Annex, where top officers considered any newcomer—anybody with an unorthodox background—to be a security risk. In a June 1945 memo, Commander J. N. Wenger wrote that he had explored the "question of employment of colored WAVES at Naval Communications Annex" and felt integrating the code-breaking unit would be too risky. He concluded that it would be "unwise to conduct an experiment of such serious implications" in a unit where security was so important, and ventured that "there are many other activities in the Navy where experiments of this sort can be carried on without so much danger in the event that difficulties arise." Black WAVES would have to take their patriotism, their intellect, and their talent elsewhere, alas.

* * *

As hard as the women worked, there were lighthearted moments even in the code rooms. One night Jane Case's unit got word that an admiral was coming to visit, and they needed to have their unit spotless by the next day. It was Jane's job to operate the buffing machine, which seemed nearly as big as a baby elephant, and harder to handle. She flipped the "on" switch and nothing happened; peering underneath a table, she saw the cord wasn't plugged in and crawled underneath to plug it. Pleased that she had solved her own problem, she backed out to see the machine flying all over the office. By the time she wrestled it into submission, the place was a disaster and they had to spend the whole night picking up messages before they could get it painted and cleaned and buffed.

Also in her office were two enlisted women whose behavior Jane observed with fascination. They sat together against a wall and did a lot of the typing. Jane liked to think of them as Myrt and Gert. One was married and one was engaged, but their husband and fiancé were away, and they used the war as an opportunity for avid extracurricular dating. "They wouldn't date anybody under a captain," she remembered. "They were very fussy about who they would date." People in the unit were pretty sure Myrt and Gert were having sex during their assignations. Jane had always been taught at the Chapin School that a woman must be properly introduced to a man before going out with him, so for the entirety of the war she did not date. "When I think of it—I could have gone out with a lot of people," she said, regretfully, later. "The rules were so set, all my life."

For many other women, their social lives were as exhausting as the code-breaking work itself. Edith Reynolds, from Vassar, found herself courted by an ardent Irish major who at one point had been in charge of mules—a bona fide muleteer—for the British Army. There was another suitor she wasn't wild about who flew his mother from Seattle to meet her. "He wanted me to know you first," the woman told her. "First what?" Edith wondered. Then she realized, with shock, that he thought they were going to get married. She broke up with him and he married her roommate.

One code breaker was standing in a movie line and realized there was a naval officer behind her. She turned to salute him, and he was so captivated by how flustered she was that they exchanged addresses and later married. At the group house where Edith Reynolds was living, they had a party and she noticed a man expertly cracking eggs into the eggnog. He was the plumber. "I came to fix a pipe, but this looked like so much fun, I stayed," he told her.

The women were working so hard that it was hardly a surprise they would blow off steam, but it did irritate some inhabitants of the neighborhood where they worked. On June 16, 1943, a Washington lawyer, James Mann, wrote a letter of complaint to Captain E. E. Stone at the Naval Annex. "I hesitate to write this letter and I sincerely hope my purpose will not be misconstrued," the lawyer began. "For some time it has been impossible for the people living on the north side of Van Ness Street, between Nebraska and Wisconsin Avenues, to sleep between eleven p.m. and 2:30 a.m. This is due to the unusual amount of noise made by the young men and women stationed at the Communications Annex." He reported that "one morning this week about eight Waves walked up the middle of Van Ness Street at 1:30 in the morning singing. In about five or ten minutes two marines came along singing at the top of their voices." The upset lawyer noted that "as the Waves and the seamen become better acquainted they are following their natural inclinations and now the street is quite a necking place."

Captain Stone wrote a polite letter thanking the lawyer and assuring him that he would endeavor "to end the unpleasant situation which you report."

While their high spirits were tolerated—up to a point—the logical consequence was not. Pregnancy was forbidden in the U.S. Navy. And yet pregnancies did occur. When Jane Case was living in a barracks, an enlisted woman got larger and larger. What struck Jane, in retrospect, was "the quiet ignoring of her. I was one of them that just would walk by and just say hello and that was it. I was stunned more than anything."

The penalty was discharge—and humiliation. Standing "captain's mast" is the traditional naval hearing to consider a possible offense, and some pregnant women were subjected to it. Jaenn Coz remembered that "when we all lived in barracks, several girls got knocked up, and it was a sad thing because she had to stand a captain's mast and watch as they ripped off her buttons and ripped off her chevrons and just humiliated her to death.... We all stood there and cried...because it was such a damned sad situation."

Nor did marriage make it okay. In late 1943, Wellesley's Bea Norton, now married, became pregnant and notified her superiors. "Horrified, they gave me three days to get out of uniform and told me I could consider it a discharge with honor." Her boss, Frank Raven, was furious at the rule and begged her to come back as a civilian, but she was too tired and angry, and she resigned in December. The same thing happened to Frances Lynd from Bryn Mawr. She was working ciphers used by rice merchants and island weathermen and thought it was the most interesting work she could imagine. When she married her college boyfriend, she tried to avoid getting pregnant, but nobody had given her good information on birth control—her mother was dead—and she didn't know how to use a diaphragm. She conceived on her honeymoon. One minute she was a respected naval officer doing work she loved and valued; the next minute she was an isolated housewife, living with her husband and several other adults. The only clothing that fit was her naval raincoat, which technically she was no longer entitled to wear. She struggled to keep house despite wartime rationing, when some days the only things she could buy were, say, bologna and canned pineapple. She tried to make a meal out of those two ingredients and was rebuked by her housemates. When her son was born, she enjoyed having an infant but later experienced profound postnatal depression.

"I felt like I had gone from being everything," she wrote in a memoir called *Saga of Myself*, "to being nothing."

* * *

The women tended to ignore the hallowed Navy rule that forbade fraternizing between officers and enlisted persons. At the boardinghouse where Suzanne Harpole was living, there was an enlisted woman named Roberta, who had attended Flora MacDonald College in North Carolina. Suzanne and Roberta worked in the same office doing the same thing, and it seemed to them ridiculous not to be friends. Also boarding at Mrs. Covington's were two women employed at Arlington Hall, and the four would take trips to Williamsburg, Luray Caverns, and New York. The two Naval Annex women and the two Arlington Hall women could not talk about what they did, of course, so even as they admired Raleigh Tavern and the House of Burgesses, they remained unaware of something even more interesting: They were all breaking codes.

For the most part, the two large cohorts of Washington-based female code breakers—Navy, Army—did not interact, or not knowingly, though they did cross paths. WAVES Barracks D, as big as it was, became overcrowded, and some Navy women went to live at Arlington Farms. Apart from that, the women code breakers likely ran into one another at all sorts of places—restaurants, movie theaters, streetcars—without realizing they were working on the same project. How could they? The work was top secret, and they couldn't talk about it.

Occasionally, though, a few did get wind of what was going on across the river. Dorothy Ramale, the would-be math teacher from Cochran's Mills, Pennsylvania, was initially hired by Arlington Hall, where she did such expert work as a "reader"—one of the most elite jobs—that she made a break into a Japanese code that led to the creation of a whole new unit. Her boss would not let her set foot in the new room, for fear he would lose her to whatever was going on in it, so she never knew what the break consisted of. After a year, she learned about the naval operation, applied to work there, and was eagerly accepted. The officer housing allowance was her incentive. She was able to live in a D.C. apartment with her sister, who was a g-girl at the Pentagon, and so she didn't have to spend her full allowance on housing. She used the stipend to buy a car: preparation for seeing the world, as she yearned to do.

Top officials at both code-breaking complexes did communicate, however. The Naval Annex had formal weekly liaison meetings with Arlington Hall, and it was a WAVES officer, Ensign Janet Burchell, who crossed the river to serve as Navy liaison for these meetings. The position required her to know about the code and cipher systems both operations were working. Ensign Burchell attended meetings where the two services discussed the forwarding of intercepts and captured materials; duplicate messages sent in different systems; reports of POW interrogations that might contain material useful to both; and other odds and ends. At one, Burchell brought a request from Frank Raven, who was trying to a break a message in Thai and knew that there was a professor at Arlington Hall who might be able to help.

* * *

The Navy women had just missed taking part in the code-breaking triumph at Midway, but ten months later they were fully embedded for, and actively engaged in, the other great code-breaking event of the

Pacific naval war. On April 13, 1943, a message came through along the E-14 channel of JN-25, addressed to "Solomons Defense Force, Air Group 204, AirFlot 26, Commander Ballale Garrison Force." The code breakers weren't able to recover the whole message right away, but the fragments they did recover suggested that the commander in chief of the combined fleet—Admiral Yamamoto himself—was headed to Ballale Island (now Balalae) on April 18. Intelligence officers concluded that this was an inspection tour.

The initial break was made in the Pacific, but Washington also got busy, recovering additives and code groups so that blanks could be filled in. More messages were intercepted, and the fast-working, far-flung teams exchanged findings. Among those digging out code recoveries was Fran Steen from Goucher. The inter-island cipher JN-20 "carried further details" about Yamamoto's upcoming trip, so Raven's crew of women were busy as well, adding facts and insights. Together the code breakers were able to reconstruct Yamamoto's precise itinerary, which called for a day of hops between Japanese bases in the Solomon Islands and New Britain. Their translation concluded that the commander would "depart RR (Rabaul) at 0600 in a medium attack plane escorted by six fighters; arrive RXZ (Ballale) at 0800"; depart at 1100 and land at RXP (Buin) at 1110; leave there at 1400 and return to Rabaul at 1540, traveling by plane and, at one point, minesweeper. He would be conducting an inspection tour and visiting the sick and wounded.

It was an extraordinary moment. The Americans knew exactly where the enemy's most valuable—and irreplaceable—naval commander would be, and when. Yamamoto was known for punctuality. Far above the pay grade of those working additive recovery, Nimitz and other top war officials decided Yamamoto would be shot down. It was not a light decision, assassinating an enemy commander, but they made it. The itinerary, as one memo later put it, signed the admiral's "death warrant."

In what was known as Operation Vengeance, sixteen U.S. Army fighter planes, Lockheed P-38s, went into the air on April 18, taking off from a Guadalcanal airfield. They knew Yamamoto would be flying in a

Japanese bomber the Americans called a Betty, escorted by Zero fighter planes. The Americans calculated their own flight plan to meet the route they anticipated Yamamoto would be taking, planning to encounter him over Bougainville. They flew for so long that the pilots were getting drowsy; the white coastline of Bougainville was racing beneath them when one of the pilots broke radio silence and shouted, "Bogeys! Eleven o'clock!" There they were, on the horizon: six Zeros, two Bettys. The Japanese did not see the Americans at first, but once they did, the escorting Zeros moved to block the U.S. fighter planes, firing so the bombers could escape. There was a hectic battle in which it never became clear who had shot down whom, but one Betty bomber plummeted into the trees, the other into the surf. Yamamoto's body was found in the Bougainville jungle, his white-gloved hand clutching his sword.

Cheering broke out at the Naval Annex when they heard the news. The architect of the Pearl Harbor attack was dead. The payback felt complete.

"Let me tell you, the day his plane went down, there was a big hoop-de-doo," recalled Myrtle Otto, the Boston-bred code breaker who had beat her own brothers in the race to enlist. "We really felt we had done something really fantastic, because that was—well, it was more than the beginning of the end. They knew it was coming down, but it was really—that was an exciting day."

CHAPTER SEVEN

The Forlorn Shoe

Undated

Arlington Hall, being more of a civilian operation, was a far cry from the Naval Annex when it came to attitude and culture. The Army's suburban Virginia code-breaking operation was equally serious when it came to work, but far more tolerant and freewheeling when it came to life. One day Dot Braden got a glimpse of just how open-minded her workplace was. Feeling nauseous while she worked at her table, Dot visited the dispensary to get something to settle her stomach. "Are you sure you're not pregnant?" the nurse asked.

"I don't think so," replied Dot, an answer that instantly struck her as preposterous. She didn't think so? Of course she was not pregnant! There was no way she could be. Why would she even have said that? Probably because she was flustered and taken aback by the question. Looking around, though, she realized how many women at Arlington Hall *were*. The nurses treated them with kind consideration and nobody expected them to quit. Some might be married; some might not be. Nobody asked. Things happened. Washington was wide-open. Soldiers and sailors were everywhere, and anything went. Men might be shipped out without warning, and couples who were about to get married didn't have a chance. Sometimes the wedding, if it happened, happened somewhat after the fact.

For her part, Dot kept her chastity intact. For her, life in Washington meant writing letters to men and having fun with other women. The same was true of her friend Crow, who was fun-loving but shy and didn't date much. Neither of them had much time to: At Arlington Hall their schedule consisted of seven days of code-breaking work, followed by an eighth day off, followed by seven more days of work. On their one day off they'd be "dead dog tired," as Dot put it, and would walk over to Columbia Pike to do errands and grocery shopping.

The adventures they enjoyed in their free time were tame and lighthearted. Once, Dot had a friend visiting from Lynchburg, and they decided to attend one of the hotel balls. As an icebreaker, the women were told to stand on one side of the dance floor and the men on the other. The women were instructed to take off a shoe and throw it onto the dance floor, and the men were to pick up a random shoe and dance with whoever owned it. But the problem with shoes was this: People didn't have many, and they couldn't get new ones often. Shoes were rationed, and they had to save up ration coupons to buy a new pair. In the interim, all anybody could do was get their existing shoes half-soled. With all the walking Dot did between her apartment and Arlington Hall—at least three miles each day—she was always wearing through the bottom of her shoes. She had purchased a pair of blue I. Miller shoes she cherished, but for the dance she had worn her other good pair, strappy white sandals, and there was a hole as big as a quarter in the sole.

Dot wasn't aware of the hole, so she lobbed the sandal onto the floor, and it flipped upside down in such a way that all the lights in the room seemed to be shining on that hole. No man grabbed it, and the shoe lay there, sad-looking, while couples danced around it. Dot didn't have a partner for that dance. Her friend was a giggly girl and they both thought that was the funniest thing they'd ever seen, Dot's forlorn shoe upturned with that awful hole in it, and no man willing to pick it up.

They did other small things that seemed daring. Dot sometimes would experiment with using carbon paper to color her hair. Hair dye was expensive, but you could sprinkle water on the carbon paper and smear it on your hair to darken it. Crow joked that she was going to tell people her roommate dyed her hair. Dot knew she wouldn't. The carbon paper didn't work all that well, but they did their best with what they had.

Not being military, Dot and Crow couldn't take the long train journeys the Navy women did; or, if they wanted to, they would have to pay full fare and risk not getting a seat. The military enjoyed fare discounts and seating priority. Even so, the two friends managed to find plenty to occupy their rare free time. They went window-shopping downtown and perfected the art of looking dressy with very little money. Dot managed to acquire a silver fox stole. They toured the museums and monuments and visited the National Cathedral, which was still under construction, but awe-inspiring even so. They took the train up to Baltimore, which had nice stores, to buy hats. In downtown Washington, they bought lipstick at Woodward & Lothrop. Back at the apartment, Crow's sister Louise—aka Sister—had a tendency toward melancholy, and Dot decided to cheer her up by throwing a party for everybody living in Fillmore Gardens, a gesture that resulted in the young women's being invited to reciprocal soirees in the apartments of neighbors, who were mostly young couples.

On their day off, Dot and Crow sometimes took a succession of buses and streetcars to do a bit of sunbathing and swimming. A popular Chesapeake Bay day resort, Beverly Beach, offered the enticements of a sandy beach area as well as a dance floor, bandstand, and slot machines. Colonial Beach, in Virginia, had a bathing area along the Potomac. Getting to either place took so long that it would nearly be time to come home by the time the women arrived, but they went anyway. The two code breakers would manage to get burned in what little time they had. When they got back they always suspected that Sister, who was fair and who wouldn't often risk going to the beach, was secretly glad to see them so red and sunburned. They thought her jealousy was funny. As they went about their travels, Dot would make tart observations about people, like, "She goes to church too much," and Crow would laugh and say, "Dot, you are an original." Dot was an entertainer and Crow was an appreciative audience. They were entirely unalike, and entirely bonded. At

Christmas, on their modest salary, Crow gave Dot a tiny pair of gold ear-rings. Dot felt closer to Crow in some ways than to her own siblings.

Some adventures transpired closer to home. There was a rather odd woman in their neighborhood who sometimes gave Dot and Crow a ride to Arlington Hall. She wore what she called "dirt-colored clothes," so she wouldn't have to wash them often. Dot and Crow appreciated the ride, so they tried to overlook her eccentricity, but when she backed into another car—to punish the driver for honking at her—they decided they'd rather walk from then on.

From time to time Dot did take the train home to Lynchburg, and sometimes she glimpsed WAVES making the same journey. Loads of girl sailors would pile on. Sitting in her seat, if she was lucky enough to get one—once, she had to make the trip standing on an outside platform, along with Crow and Liz and Louise, getting covered with smoke—she reflected enviously that the Navy women were cute girls and their naval uniforms looked very smart. Unlike the WAVES, she and her Arlington Hall colleagues were largely unrecognized for their war service. They were not feted or celebrated, and nobody asked them to model in fashion shows. People in her family knew Dot was doing something for the war, but they assumed it was secretarial and low-level. She could not even tell her mother. But even as she admired the Navy women's outfits, it never occurred to Dot that the WAVES might be engaged in the same war work that she was, endeavoring—just as she was—to beat back the fascist menace and break the codes that would bring the boys home. The very thought that so many young women were all working the same top secret job, Dot and Crow and those distant, glamorous-looking WAVES, never crossed her mind. Nor was she remotely aware that the long-simmering competition between the U.S. Navy and the U.S. Army had come to a head as the Army struggled to match the Navy's efforts in the Pacific Ocean.

CHAPTER EIGHT

"Hell's Half-Acre"

April 1943

Young Annie Caracristi washed her hair with laundry soap. Observing her, Wilma Berryman felt convinced of it. Fels-Naptha, most likely: the strong-smelling bar soap meant for treating stains. You weren't supposed to use Fels-Naptha on your skin unless you had something dire like poison ivy—certainly not on your hair—but some people did, these days. Shampoo, like so many items, was not always easy to come by. The results were not ideal: Annie's hair was thick and curly and flew everywhere. But a tendency to dishevelment only increased Wilma's fondness for her.

Blue-eyed, blond, and good-natured, Ann Caracristi came to work at Arlington Hall each day wearing bobby socks, flat shoes, and a pleated skirt that billowed and swung. She looked like a bobby-soxer, the kind of carefree and heedless college girl who lived for boyfriends and swing dances. But appearances were deceiving. What hidden depths Ann Caracristi had. What capabilities. General Douglas MacArthur did not know it, but his secret weapon—or one of them—was this affable and somewhat cosseted twenty-three-year-old from the upper middle classes of exurban Bronxville, New York. Intellectually ferocious, Annie worked twelve-hour shifts, day after day. The only time she missed work for any time at all was when she came down with chicken pox. She phoned apologetically to say in a tiny, pitiful voice that she could not come in. Wilma Berryman took her some soup.

Annie Caracristi surprised everybody, most of all herself, with her cryptanalytic feats. Though she had been an English major in college, she possessed the mind of an engineer. It was fascinating for Wilma Berryman—the West Virginia schoolteacher who had been one of William Friedman's early Munitions Building hires, now supervising a major unit at Arlington Hall—to see what Annie could do. Nothing the Japanese did could shake her off. Conversion squares, encipherment tables, cleverly cannibalized additive books—Annie was onto all their ruses. So gifted was she that Wilma made Ann the head of her research group. At Arlington Hall, to have a recently graduated female in charge of a key unit was not unusual. It was normal.

Unlike the Navy operation, the Army's code-breaking operation at Arlington Hall was polyglot, open-minded, and nonhierarchical. Anybody could be in charge of anything. There was a wide assortment of ages and backgrounds working at its wooden tables. Bespectacled middle-aged men labored alongside pin-curled young women with names like Emerald and Velvet. Which is not to say that there wasn't sexist condescension: One of the bookish men, a New York editor named William Smith, referred to Arlington Hall's contingent of female southern workers as the "Jewels." It was a lofty and rather snide reference to the number of women working there whose parents had seen fit to name them after precious stones. He wasn't wrong: In addition to a profusion of Opals and Pearls, the workforce included a real Jewel—Jewel Hogan—who worked in the machine section. And there was Jeuel Bannister, the band director recruited out of South Carolina.

At Arlington Hall there also were "BIJs," or born-in-Japans, the term for people who grew up in missionary families and worked in the translating section. There was the actor Tony Randall—later famous as Felix Unger in *The Odd Couple*—clowning around (at one point he danced on a table) as he waited for the intelligence summary to be taken to the Pentagon. There was an extended group of siblings and cousins—the

Erskines—who had relocated as a family unit from Ohio. There was Sumner Redstone, the future billionaire media magnate, now a young officer in the translating unit. There was Julia Ward, former dean of students at Bryn Mawr, czar of a well-run library unit. There were nannies, beauticians, secretaries, restaurant hostesses. Josephine Palumbo at eighteen was virtually running the personnel unit, plucked out of McKinley High School in Washington. Tiny Jo Palumbo, daughter of an Italian immigrant laborer, was the person who swore in newcomers, and the sight of her administering the grave secrecy oath had inspired one code breaker to write a lyrical poem in her honor.

Unlike the Navy, Arlington Hall also had an African American codebreaking unit. This was not so much because the place was unusually liberal-minded, but rather because Eleanor Roosevelt—or somebody at the top—had declared that 12 to 15 percent of the Arlington Hall workforce should be black. It was poor recompense for the fact that many of Arlington's black residents had been pushed out of their homes by the construction of the Pentagon and other military edifices, but work was welcome and this was better than nothing. Arlington Hall's African American workers had to take segregated transport to get there, and many, even those who were college graduates, were given menial jobs as janitors and messengers. But there also was a special code-breaking unit whose existence was unknown to many of the white workers. The African American unit monitored the enciphered communications of companies and banks to see what was being transmitted in the global private sector and who was doing business with Hitler or Mitsubishi. They kept a library of 150 systems, with careful files of addresses and characteristics of all the world's main commercial codes. There was no shortage of qualified people to staff it: Despite its segregated school system and the inequality of resources that accompanied segregation, the city of Washington had a number of highly regarded black public schools, as well as Howard, one of the country's premier historically black universities. One of the team members, Annie Briggs, started out as a secretary and rose to head the production unit. Another, Ethel Just, led the expert translators.

The team was led by a black man, William Coffee, who studied English at Knoxville College in Tennessee, started out as a janitor and waiter at Arlington Hall, and rose to this position.

In short, in its eclecticism and, often, its eccentricity, the atmosphere at Arlington Hall was unlike anything the U.S. military had ever produced. When Juanita Morris, a college student fresh from North Carolina, showed up on her first day, she was directed to a dim room where she saw a woman holding an ice pack to her head, another wearing a sun visor, a man walking around in his underwear (he had been caught in the rain and had hung his clothes up to dry), and another who was barefoot. "This is the German section," someone told her, and left. Her father had told her to come back home if things didn't work out, and for the first couple of weeks, she was tempted.

Arlington Hall was a military operation, but only in a nominal sense. An Army colonel named Preston Corderman had taken over as head, but Corderman was hardly a rigid disciplinarian. He knew whom he was dealing with. When the code breakers figured out how to rig a Coke machine to make it run continuously and spew free liquid— they achieved this by inserting a coin and unplugging the wall plug— Corderman sent around a memo drily congratulating them for having "solved the machine" and suggesting that it was time to pay the requisite nickel per cup.

Physically, the place was a hodgepodge. The main schoolhouse retained its languid Old Virginia finishing-school flavor, but the new temporary buildings were purely functional. Building A—where the Purple cipher was attacked along with other ciphers—consisted of two floors and a basement with a fireproof vault. Designed to hold 2,200 people, it quickly proved inadequate, and so B was built. Soon the campus boasted a beauty parlor, a tailor, a barbershop, a dispensary with fourteen beds, a 620-seat auditorium and theater, a mess hall, a car repair shop, a warehouse, and a recreation building. The two main buildings were poorly insulated, chilly in winter and sweltering in summer. There was much fan drama, what one worker called "a constant war of direction and re-direction," in which the goal was to get the full force of the fan directed at somebody else while the gentle wafting breezes floated toward you.

Feeding the workforce was a perpetual headache. In this sparse part of Arlington County there existed one restaurant and two drugstores with lunch counters, but the food was bad and so was the service. Box lunches were attempted and abandoned. A cafeteria was created, soon to be replaced by a bigger one. The bigger one was inundated and the first was reactivated, and both were kept up and running. The snack bar stayed open most of the night, so people on the graveyard shift could eat. At night the smell of brewing coffee pervaded the premises.

Transportation was similarly a challenge. Bus drivers were in short supply and so were buses. Entrance was gained through one of four gates, where employees had to display badges with photos and a color denoting their clearance level. If a code breaker forgot her badge, she had to wear a "forgot my badge" badge designed to humiliate. One version displayed the cartoon face of a donkey. Often, enlisted men recruited for their intelligence—high-IQ GIs, they were called—were called away from code breaking and assigned to guard the post. Many had never held a firearm, and there were rumors of accidental gun discharges.

Still, there was plenty of evidence that a real war was going on. Military officers headed many units, and enlisted men rotated in and out. Even among the soldiers, though, rank had little meaning. A lieutenant might report to a sergeant or a civilian or even to a private first class. If the officer objected, he was sent overseas. "You didn't go by rank," said Solomon Kullback, another early Friedman hire. "You went by what people knew."

The same was true of women. It would be an exaggeration to say women enjoyed true parity in the Arlington Hall workforce: Among the early William Friedman hires, men like Frank Rowlett, Abraham Sinkov, and Solomon Kullback were awarded military commissions to put them on an even footing with the Army brass, but the veteran women were

Agnes Meyer Driscoll, a former Texas high school math teacher, became one of the great cryptanalysts of all time, cracking Japanese Navy fleet codes during the 1920s and '30s. *Courtesy of National Security Agency*

Elizebeth Smith Friedman, another ex-schoolteacher, took a job in 1916 at an eccentric Illinois estate called Riverbank, where she helped found the U.S. government's first code-breaking bureau. She later broke the codes of rumrunners during Prohibition. *Courtesy of the George C. Marshall Foundation, Lexington, Virginia*

Genevieve Grotjan aspired to be a math professor but couldn't find a university willing to hire a woman. In September 1940, after less than a year as a civilian Army code breaker, she made a key break that enabled the Allies to eavesdrop on Japanese diplomatic communications for the entirety of World War II. *Courtesy of National Security Agency*

Japan's ambassador to Nazi Germany, Hiroshi Oshima, was a confidant of Adolf Hitler. Oshima communicated with Tokyo using an enciphering machine the Allies called "Purple." Grotjan's break enabled the United States to monitor these missives, yielding some of the best wartime intelligence out of Europe. The Allies called it "Magic." *Courtesy of National Security Agency*

The German military forces used their own enciphering machine, the portable Enigma. *Courtesy of National Security Agency*

America in the Depression was still a rural country. By 1940 only about 4 percent of women had completed four years of college, in part because many colleges would not admit them. Dot Braden, shown here as a girl with her nanny and brothers, was steered to Randolph-Macon Woman's College by her spirited and determined mother. *Courtesy of Dorothy Braden Bruce*

Dorothy Ramale, shown here in her 1943 yearbook photo, grew up on a farm in Pennsylvania and wanted to be a math teacher. But the dean of women at Indiana State Teachers College called her in and told her the U.S. Army had another idea for her. *Courtesy of Indiana University of Pennsylvania Special Collections*

Women's colleges in the 1940s were a mix of cerebral inquiry, marital ambition, and hallowed rituals. The 1942 May Court at Goucher College included Jacqueline Jenkins (fourth from left) and Gwynneth Gminder (second from right). Their lives changed when they received a secret summons from the U.S. Navy, as did Fran Steen, shown in the photo on the left. *Courtesy of Goucher College Archives*

Unbeknownst to these young women, the Army and Navy were hotly competing for their talents. The competition began when Ada Comstock, president of Radcliffe, was asked to suggest undergraduates who could be trained by the Navy in cryptanalysis. The surprise attack at Pearl Harbor had exposed the country's intelligence deficit and created a new demand for educated women. *Courtesy of Schlesinger Library, Radcliffe Institute, Harvard University*

Als
Afri
female, a
workers–
keepi
were
M

As war progressed, the demand grew. In 1942 the United States grudgingly decided to admit women into military service, a wildly controversial measure. "Only bad women join the service," one code breaker's mother told her, but soon recruiting was in full swing and the mother was proudly snapping photos of her daughter in uniform.

Women rushed to enlist; one reporter described a WAAC recruiting station as a "tidal wave of patriotic pulchritude." Women who tested high for intelligence and aptitude—and passed background checks—were routed into code-breaking service. *Courtesy of U.S. Army INSCOM*

Wome
Germa
that
convoys,

The U.S. Army recruited schoolteachers in the South. Handsome young officers were dispatched to lure them. When Dot Braden approached recruiters at the Virginian Hotel in Lynchburg, they did not tell her the nature of the work, and probably did not know themselves. *Copyright: Nancy B. Marion; Credit: Courtesy of Lynchburg History*

During their off hours, the women wrote letters incessantly, sending the missives—often with an enclosed snapshot—to soldiers and sailors. One g-girl was writing twelve different men. *Courtesy of Library of Congress*

The women worked around the clock and often didn't know whether to eat breakfast or dinner when they finished a shift. Navy code breaker Edith Reynolds (center), recruited from Vassar, relaxes with colleagues. *Courtesy of Edith Reynolds White*

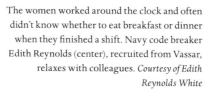

The women enjoyed their freedom. Dot Braden (right) and her best friend, code breaker Ruth "Crow" Weston from Mississippi (left), were glad to get away from crowded Arlington Farms and rent an apartment on their own. *Courtesy of family of Ruth "Crow" Weston*

On their rare days off, the women would ride buses and streetcars to the beaches of Virginia and Maryland. This group of Arlington Hall code breakers includes Crow (far left), and Dot (peeking out from behind the pole). *Courtesy of Dorothy Braden Bruce*

When, in the spring of 1943, a group of members of the WAVES were loaded onto a troop train and sent "west" under sealed orders, they hoped they might be headed to California. They emerged to find they were in Dayton, Ohio. *From the collections of Dayton History*

At a top-secret building on the grounds of the National Cash Register Company (headquarters shown here), the women helped build more than one hundred "bombe" machines designed to break the German naval Enigma cipher. *From the collections of Dayton History*

Their bucolic campus was called Sugar Camp, named for the maple trees on the grounds. The women were warned to never go into town alone, for fear of being kidnapped by spies. *From the collections of Dayton History*

The women at Sugar Camp enjoyed great camaraderie even as they felt responsible for the welfare of brothers, husbands, and fiancés serving overseas. For some, the psychological strain would not abate even after the war was won. *Courtesy of Deborah Anderson personal archives*

After World War II ended, Stephen Chamberlin, operations officer for General Douglas MacArthur, declared that code breakers shortened the war and helped save thousands of lives. The postwar accolades did not mention that more than ten thousand U.S. code breakers were women. *Courtesy of U.S. Army INSCOM*

Many women stayed friends for decades after their service. A group of Navy women kept a "round-robin" letter going for seventy years. The group included Elizabeth Allen Butler (front far left), Ruth Schoen Mirsky (front, second from right), and Georgia O'Connor Ludington (back, second from right). *Courtesy of Ruth Mirsky*

Dot and Crow's friendship remained so strong that Crow insisted Dot and her husband, Jim Bruce, come to Washington to meet her own fiancé, Bill Cable, and give their stamp of approval before she would marry him. Jim Bruce is at right and Bill Cable is at left. *Courtesy of Dorothy Braden Bruce*

The women took their oath of secrecy so seriously that even now, at age ninety-seven, Dorothy Braden Bruce (shown here at her birthday party with her grandchildren) has trouble bringing herself to utter certain words she was told never to say outside the grounds of Arlington Hall. *Courtesy of Dorothy Braden Bruce*

not similarly promoted. Even so, all of the early women—Wilma Berryman, Delia Taylor, Genevieve Grotjan—found themselves running top units. To a striking degree, Arlington Hall was what would nowadays be called a "flat" organization, an egalitarian work culture in which good ideas could emerge from any quarter and be taken seriously.

In part this was thanks to the open-mindedness of the people in charge, but it was also thanks to their desperation. In the early months of 1943, Arlington Hall struggled to make headway in one of the Pacific War's most urgent challenges: breaking the codes of the Imperial Japanese Army. In this effort, it was Wilma Berryman and her bobby-soxer protégée, Ann Caracristi, who would make the first significant break.

* * *

During 1942 the U.S. Army and Navy had hammered out a sane division of code-breaking duties, abandoning the odd-even rivalry dating back to the Purple break. The U.S. Navy took responsibility for breaking Japanese naval codes as well as helping with the German naval Enigma. Swamped by the demands of these major enemy systems, the Navy ceded to the U.S. Army responsibility for keeping up with Purple, as well as the codes and ciphers of many enemy and neutral nations. But the Army's toughest assignment was breaking a fiendish tangle of Imperial Japanese Army codes, which were separate from those of the Imperial Japanese Navy and—this long into the war—remained unbroken. Tackling them was a monumental task, one that for some time appeared beyond even William Friedman's talented and experienced group.

Part of the problem, at first, had been a lack of message traffic. Prior to the war, the Imperial Japanese Army had some 2 million troops stationed in China and on the Manchurian border, but these units were close enough in proximity that they could use low-frequency, low-power transmissions, and the U.S. Army had a hard time getting radio intercepts. To break a complex enciphered code like the ones used by the Japanese military, it's essential to have what's known as "depth": lots and lots of intercepted messages that can be lined up and compared. In the panicked atmosphere after Pearl Harbor, the group made a stab at a solution nonetheless. In 1941, a British colleague brought some Japanese Army intercepts to Friedman's operation in the old Munitions Building. Friedman shut four of his staffers—Solomon Kullback, Wilma Berryman, Delia Taylor, and Abraham Sinkov—in a room, telling them not to emerge until they had broken something. The job was just too big. After about three months, Solomon Kullback stood up and shoved his desk back into the German section. "I've had it," he said. The only positive outcome was that Delia Taylor and Abe Sinkov fell in love, got married, and moved into a houseboat at the wharf. Other than that, the first attempt was dispiriting and the winter of 1942 was a gloomy one.

In many ways, however, great success was the Japanese Army's great undoing. After its stunning victories in the first half of 1942, the Japanese Army began to spread out. Millions more troops now occupied a greatly enlarged amount of territory. Japanese units fanned out over Asia and the Pacific archipelagos: China, Hong Kong, the Philippines, Thailand, Burma, Malaya, the Dutch East Indies. The Eighth Area Army concentrated around the stronghold of Rabaul, on the island of New Britain. Each unit remained tied to its home base in Japan and was obliged to send back reports on things like casualties and the need for reinforcements. As the Japanese Army got farther from Japan, radiomen increased the power of their transmissions, and this made it much more feasible to intercept them.

Soon enough, the problem at Arlington Hall was not a dearth of intercepts. Japanese Army messages began pouring in by the tens of thousands—by airmail, by cable, by teletype. The problem was that there were many systems, each complicated in its own way. In the final months of 1942 and into early 1943, the code breakers of Arlington Hall worked in a fever—painfully aware that their colleagues in the U.S. Navy had broken JN-25 and were laughing into their gold-braided sleeves at the Army's galling lack of progress.

* * *

The Arlington Hall code breakers also knew that the situation in the Pacific was at a tipping point. In the wake of Midway, an American pushback plan was being formulated, as the fighting forces of the U.S. Army and Navy hammered out their own division of labor in the world's biggest ocean. After much sparring, the services agreed that Admiral Chester Nimitz would be the Pacific theater commander and that the U.S. Navy under Nimitz would deploy mostly in the central and North Pacific, a huge blue-water arena where the only wisps of land were tiny atolls. Down in the southwestern part of the hemisphere, where there was more terra firma, the U.S. Army—led by MacArthur—would fight along the islands, adopting an "island hopping" strategy in which some Japanese-held islands would be assaulted with the intent of seizing airstrips and establishing overlapping zones of air power. Cut off from supplies and reinforcements, other islands would be bypassed, isolated, and allowed to wither.

MacArthur's goal was to retake the Philippines and invade Japan. Operation Cartwheel called for the U.S. Army and Navy to work together in this region. As MacArthur's troops made beach landings and fought a fierce, dug-in island adversary, Admiral William "Bull" Halsey would support them from ships in the South Pacific. The U.S. Army at this juncture had more troops fighting in the Pacific than it did in the Atlantic war theater. Those soldiers faced months, maybe years, of beach and jungle warfare. They needed intelligence to tell them where the enemy's land troops were and what they were getting into. Even with JN-25 being read, the United States needed access to Japanese Army communications.

For a code breaker, no situation is more stressful than knowing lives depend on your own success or failure. If you crack the code, people will live. If you don't, they may die. Once more a unit was set up to break the Japanese Army systems, led by Frank Lewis, one of Friedman's most brilliant—if unlikely—hires. Lewis, the Utah-bred son of an Englishman turned cowboy, was a gifted musician, a lover of puns, and a puzzle enthusiast who would later write the cryptic crossword for *The Nation*. Before the war, he had been languishing, bored out of his mind, as a secretary in the death benefits section of the civil service. Friedman snatched him up. Friedman's top women also worked the Japanese Army problem; after the move to Arlington Hall, all found themselves sweating it out on the hot upper floors of the old schoolhouse. The rooms had a country club flavor that would have been lovely and relaxing, had they not been so crowded and had the task not been so daunting. It was not unusual to find an exhausted code breaker napping in a tub.

Matching wits against an unseen enemy, their little band was vastly outnumbered.

"Visualize, if you will, the entire communications set-up of the Japanese military forces, with many thousands of men whose entire existence centered around the preparation and transmission of messages," Lewis later wrote, "and then contrast that huge organization with the small group of technicians at our own establishment."

Much of their challenge had to do with the island environment in which the foe was operating. As the Japanese Army spread out around the Pacific, its cryptographers had to create new code systems and subdivide old ones. The Japanese devised a host of minor codes and at least four major four-digit systems: one for ground forces, another for air forces, another for high-level administrators, and another for the "water transport organization," which was a vital lifeline of marus, or commercial merchant ships, commandeered by the military to carry resources, including oil, food, and equipment. Each system was identified by a discriminant—an unenciphered four-digit code group at the beginning. It was a huge tangle, and as of January 1943 there were just fifteen American civilians, twenty-three officers, and twenty-eight enlisted men working on breaking all of them.

* * *

One of the civilians was Ann Caracristi, the twenty-three-year-old who washed her hair with laundry soap. She had been recruited out of

Russell Sage, a women's college in Troy, New York. Ann had grown up in Bronxville, in a middle-class family of Italian-Austrian heritage. She had two older brothers, one of whom was serving in India. Her father, a businessman and inventor, had permitted the older brother to attend college, but the younger wanted to be a liberal arts major and that was not considered sufficiently serious-minded, so he could not go. Her father died while Ann was in her teens, and Ann herself was able to attend college thanks to a friend of her mother, who saw her intellectual promise and helped persuade a group of Bronxville women to finance her tuition. She played basketball and edited both the campus newspaper and the literary magazine. She did not see herself as smart, but those who instructed her saw differently.

In May 1942, the Army Signal Corps had met with emissaries from twenty colleges gathered at the Mayflower Hotel, including Dr. Bernice Smith from Russell Sage. The Signal Corps asked Dr. Smith to handpick a few of her top students. As a result of these faraway high-level meetings, Ann Caracristi found herself sitting in the office of a Russell Sage dean she knew well. The dean told Ann and two classmates that there were jobs in Washington for women with brains and imagination. Thinking it a bit of a lark, the three friends traveled down after graduation, taking rooms in a Wyoming Avenue boardinghouse that had once housed the Armenian embassy. Soon enough, Ann too found herself laboring under the eaves in the attic of the Arlington Hall school building. She started by learning to edit traffic, sorting intercepts to be typed up and put onto IBM punch cards so that code groups could be compared. Everybody was playing it by ear, and newcomers could—and did—innovate techniques. Another female newcomer suggested they first edit by date and time, as a way to identify duplicate messages. The suggestion—it came to be known as de-duping—revolutionized the process.

Progress was slow, however, even with the influx of helpers. To the naked eye, the major Japanese Army code systems consisted of an unbroken string of four-digit numbers: 5678 8757 0960 0221 2469 2808 4877 5581 1646 8464 8634 7769 3292 4536 0684 7188 2805 8919 3733 9344, say.

The code breakers knew only that the systems were enciphered codes, somewhat like JN-25, involving both a code and an additive book. But the Arlington Hall team could not figure out the encipherment method; despite attack after attack, they were stuck.

At the suggestion of a visiting Bletchley Park colleague, the brilliant John Tiltman, they decided to break the job down to see if they could manage to crack even just the address that began every message. This was a series of only a half dozen or so code groups, but they were important ones. After a Japanese Army message was enciphered, it passed to a radioman who attached an address specifying who the message was from (the *hatsu*), whom it was going to (the *chiya*), and where those people and their units were located (the *ate*, or address). The address designated the army command, installation, or officer for whom the message was intended.

Details like that may sound dull and ordinary, but they provide a lot of eye-opening intelligence about the makeup and location of enemy forces. The address was in its own code system, used only by the radiomen and distinct from that of the message itself. If Arlington Hall could break the address code, it would tell them who in the Japanese Army was where, but it also might provide clues that could help get into the messages themselves.

Wilma Berryman was assigned to the address problem in April 1942, and in June, as they were plunging into work in earnest, Ann Caracristi joined the group. As they labored and strategized, Ann and Wilma formed an immediate bond. It was an unlikely pairing: Wilma was a down-home, dyed-in-the-wool West Virginian, a shrewd, world-wise, tall, big-boned, genial southerner who liked to call people "honey" and to use adjectives like "stinkin'." Her husband had died soon after they moved to Washington—she would marry three times—and her colleagues got her through her grief, gathering "round like family and it made life really worth living and going on," as she later said. Ann was ten years younger, a northerner, new, well-bred, uncertain yet of her own abilities, inexperienced in the workplace, barely initiated into the tight fraternity of Arlington Hall cryptanalysts. But both had an appreciation for humor: Wilma was "a fun person," and Ann liked working for her. The two women shared high spirits, humor, brains, imagination, and a relentless determination to prevail.

Both enjoyed the work. Ann would later say that she felt her time at Arlington Hall was "sport," rather than labor. The women tried a number of approaches. At the suggestion of one military officer working with their unit, Ann was assigned to do something called "chaining" differences, which is a long, agonizing process that involves subtracting one code group from another with the hope that two code groups might have been enciphered with the same additive. Chaining differences was a task of the most routine and time-consuming order. It is code breaking the hard way, a brute-force method used when there are no other clues or ways to get a start.

They had caught a small break, though. When a Japanese plane crashed in Burma, the British captured some message templates—a template is a blank form with some code groups filled in to speed the process—and sent them to Arlington Hall, which had them in its possession by late January 1943. Wilma and Ann began having quiet conversations about how best to use this material. They agreed that chaining differences was "silly" and that it made more sense to use the templates and the bits of information they contained.

They had a couple of other tools as well. They had received a report from Australia with some names and ID numbers for Japanese Army units in the southwest Pacific. In addition, the U.S. Navy had sent over some cribs. At times, the Japanese Army was obliged to send messages over Navy radio circuits. In that event, a naval address code was appended. The naval address was in a simple code that had been solved, and it provided the basic pattern of Japanese military addresses; the unit, followed by two numbers, then an honorific such as *butaicho,* or "squad leader."

The Arlington Hall address team had well-kept records, and Wilma Berryman remembered some Japanese Army messages with addresses that might be the same as these Navy ones. She found them and began fiddling around, writing the plain words from the Navy address codes below the Army ciphers. "I sort of remembered having seen something in that file and I went back to the file and found it," was how she put it later. "I found what I thought looked like it ought to be that, the same thing. I had it on my desk and I just wasn't positive."

Suddenly she noticed one of the men who worked with their unit, Al Small, standing behind her. He stood there for a while, watching.

"Wilma, what are you doing?" Small asked. She showed him how she was lining up the Army messages against the Navy cribs. It seemed to be working. The captured message blanks gave her the underlying code group, and the Navy crib provided its likely meaning. If she saw, say, the enciphered group 8970, and knew from the captured blank that the basic code group was 1720, she could figure out that the additive was 7250. The Navy crib told her this was the syllable *mo*. She could hardly believe it. Was she imagining things? Al Small stood and looked at it for a long time. "You've got it," he said finally. "That's it! That's it!"

"I'm afraid I'm forcing it," replied Wilma, fearful that the juxtaposition was an illusion. "I'm pushing too hard. I want it to work."

"No—that's it!" Small repeated. "You've got it!"

Her solution was Arlington Hall's first real break into any Japanese Army system. In early February 1943, a memo reported that "with the aid of the captured messages it has been possible to read the first encoded and enciphered addresses." It would be followed by months of what amounted to the most laborious kind of coal-mining extraction. The problem with enciphered codes is that they do not yield to immediate solution like machine ciphers do. Even after a basic understanding is achieved, there remains a great deal of work. The code breakers had to build a bank of additives and figure out what each code group stood for. It had taken Agnes Driscoll and her team years to master the Imperial Japanese Navy code systems, but the Arlington Hall team did not have years. Deriving the meaning of each code group is like extracting ore, chunk by hard-won chunk. Ann Caracristi dove in, blissfully at home in an environment where there was very little oversight and "you assumed that you were going to have to figure your way out of most problems."

Arlington Hall began producing weekly memos to keep the Pentagon and others apprised of their progress. On March 15, 1943, a memo reported that more address code values were emerging. Ann, Wilma, and their few colleagues had ascertained that 6972 meant *i*, 6163 meant *aka*, 4262 meant *tuki*, 3801 meant *si*, 0088 meant *u*, and 9009 meant *dan*. They were beginning to recover additives and to understand nuances such as the system's unique sum check. Like the Japanese Navy, the Japanese Army cryptographers were fond of sum checks as a guard against garble, but they had devised a method different from the "divisible by three" sum check of the Japanese Navy. In this system, the code breakers discovered, the four-digit code groups were actually three digits, plus a fourth digit that served as a sum check. If a code group was 0987, for example, 098 was the actual code group, and 7 (using false math) was the sum check: $0 + 9 + 8 = 7$.

Reconstructing an enemy codebook is called "book-breaking," and Ann Caracristi, working with the linguists, turned out to be stunningly good at it. By summer 1943, they knew that 1113 stood for *shibucho*, 1292 for *taichoo*, 1405 for *butai*, and 3957 for *bukkan*. They dug out the code groups for Hiroshima, Singapore, Kupang, and Tokyo.

The address codes carried a bounty of operational military information. The work that Ann Caracristi and Wilma Berryman were doing enabled U.S. military intelligence to construct what is called the order of battle: an accounting of the strength, equipment, kind, location, and disposition of Japanese Army troops. They were able to pinpoint where the enemy was quartered and headed. Soon "MacArthur's headquarters had as good a picture of the Japanese military set-up as he had of his own," as Solomon Kullback put it. As it grew in importance, the address unit grew in number. In addition to Ann and Wilma, it included women with names like Olga Brod, Bessie E. Grubb, Edna Kate Hale, Mildred W. Lewis, Esther A. Sweeney, Lottie E. Miller, Bessie D. Wall, Violet E. Bennett, Goldie M. Purks, Mabel J. Pugh—and, yes, a jewel, Ruby C. Jones.

"That outfit was 100 percent female," Solomon Kullback said later. "There were only girls working on that address system." There were two military men assigned to the unit—Reuben Weiss and Mort Barrow— and the pair used to circulate, joking and stoking morale, showing up in the middle of the night to raise the spirits of the graveyard shift and jolly the women along. There were "only two officers," Kullback recollected, "and at least a hundred girls."

Wilma's team worked alongside a unit called "traffic analysis," also mostly female. It was the job of traffic analysis to follow fluctuations in Japanese Army message traffic, without worrying about the actual message content. External information helped in developing the order of battle. If a flurry of radio messages began going back and forth to a new location, this meant somebody was on the move. Before long, Wilma would find a representative of G-2—military intelligence—standing behind her daily, urging her on and telling her which addresses to concentrate on. Arlington Hall began to produce a daily order-of-battle summary. Every morning at five o'clock there was a "black book session" discussing the report that would be compiled each evening and taken to the Pentagon.

For the women code breakers, it was exhilarating. The same address code was appended to all the different Japanese Army systems, so soon enough every kind of message—air force, administrative, water transport, ground troops—was passing through their hands. "I think that's one of the things that made it so much fun. We saw everything, everything had to come through us," Wilma later said. "You had cryptanalysis, you had traffic analysis, and you had order of battle, and you could hardly ask for a better job. You were sitting right on top of the world and you were following the Japs all over hell's half-acre, all the time. You weren't just reading something about, 'Send me three pounds of sugar and ten pounds of rice.' Or something like that. I thought it was the most exciting job in the place."

* * *

Staying current was a never-ending task, though. As with the Navy, the Japanese Army routinely changed both its books and its methods.

There was a grim period when the radiomen began using one method to encipher addresses that had an odd number of code groups, and another method to encipher those with an even number. The address unit seemed on the verge of collapse.

"We were in an awful pickle, because it was war," Wilma Berryman later recalled. "They had been getting such beautiful order of battle, and all of a sudden, just like that, it was wiped out. And so we had to do something about it." Wilma and Ann began working on traffic that had an even number of group counts, and they were on the verge of being able to read it when Solomon Kullback came by and said he would help. Their boss told them to gather up half of the messages and bring them into his office.

"I'll take care of the evens, you take care of the odds," he told them.

As grave as the situation was, the two women thought this was extremely funny. Given all the work they'd just done on the even-numbered code groups, their boss was getting by far the easier job. They had to dash away so they wouldn't lose their composure. "Annie and I just ran, it was so funny," said Wilma. For years after that, all Wilma had to say was, "You take care of the odds, I'll take care of the evens," and Ann would crack up.

Nothing could shake loose Ann Caracristi, who, even in this elite group, was in a class by herself. "My capabilities compared with Ann's were nothing," Wilma said. Wilma put Ann in charge of a small research team that included a mathematician named Anne Solomon and a male Harvard graduate, Ben Hazard, who had a physical condition that disqualified him from service. Naturally, they nicknamed him Hap. They were young but formidable. As the Americans began to retake Pacific islands, the Japanese had a hard time distributing new codebooks to isolated outposts. Sometimes they could sneak the codebooks in by submarine, but often the marooned cryptographers had to devise a jury-rigged solution, using old books in a new way. The Japanese would make squares, in which they would take additives from the old book, and—rather than adding them—run them vertically and horizontally to construct a table. Squares are hard to break for a number of reasons, among them the fact that there is no mathematical relationship between the underlying code group and the enciphered group that gets transmitted, because no addition has been done.

But group counts and sum checks do create patterns, and Ann's scanty training included some preparation for breaking squares. The first time she encountered this kind of enemy innovation, she sat puzzling over it. Solomon Kullback was instantly at her side, urging her on as the solution came into view. "It was fascinating, actually, to work in the world, become aware of it, and find you can really do something useful, without being a mathematician or linguist," Ann recalled many years later.

It was far more than that. What the Japanese were doing was encryption, and what Ann and her team were able to do, over and over, was unlock the encryption. They would work through the night to recover a key. They felt grudging admiration for how the enemy, even under attack, was able to cope with systems that demanded so much handwork, flipping through volumes and consulting tables. When Solomon Kullback received visitors, he liked to take them into a room with a table piled high with a sheaf of materials—codebooks, additive books, stacks of papers—and say, "These are the current materials which are being used by the Japanese."

As the Arlington Hall veterans refined their training, they tried to understand what to expect from a newcomer. Wilma Berryman would give Annie, as she called her, a smattering of code groups—without telling her they came from a captured book—to see how many she could book-break. She was like a living cryptanalytic experiment. One year into her tenure, Ann began training enlisted men. Years later, when Solomon Kullback was asked whom he would want if he were stranded on a desert island and only one person could crack the message that would get him home, he didn't hesitate: "Ann Caracristi." The address unit also did its bit to shoot down Yamamoto. During that tense period after the first itinerary-related message was received, the U.S. Navy sent over for help on the address codes. Wilma recalled the episode as one of her team's "biggest achievements," though afterward, she said, the Navy did not like to admit the Army had helped. "The Navy took a lot of credit," she said. "I was kind of unhappy."

* * *

In the long-running beef between the U.S. Army and Navy, one of the Navy's objections was that, by engaging a civilian workforce, the Army was likely to hire reckless, undisciplined people who could not keep a secret. This could not have been further from the truth. Sure, some people at Arlington Hall joked about what would happen if they blabbed. Signing the secrecy oath made them liable for prosecution, and a violation carried a fine of $10,000 or ten years in prison. Wilma liked to say nobody working on a civil servant's salary "could ever have found ten thousand dollars," so they'd just have to go to jail. But they took the secrecy oath in earnest. When Ann was asked what she did at Arlington Hall, she spoke in airy terms about clerical work.

They preferred not to have to explain things at all. The easiest way was to hang out with one another. The top Japanese Army code breakers ate lunch together and dined in the few local restaurants. Ann and Wilma and a few others even bought a sailboat together, pooling their funds to raise what was likely just a few hundred dollars, and struggling to sail in the wind-challenged Potomac River, where once they were almost mowed down by the night boat to Norfolk. It was dark, their sailboat was becalmed, and there was much horn blowing and hilarity. The code breakers formed a glee club and a theater group, played tennis, set up duckpin bowling in the alleys of Clarendon and Colonial Village, two nearby neighborhoods. There was a lot of drinking. People played the piano and sometimes slipped beneath it, passed out from alcohol and exhaustion.

There were love affairs, and there were deep and abiding friendships. Ann became close friends with a former Kansas schoolteacher and aspiring writer named Gertrude Kirtland. Gert was gregarious and more extroverted than Ann, also sixteen years older, fiercely literary, and erudite. She had been recruited out of the University of North Carolina at Chapel Hill. Gert, a people person, soon transferred into the personnel operation, where she provided her friends with a direct conduit to the very top brass. Gert, Ann, and Wilma would save up their gas coupons and drive Wilma's old car out for weekends in the rolling Virginia countryside around Leesburg.

But restorative weekends did not happen often. Tooth and nail they worked. No one jostled for promotion. All this, they knew, was temporary. The point was to win the war and get back to their regularly scheduled lives. One civilian woman complained often about her pay grade, and Ann thought her careerism was appalling. The goal was not to seek advancement. The goal was to serve the war effort. There was competition—with the Navy, with the British, with colleagues in Australia, with one another—but the point was just to get a solution first. They couldn't get people to stop working. Linguists, if they had spare time, would help with logging. There was a huge snowstorm, and everybody managed to walk to work. There was a spirit that prevailed of collegiality, fun, dead earnestness, and intellectual gamesmanship that led to the next major break, one that would have a consequence on the war as important as—if less publicly celebrated than—the Battle of Midway.

* * *

Breaking the address code was a vital achievement, but it left unsolved the main Japanese Army codes. By the spring of 1943, the situation was dire. The staff at Arlington Hall, along with a companion unit in Australia, began concentrating on the system known as 2468, the "water-transport code" used by the Japanese Army to route its supply ships, or marus. The team knew that 2468 was an enciphered code and that buried somewhere were two four-digit groups that comprised the indicator, an embedded item that told which part of the additive book had been consulted. The indicator was the central, elusive clue, and people all over the world were working to find it. As late as March 1943, the effort felt hopeless. The team tried brute-force attacks, punching cards and doing

IBM run after IBM run, coming up with hypotheses, raising their own hopes only to see them dashed.

"The mere statement of facts and figures can convey little idea of the more exciting and well-nigh incredible aspects of the work—all-night sessions, when each message was pounced upon as being perhaps the final link in a chain of cryptanalytic attack; the periods when the borderline between complete blankness and complete readability seemed dangerously close; the narrow margins by which the cryptanalysts, working in close collaboration with the traffic analysts, sometimes just 'squeaked by!'" Frank Lewis would later write—romantically but accurately. "An account of these aspects would put any mere 'spy thriller' to shame."

Then in April 1943, a couple of things happened. The Arlington Hall unit received a telegram from England that mentioned a peculiar aspect of 2468: In the second code group of each message, the first digit did not seem random. A second telegram, from Australia, confirmed and elaborated on that finding. A young American officer, Joe Richard, had been assigned the routine job of sorting traffic at the facility in Brisbane, working under a "weak droplight" in a two-story suburban house with its blackout curtains perpetually drawn. He too had spotted some non-random behavior that suggested some sort of mutual affinity, some tie between certain early groups in a 2468 message. He saw that any digit in the first position of the third code group shared a relationship with the corresponding digit of the second group: if the latter was 0, then the other would be 2, 4, or 9.

With just these tiny clues, the Arlington Hall code breakers began to look at the code in a new way, focusing on the second and third groups. Around midnight of April 6 and going into the wee hours of April 7, an elite team including three men and four women—Delia Taylor (now Sinkov), Mary Jo Dunning, Louise Lewis, and Nancy Coleman—shut themselves in a room and put a sign on the door barring anybody else from entering. They understood now that some digits in early groups seemed to be exerting control over other digits, and they saw that if two-number pairs, such as 11 or 77, appeared in a certain place, another pair appeared elsewhere. It was like a call and response between groups. The numbers were controlling one another's behavior in a way that suggested they were interdependent. Around midnight, the team began to see what was going on. The digits, they realized, lined up in a revelatory pattern.

They had broken a complex and crucial system. The Japanese water-transport organization used three additive books, each page of which had one hundred four-digit additives arranged in a ten-by-ten square, with randomly ordered numbers denoting the line and column. The indicator consisted of two four-digit code groups. The first indicator group provided, in its first digit, the number of the additive book used. The second two digits gave the page number, and the fourth digit was a sum check. They called this pattern BPPS, or "book page page sum check." The second indicator group gave the coordinates of the row and column. The full indicator pattern was BPPS RRCC.

But here was the truly diabolical part. These two indicator groups were placed between two early code groups of the message. Only then was the indicator enciphered, using the first two enciphered groups from the message. The point being: The message was enciphered, and then used to further encipher the indicators. It was a horribly interlaced system, a sort of Russian nesting doll involving layers upon layers of disguise. When the team told their boss, Preston Corderman, what they had achieved, his instinct was to pull down the shades, as if the enemy might be outside. They considered telling no one in the world, then gathered their wits and notified Australia. By a remarkable coincidence, Brisbane had gotten the solution at the same time.

The feeling inside Arlington Hall was electric. "New life has been given to the entire section," read an April memo, "and several problems heretofore seemingly impregnable are being attacked in the light of what has been proved in 2468." They were able to use the word "maru," which appeared often, to recover additives. They proceeded to break other systems and see how the layout of the Japanese Army could help them. As the Japanese Eighth Area Army split around Rabaul, it began sending identical messages, enciphered by the same additive but using different squares, to different units. The code breakers could compare duplicates—they called them cross-dupes—to tease out additives.

It didn't take long to grasp the import of what they had done. In July 1943, one of the first 2468 messages broken by Arlington Hall revealed that there would soon be four marus in Wewak Harbor. Wewak, on the island of New Guinea, was the site of a major Japanese airbase. The code breakers passed this information to military intelligence. Not long after, Solomon Kullback heard over the radio that the U.S. Navy had sunk four marus in Wewak Harbor. He listened with satisfaction. There was no sympathy for the drowned enemy sailors and soldiers, not in wartime.

The break into 2468 was one of the most important of the war. It was every bit as vital as the breaking of Enigma or the Midway triumph. The 2468 code routed nearly every single maru making its way around the Pacific to supply the Japanese Army. As with Japanese Navy vessels, many marus sent a daily message giving the exact location where they would be at noon. The information would be turned over to American sub commanders. "What nicer bit of information would be necessary for a submarine than to know that the ship was supposed to be at a certain spot at a certain day?" Kullback said later. The U.S. military employed ruses so the Japanese wouldn't know the maru sinkings were the result of a broken code. American planes would be sent up so it would look as though they had spotted the maru from the air. The Japanese sent messages saying they thought coast watchers—spies along the island coasts—were to blame, which the Arlington Hall code breakers read with glee.

Buoyed and elated, Arlington Hall became ambitious. They wanted mastery of every Japanese Army system. The solution of 2468 led to breaks in code systems called 5678, 2345, 6666, 7777. They solved 3366, an aviation code, and 6789, which dealt with promotions and transfers, pay and requests for funding, troop movements, and reports from the "hygiene bureau," telling how many Japanese soldiers were dead and wounded, how many sick from typhus and other diseases. They knew not only the enemy's location and pay, but also his state of health.

They attacked a major administrative code, 7890, an effort that shows what a group achievement this was—the way the Arlington Hall code-breaking unit had become one big communal brain. For weeks the administrative code seemed to defy their efforts. Then one day, a lieutenant working on the team came to pick up Frank Lewis; the two men had a fencing engagement in the nearby gym. Getting ready to go, they started chatting about the administrative code. The lieutenant wondered whether it might be enciphered using the kind of square that some other systems employed. Lewis thought not. The lieutenant asked how, if it were a square, one might tell, and Lewis went into a windy discourse about repetitions of ciphers within square periods but never crossing them, limitations of plain text, limitations of key, and so on. In the unlikely event that all of these conditions were present, he added, they could expect certain numbers to start showing up, like 9939.

Delia Taylor Sinkov overheard them. She pointed out that 9939 was the most frequent group in the 5678 system. It was this chance conversation—the men's casual hypothesizing and Delia Sinkov's steel-trap recall—that broke the administrative code, yielding intelligence including the numbers of Japanese soldiers killed and wounded, and, at least once, tactical information about a major planned attack.

Arlington Hall broke everything. "There wasn't a damn thing that the Japanese transmitted that we weren't able to read," said Solomon Kull-back. They read messages before the intended Japanese recipients did. They also were privy to the unintended consequences of Japan's strict attitude toward code security. In the Japanese Army, punishment for a lost or captured codebook was so severe that soldiers often would not admit a codebook had gone missing. In January 1944, Australian soldiers captured the entire cryptographic library of the Twentieth Division in New Guinea, which was found in a deep, water-filled pit. The codebooks were passed to Arlington Hall, where the code breakers used the materials to read messages a soldier sent to his superiors, assuring them of just how thoroughly he had destroyed the very books they were holding. The books provided intelligence that played a major role in MacArthur's campaigns in New Guinea and the Solomon Islands.

The address codes—whose solution had kicked everything off—continued to be integral. All intercepts would start out in Wilma Berryman's unit, where the address would be solved and appended to the intercept. Often, a single Japanese Army message was sent in eight or ten sections. The serial numbers, which were part of the address preamble, helped the staff members who were doing the message sorting to reassemble the pieces. Then it was on to the code-breaking unit. "Our job was to get the address solved so that it could be passed to the next wing, then people could line it up and actually tackle the text," said Ann Caracristi.

But it was still a small group of people tasked with reading the mail of the entire Japanese Army. What Arlington Hall had by mid-1943 was a select group of Munitions Building veterans, some promising civilians including Ann Caracristi, and a handful of military men. To assist the U.S. military in its hard-fought and arduous Pacific campaign, they had to "build an organization that would produce results as rapidly as possible," as one memo put it. That meant breaking down the work into smaller elements, developing a well-oiled assembly line—and hiring many more Jewels. It was the 2468 breakthrough and the floodgates it opened that led to the recruitment of Dot Braden and Ruth Weston.

CHAPTER NINE

"It Was Only Human to Complain"

August 1943

Arlington Hall in the summer of 1943 was like a start-up that had received a big dose of venture capital funding and needed to scale up overnight. The breaking of 2468—the all-important water-transport code—created a need for thousands more workers to decode the stream of maru-related messages. By this point in the war there were major recruiting challenges. The Navy was picking off women right and left, as were federal agencies, including the OSS and the FBI, as well as factories, defense companies, and other private firms. The Army did not pay as well as private industry, but it could appeal to the patriotism of workers—and it paid more than teaching school.

That's when Arlington Hall decided to lure schoolteachers, concentrating its efforts in the South. Targeting the southern states was not exactly a preferred strategy: In recruiting a civilian workforce, the Army had to abide by the mind-boggling bureaucratic rules of the Civil Service Commission, which required that for its Washington workers the Signal Corps had to recruit from something called the Fourth Civil Service District, a region that included Maryland, West Virginia, Virginia, and North Carolina. Owing to the obsession with secrecy, recruiters were not told what job they were recruiting for. This—plus quota pressures—created fertile ground for exaggeration. The void of information led recruiting officers "on occasion to make invalid assumptions and inaccurate and misleading statements to candidates for positions," one wartime report acknowledged. They issued "tempting promises."

In other words: They lied.

And because they were going after southerners, the Army devised tactics based on stereotypes about southern women—namely, that they were more man crazy, more sentimental, more emotionally gullible, and more hell-bent on marriage than women from elsewhere. It did not seem to occur to them that some of the women, like Dot Braden, might be seeking to disentangle themselves from marital engagements rather than enter into them. Arlington Hall selected good-looking officers to do the recruiting. One officer of Finnish extraction, Paavo Carlson, was considered particularly handsome. Responsible for Richmond, Virginia, and its vicinity, he may well have been the officer at the Virginian Hotel who recruited Dot Braden.

"Young Army boys we used because then the girls that we recruited would come back to Washington and think they were all going to get husbands that way," said Solomon Kullback years later, still proud of this ruse. "Literally in a lot of cases these barefooted girls from the hills of West Virginia were brought in and given some training.... We would get these new people in and teach them about the Japanese system and a little bit about some of the basic words, the common words, and turn them loose on the overlaps." There was a bit of condescension exhibited by more experienced workers: As the new women came in, Ann Caracristi admitted, "I think the northern members of our community were not as generous about their southern recruits as they might have been."

Arlington Hall added hundreds of civilian women a month. There was a major recruiting wave in the summer of 1943—the one that brought in Dot Braden, from Lynchburg, Virginia, and Ruth "Crow" Weston, from Bourbon, Mississippi—and another in February 1944. Every time there was a new break, or a big military assault in the Pacific, the Signal Corps went out and scooped up more women. By 1944, recruiters were allowed to expand into the Midwest and Northeast: Maine, Michigan,

Wisconsin, Illinois, Indiana, Minnesota, Iowa, Nebraska, Kansas, Oklahoma, Missouri, and others. The Army decided to pay the women's way to Washington as an added inducement. Eventually recruiters were told what they were recruiting for, but because of the changing nature of the work—new codes were always being broken, new techniques developed—they had little idea what skills were needed.

So they just vacuumed up women.

"All the branches needed so many new employees that no attempt was made to designate what kind or quality of employee a branch desired," a memo noted. "Branches were constantly requesting personnel in large numbers. It was not unusual to receive a request for 200 clerks with no stipulation as to classification, age, education or experience."

The Army began running ads, pasting flyers and posters in public places, and issuing press releases. An article in a Minnesota newspaper noted the recruiting effort and reported that "more than 100 Twin Cities girls have gone. Phyllis La Due is one; she wrote to her parents that she has found many St. Paul girls at Arlington Farms and the work is 'exciting.'"

The Army launched training courses at cooperating colleges, including Winthrop College in Rock Hill, South Carolina. The instructor, Ruth W. Stokes, head of the Math and Astronomy Department, wrote her superiors urging them to budget another math teacher, to help meet the demands of the "secret war course." Stokes's letter gives a sense of the frenzied nationwide competition for female math majors:

The Ballistics Laboratory at Aberdeen Proving Ground asked for [Winthrop's] entire class of 14 majors. David Taylor Model Basin Engineering Laboratory asked for all the class and further suggested that the U.S. Bureau of Ships could employ all the girls Winthrop could train, for engineering assistants. Langley Field offered employment at $2400 to the first 6 of my students who applied....I had a telegram and a letter last week from Rear Admiral H. S. Howard, literally begging me for as many as six girls with majors in mathematics....Of the 34 senior students I had in the Cryptography class last spring, 33 were offered employment with the

Signal Corps, at the initial salary of $1970. The one not accepted was of foreign parentage, Syrian. In the last year the mathematics department at Winthrop College has trained and placed in essential war work more than 50 young women.

Arlington Hall also did its best to lure WACs from other Army units. A pamphlet was printed for military women, which made working at Arlington Hall sound like a spa vacation. Titled *Private Smith Goes to Washington*—the popular James Stewart movie had premiered in 1939—the cover featured a young woman in an Army uniform with a handbag and suitcase and, in the background, the Washington Monument and Tidal Basin with cherry blossoms in bloom. Inside was a photo of Arlington Hall with its lovely facade, an American flag flying high. "Arlington Hall has been acclaimed as one of the most beautiful buildings in the South," the pamphlet enthused, describing it as being "five miles down a beautiful, swift, tree-lined road from Washington, the Nation's Capital."

Emphasizing the glamour of the work, the pamphlet traced Private Smith's journey, saying that she "alights from the bus, crosses the lovely stretch of walk, opens the great door and walks into the subdued bustle of Arlington Hall. This is no ordinary Army post. There is an atmosphere of excitement and mystery here which starts her blood tingling even as she crosses the threshold."

More photos were scattered enticingly through the pamphlet, which touted the barracks ("bathtubs and showers"), the food ("tempting, nutritious dishes"), the clothing ("smartly tailored dress uniforms"), and the career opportunities. It praised the PX, the beauty parlor, the exciting moment of mail call. It noted that a WAC's basic pay was the same as a man's: $50 a month plus food, lodging, clothing, medical and dental care, and benefits including life insurance. There was a quote from a private. "The $50 I get every month is all velvet—all mine! When I was a civilian, I never had that much left after all my bills were paid."

The pamphlet attracted about a thousand WACs to Arlington Hall—where they got the "stinkinest jobs that there were to have," in Wilma

Berryman's view. (Women made up most of the seven thousand civilian employees at Arlington Hall, for a civilian-to-military ratio of about seven to one.) The WAC barracks were so primitive that women had to shovel coal into potbellied stoves to heat them. Some WACs were billeted in horses' stables at a nearby Army post. The women living in the stables were called "Hobby's Horses," in homage to Colonel Oveta Culp Hobby, the Texas newswoman who headed the WACs. Many were put to work in the Arlington Hall machine room. Some were placed on security duty, made to sit in chairs and guard doors. One WAC took her job so seriously that she barred a general from entering Wilma Berryman's office on official business. When he protested, telling the WAC who he was, she shouted, "I don't care if you're a colonel; you can't go in there!"

Stinking though some of the jobs may have been, the U.S. Army—unlike the Navy—did allow its military women to be deployed overseas. Some WACs were trained in cryptography and sent to the war theater to encode American messages. They went to France, Australia, and New Guinea, where they worked in bunkers, basements, and fenced-in compounds.

Others were trained as radio intercept operators, manning clandestine receiving posts at Two Rock Ranch, a coastal station in California, and Vint Hill Farms, the intercept station set up in a red barn at a farm in northern Virginia. It was a welcome career opportunity, but one that, for some, came at a psychic cost. Norma Martell, one of the WACs assigned to Vint Hill, had grown up in West Virginia, the eleventh child of a subsistence farmer. She won a full scholarship to a nearby college, but her family could not raise the $7 bus fare, so she did not go.

After joining the WACs, Norma found herself arriving at Vint Hill on the same day the men working there were departing. "All the men in that unit went overseas and died on the beaches within a month," she recalled in a 1999 interview. Traumatized by what became of the men she helped free for service, she embraced pacifism and became a Quaker. Her work was so secret that she couldn't confide her guilt to anyone—not her parents, not her friends, not a therapist, not a pastor, not then and not ever.

* * *

Some WACs got assignments that were challenging and important, but in a much less painful way. In May 1945, two WACs working at the Vint Hill intercept station—their last names were Regan and Solek—were assigned to test Arlington Hall security by seeing if they could penetrate the code-breaking compound and steal classified information. The two resourceful women were put up at a nearby hotel and told to apply for a job. They did not know anything about the layout, nor how the system worked. They came to the front gate wearing civilian clothes, said they wished to apply for work, and were admitted and given visitors' badges. They acquainted themselves with the compound, went to the PX and started chatting, learned what badges gave entry into certain buildings, altered their visitors' badges accordingly, stole more badges from some coats lying around, and proceeded to wander around taking classified documents from desks and drawers. They turned in everything to an intelligence officer at the end of the day. The next day they did exactly the same thing. Nobody reported the badges or documents missing. The escapade found its way into the *Washington Post*, which noted the breach in a gossip item headlined SECRET WAR DOCUMENTS PROVE ABOUT AS HARD TO GET AS A COLD.

* * *

Civilians at Arlington Hall soon learned that conditions at the top secret facility differed from the fetching language of the recruiting posters and the honeyed promises of handsome Army officers. The place was chaos. Productive chaos, but still chaos. Many women were excited to move to Washington, the beating heart of the free world, whose population had swelled from some six hundred thousand to almost nine hundred thousand. But once they became bona fide civil servants, the women— and men—proceeded to do what civil servants have done from time immemorial: complain.

Morale might have been high among the elite code breakers whom

Ann Caracristi hung out with, but among the rank and file at Arlington Hall Station, complaints included everything from whirring fans to coworkers snapping gum. Bad insulation, unpleasant bosses, tablemates who smoked and lollygagged—all these were grounds for bitter complaint. It was the first time many of the women had spent time in a bona fide workplace—apart from a classroom—and they discovered what workplaces are and have been since the dawn of time: places where one is annoyed and thwarted and underpaid and interrupted and underappreciated.

A report conducted in 1943 concluded that at any given point 30 to 35 percent of the staff at Arlington Hall were "more or less openly dissatisfied" with their working conditions, job, supervisor, or pay. The report noted that even a small group of unhappy workers could "kindle a flame into an uncontrollable holocaust" and that this sort of infectious discontent was what led to the rise of labor unions. A campaign was mounted to keep code breakers happy, involving things like inspirational posters and movies. In the early fall of 1943, a "morale survey" was conducted to give the workforce an opportunity to vent.

Many complaints doubtless sprang from the fact that the place was growing so fast. Others were the natural result of generally contented people being invited to name the aspects of their workday they objected to. One female worker said she thought most people liked the work, very much, but that "it was only human to complain." And now they had their chance. The survey compiler, Rhea Smith, a professor at Rollins College, wrote an introduction laying out the factors that conspired to make 1943, with all its achievements and breakthroughs, the summer of so much worker discontent.

These included the heat, which, Smith ventured, played on the nerves of people with "high strung temperaments" and "overwrought imaginations." (It is hard not to read these phrases as synonyms for "women.") He faulted the rush to fill quotas, resulting in selectees "improperly advised" and "often grasping at what appeared a glamorous lark." There was also the problem of pay. Unlike Ann Caracristi and other elite workers, many women did care about their salaries and pay grades. Some wanted to pursue careers in stenography. Typists complained about losing speed while they were stuck in training. Others worried that, competing for jobs in a postwar workplace, they would not be able to tell prospective employers what, exactly, they had spent the war doing.

There was gender-related conflict, but it was not between women and men. Rather, it was between male civilians and male military officers. The civilian men were mostly professors who weren't of military age, and men with conditions that disqualified them from serving. These men had a legitimate reason for not being in the shooting war, but they were touchy even so. The close juxtaposition of civilian and military made for a combustible masculine rivalry. Leslie Rutledge, a Harvard graduate who had been classified 4-F, said there was resentment at the "pretentious GI bearing."

The angriest man was one William Seaman, who felt the civilian men were treated much worse than the soldiers. It was his view that soldiers were saluted when they entered the gates, while civilian men were "bawled out because the badge is in the wrong place." Civilian men, he claimed, were the only men who did good work. Officers enjoyed too many privileges; enlisted men were often called away for other duties; and the situation was "not going to be helped by the introduction of the WACS."

Soldiers had their own issues. "At first, the soldiers had a bad feeling," said Captain Javier Cerecedo. "They felt they were being run by women or civilians. This was corrected."

Each section bred its own dissatisfactions. In the message-routing section, it was noted that "some of the college girls look down on the non-college people." In the indexing and sorting section, a worker named Bernice Phillips complained that "some people sat around and did absolutely nothing" and "do not appear to know a war is on." Olive Mickle ventured that chatterers and idlers wasted time visiting at tables.

In a section devoted to compiling internal reports, complaints included the fact that "many have been employed under the illusion that they would be engaged in thrilling work full of adventure" only to find that the job was "dull and routine." The information section was run by a professor named John Coddington, who complained that he needed women of high caliber, "girls who not only graduated from college but did well in college, girls who went to a good college." He wanted girls with a "wide background of reading and culture, some linguistic ability. Some people who know geography."

Coddington also noted that the "girls" in his unit had to fight for typewriters—there was a typewriter shortage throughout the Washington area—and that they complained about one employee who "smokes cigars constantly." The information section was constantly fielding requests. Kay Camp, a graduate of Swarthmore, handled the geographic unit; Alene Erlanger, a Smith graduate, was building a file on Japanese shipping; Anna Chaffin was compiling Japanese place names. Like other units, the information section worked all night. The women resented having to walk at midnight all the way to the Buckingham area, a good half mile away, to catch a bus to Eleventh and E downtown, when women living at Arlington Farms had a night bus that came right up to the gate.

Other rivalries emerged. The units working codes other than Japanese felt undervalued. In the section responsible for Near Eastern, Turkish, Persian, Egyptian, Afghan, and Arabic systems, Lieutenant Cyrus Gordon pointed out that a global war "demands that all parts of the world should be covered" and that the Japanese section "should not gobble up all the trained personnel." In the German section, one lieutenant complained that some people wasted so much time that they were "little better than saboteurs"; that there was "appalling inefficiency"; that one girl was exceedingly inaccurate but "wept when corrected" and so was not corrected.

In the Portuguese and Brazilian diplomatic section, one woman was unhappy that work sometimes poured in, and that often—after they had broken the messages—they realized that the spike in South American traffic had nothing to do with the war but was the result of "ambassadorial weekends," which is to say, big diplomatic parties.

In the Italian section, Harold Dale Gunn reported that people became irritable in hot weather. "The recent directive on venetian blinds upset a number of people," he said, bringing up what many cited as the chief workplace outrage during the summer of 1943: a directive that all window blinds had to be lowered to the same position. It was a classic Washington move, overbroad and ill thought out and made more classic by the fact that it was soon withdrawn. "The venetian blind directive almost caused the resignation of three people," another worker noted. The biggest complainer was the perpetually aggrieved William Seaman, who declared that "one would gather that officers have nothing to do but think up annoying orders."

There was some rivalry between women—young and old, married and single. Mrs. Ruth M. Miller ventured that many young women came to Arlington Hall with a "rosy picture of conditions and an anticipation of romantic adventure." Mrs. Miller ventured smugly that married women like her worked harder than single women did, in that they had "someone to work for and something to defend."

The perennially discontented William Seaman seems to have been pushed around by a clique of what would now be called mean girls. "When I came into our section it was controlled by a small group of young girls, who made it difficult for new men by assigning us disagreeable jobs and by preventing us from learning how to do the more technical work. They are not in control now, but it is generally known that they pass judgment on new people and are responsible for some being transferred."

Food! Orders! Window blinds! Men! Women! Coworkers! Young people! It was a cascade of standard workplace discontent. There were some inconveniences unique to women, who often had domestic chores on top of their code-breaking shift work and who struggled to get by in an expensive city. Ruth Scharf was divorced from her husband, who was in the Army and refused to pay child support. Miss Lucille Hall "has tried to save money, but she gave up in despair." Jane Pulliam's mother had to send her money.

But many others were fulfilled, happy, and thriving. Doris Johnson, from North Carolina, said the work was interesting. A former teacher, Lillian Parmley, said that "her job here has not been as nerve wracking as handling forty school children." Lena Brown "got discouraged when she has been unable to dig anything out of a message," but when she did get something, she "has become elated, and has worked overtime." Lillian Wall, supervisor of the stenographic unit, said that the women in her unit were "pleased with the prospect of learning a profession."

Lillian Davis, in charge of logging in the traffic analysis section, had her unit running like a high-end race car. She permitted no gossiping, no gum chewing, no backbiting or nagging. People who were noncooperative were transferred.

Another young woman flexed her managerial talents. Jane B. Park was a recent graduate of the University of Maryland and now ran the cryptographic training unit. There had been a snafu when Dr. Harold Briggs, a history professor, had been assigned under the misunderstanding that he was to supervise training, but everybody—including Briggs—agreed that Jane Park was better qualified. The report noted that Jane Park was "a bright, vivacious young lady who stood high in her class, although her major was Home Economics." At twenty-three, it was her job to train workers in cryptographic security—encoding American messages—as well as procedure and systems analysis. Today, what she was in charge of would be called cybersecurity.

Perhaps, though, chaos and a certain degree of boredom had their uses. During down periods, the well-educated literary mind turned itself toward ways of commemorating its own secret efforts. In April 1944, two code breakers identified only as M. Miller and A. August—likely Marjorie A. Miller and Ann R. August, who worked on a team together—wrote a poem in celebration of the first anniversary of the breaking of 2468. The poem would not be declassified for nearly seventy years. It was written "with apologies" to Edgar Allan Poe's "The Raven," with a nod to a popular song titled "Pistol Packin' Mama." The first three stanzas described the preparations for the 2468 breakthrough. The next ones described the break itself, Frank Lewis's shaving off his Van Dyke beard in celebration, and the recruiting aftermath.

In June the state of Carolina, and missionaries back from China
Put on their shoes, packed up their Bibles, swarmed like locusts to the
    Hall,
And school kids all across the nation got a permanent vacation,
Cause teachers headed for the station, heeding the recruiters' call,
M.A.'s, B.A.'s, PH.D's, candidates for Sp-4
(They still are this and nothing more.)

Then along came period 7, everybody was in heaven,
Pencil-pushing-mammas sank the shipping of Japan,
But then the nasty Nip did dare, in Period 8 to change the square,
Making it a vigenere production; charts went down again,
But Seidenglanz knew the solution that would our confidence restore,
"Boys, move the furniture once more."

Now April 6th we celebrate your birthday, dear 2468,
Even though you're growing tougher, tougher with each passing year,
Though our overlaps are stalling, and production charts are falling,
Cryptanalysis is still our calling, it's got to be we're frozen here!
And we'll reply when our children query what did you do in the war,
I bought red tape for the Signal Corps.
(Praise the Lord there is no more.)

CHAPTER TEN

Pencil-Pushing Mamas Sink the Shipping of Japan

1943–1944

Ambon. Canton. Davao. Haiphong. Hankow. Kiska. Kobe. Kuching. Kupang. Osaka. Palembang. Rabaul. Saigon. Takao. Wewak. Dot Braden until a few months earlier had never heard of most of these places. Now they ruled her life. They kept her running from the big table where she worked, over to the overlapper's console, then back again to her spot at the big table. These were the names of places, somewhere in Asia or the South Pacific, likely to be mentioned toward the beginning of messages coded in 2468, the main Japanese water-transport code, or one of the other, smaller transport codes.

Or rather, they were some of the places. Transport code 2468 was massive; 2468 was everywhere; 2468 dominated the Pacific Ocean. Anything anybody needed was sent by water. Water was how the rice was transported, and the soldiers, and the spare airplane parts. To move the goods the Japanese Army needed, the marus were always sailing. Always leaving and arriving. A maru could be a tanker, a freighter, a cargo ship, a barge, a cable layer, a motor transport. They plied between Hiroshima, Yokohama, Wewak, Saipan, Tokyo, Manila, the Truk Lagoon. Exotic places. It was not necessary for Dot to know how to pronounce the cities and ports, but it was helpful to know the four-digit code groups that stood for them. Code system 2468 commanded Dot's attention, controlled Dot's movements. It filled her brain.

A job more unlike teaching Virginia schoolchildren would be hard to imagine. No longer was Dot Braden standing at a chalkboard, explaining physics formulas to eye-rolling teenagers, or ordering senior girls to march and salute. Instead, she was sitting head down at a table puzzling over words she had never heard before she came to Arlington Hall. "Sono." "Indicator." "Discriminant." "GAT." The sono was the number appended to messages that had been divided into parts before being transmitted. Sono #1 was the first part, Sono #2 was the second part, and so on. The discriminant was the number that identified the system—for instance, 2468. The indicator was the tiny clue that told you what book to look in. GAT stood for "group as transmitted": the code group plus the cipher. The GATs were what you saw when you looked at the message for the first time.

Dot, of course, was not to utter any of these words outside the high wire double fences of the Arlington Hall compound. People were warned never to use, outside the building, the words they used inside it. "This material is extremely secret and must be treated with the utmost care," one training document said. "Some of the words which you will consider elementary have been used only in this code, eg KAIBOTSU SU 'to sink a ship'. If you should mention this word to any one connected with the Axis or in some way succeed in letting it get into improper hands, this one fact alone would betray to the Japanese that we are reading their most recent transport code."

Dot did not know anyone "connected with the Axis." Even so, she never mentioned anything to anybody. On the bus, she kept mum. She and Crow still never discussed their work, even though they lived together, ate together, and shared a bed. When she wrote letters to her brothers, or to Jim Bruce or George Rush, she did not tell them what she did. She talked about eating red beans and rice and frozen peaches and going to the beach on the bus and the streetcar. Just the silly things she

244 • CODE GIRLS

and Crow did in their time off. She liked the work at Arlington Hall and, unlike some of her colleagues, had few complaints, apart from the fact that northerners thought southerners were backward and stupid.

At Arlington Hall, Dot worked at a wooden table with other women, all of them sitting together in a big room in Building B. She was given cards with series of four-digit GATs, and it was her job to run the GATs against a bank of code groups she had memorized. The messages she got were the urgent ones. Routine 2468 messages were sent to the punch-card room, to be processed by machines. It was a tenet at Arlington Hall that every message had to be processed, no matter how insignificant or routine. No message was devoid of intelligence use.

The ones that must be decoded quickly—the ones that might require action—had to be deciphered by hand, and these were the messages Dot got. She would scan each one and compare the groups on the page to the code groups she kept stored in her head. She would look for a group whose position suggested that it likely meant "maru," or—this was always exciting—"embarking" or "debarking." Dot sat near a pole, and when she saw a code group that seemed important, she would jump up and almost hit her head on the pole as she ran to take the message to the overlappers' unit, a group of women who worked in another room. They would take her message and place it on a big piece of paper with other messages encoded using the same additives.

Often, a young woman named Miriam was the overlapper waiting to receive Dot's breathlessly delivered handoff. Miriam came from New York City, and she was one of the most condescending northerners Dot had ever known, and that was saying something. One day, over lunch in the cafeteria, Miriam said, "I have never yet met a southerner who can speak proper English." This offended Dot, as it was intended to do. "Another smart aleck New Yorker," she thought, but she did not say it. She comforted herself by assessing the so-called yellow diamond on Miriam's ring finger—Miriam had a fiancé, or claimed to have one—and reflecting that both the diamond and the fiancé were likely fakes.

Despite their low opinions of each other, Dot and Miriam had to work seamlessly together, and they did. Dot would get the messages started by identifying some of the code groups, and Miriam would place the message Dot brought her. From the overlapping station, the work sheet went to a reader, who would decipher the meaning. Information from the finished translations would make its way to the staff of General MacArthur or to a submarine captain who would do what needed to be done.

The language of the 2468 messages was telegraphic in style. Short, straightforward, and no-nonsense, the messages consisted of sailing schedules, harbormaster reports, reports on the water levels of ports and transportation of cargo. Sailing schedules were the simplest. These included the transport number, the date, the time the maru would be arriving or leaving, and its destination. Others concerned the movement of troops or equipment. A few dealt with transportation of the wounded or ashes of the dead. The marus out there in the Pacific Ocean carried everything: food, oil, supplies, human remains.

When a new message arrived, Dot looked for stereotypes, which were words that occurred frequently in the same place. "Maru" was a common one, but there were others as well, depending on the origin and the goods being transported. For example, one station transmitting from Singapore—the #3 Sen San Yusoo—sent a regular report on the shipping of oil to Hiroshima, Manila, and Tokyo. Stereotyped words might include ship names and numbers; the number of kiloliters of light oil, crude oil, heavy oil, aviation gasoline, or other gasoline aboard; how many trips each ship would take, and when. Another Singapore station transmitted to Hiroshima, Tokyo, and Moji a report of ships leaving for Palembang. Stereotypes might include the ship number or name, the date and hour of departure, the speed, the course, and the date and hour of scheduled arrival at the mouth of the Musi River.

Another transmitted a daily weather report with data including wind velocity and direction, temperature, and condition of the surface of the Andaman Sea, the South China Sea, the Yellow Sea, and other faraway bodies of water. Dot handled a lot of weather reports. Sitting at her table in Arlington, Virginia, Dot was amused at how many bits and pieces of information she knew about what the weather was like eight thousand miles away.

Another station transmitted a report on small boats available for supply services. Mentioned might be steel barges, wooden barges, special boats, small boats, twenty-metric-ton boats, plywood barges, and cargo submarines.

A station at Surabaya originated a report on the departure of ships escorted by a single Navy plane, including the names and types of ships (motor, sail, fishing), number of barges being towed, date of departure and destination, speed, scheduled date of arrival, route, and daily position of ship on consecutive days at given hours. A report from Shanghai about a supply ship might include a message spelling out that the probable route was "from Shanghai along the coast to the Yangtze River up the Yangtze River to WUU (Buko) down the river to Nanking and finally across the East China Sea to Moji."

Here is how Dot did her work: Let's say she knew that the code group for "arriving" was 6286 and she knew where this word was likely to appear. She would find that place in the message and look at the GAT before her. Books at Arlington Hall listed common code words as well as possible enciphered versions. She would look for a match, or she could do the math in her head and strip out the additive herself. Sometimes—when they were desperate—the code breakers would take the code groups and encipher them with every possible additive. A smattering of 2468 code groups included:

4333 *hassoo*—to send things
4362 *jinin*—personnel
4400 *kaisi*—beginning, commencing
4277 *kookoo*—navigate, to sail
4237 *toochaku yotei*—scheduled to arrive
4273 *hatsu yotei*—scheduled to leave

There were vocabulary words associated with sailing schedules. According to training materials compiled at Arlington Hall, *atesaki* was "destination" or "address"; *chaku* was "arriving"; *dai ichi* was "first"; *honjitsu* was "today." *Maru* was "commercial ship"; *sempakutu* was "ship"; *sempakutai* was "convoy unit"; *teihaku* was "anchoring"; *yori* was "from"; *yotei* was "schedule." *Gunkan* was "warship." *Chu* was "now." *Hatsusen* was "ship leaving." *Hi* was "day"; *hongetsu* was "this month"; *senghu* was "onboard ship"; *shuzensen* was "ship being repaired"; *tosai sen* was "ship loading."

Dot's workday consisted of messages that, once deciphered, said things like "PALAU DENDAI/ 2/ 43/ T.B./ TRANSPORT/ 918/ (/878/)/ 20th/ 18/ JI/ CHAKU/ ATESAKI/ DAVAO/ SEMPAKUTAI/ 4/ CEBU/ E.T./"

If it sounds hard and exhausting, it was. The administrators at Arlington Hall concluded that Section B-II—the Japanese Army section—had the most complex mission in the place. This was owing to the intricacy of the Japanese Army's cryptologic system and periodic major changes in how the systems worked.

At first, the 2468 system was enciphered by false math, but in February 1944, a few months after both Dot and Crow arrived, the Japanese began using squares. As the Japanese coped with their changing island situation, the ex-schoolteachers had to cope with Japanese cryptanalytic changes. During 1944, there were thirty thousand water-transport-code messages received each month. This meant breaking a thousand messages a day. In August 1944, the Japanese began using a new additive, new codebooks, a new square, and new indicator patterns. Staying abreast of these was Crow's department. Crow with her math skills had been assigned to the "research unit," which did the ongoing analysis that enabled Dot to do the active processing work. Dot didn't know that, nor did Crow.

The schoolteachers working 2468 got a bit of specialized training. The course was developed by an all-female committee—Evelyn Akeley from Skidmore, Alice Beardwood from Bryn Mawr, Elizabeth Hudson,

Juanita Schroeder, Mildred Lawrence, and Olivia Fulghum. The instructors were likewise female. Among them were Lois Harer, Lenore Franklin, Margaret Ludwig, Margaret Calhoun, and Alice Goodson.

The team of women came up with a ten-day course explaining controls and indicators, decipherment using squares and charts, traffic procedures, preambles, and message analysis. Students studied the mathematical recovery of additives, the patterns of Japanese texts, the reconstruction of squares, and the recovery of indicator keys. Women learned to compare messages such as

4 Oct 1944: 8537 1129 0316 0680 1548 2933 4860 9258 4075 4062 0465

6 Feb 1945: 5960 1129 1718 6546 1548 3171 0889 9258 4075 4062 0465

6 Mar 1945: 7332 1129 1718 3115 1548 8897 7404 9258 4075 4062 0519

and to see that these messages were relayed about the same day every month and that certain code groups recurred in the same place and probably represented a stereotype word.

The workers breaking 2468 received training in basic Japanese vocabulary, focusing on words found in shipping reports. The course concentrated on kana, which are syllables that amount to a phonetic rendering. They learned that the "typical Japanese syllable consists of one consonant followed by one vowel, eg HI-RO-HI-TO. YO-KO-HA-MA. The verb is at the end. Nouns have no singular or plural." The women were given vocational aptitude tests and rated as "clerical," "technical," or "analytic." Analytic work was the most difficult, and it was the category both Dot and Crow had been chosen for.

When Dot was hired, Department K—the unit that decoded 2468 messages—was doubling in size, from 100 workers in July 1943 to 217 in 1944, and it was steadily growing more expert and efficient. "The history of the department during the past year is one of increasing production with a continuous decrease in the time necessary for solution," noted a 1944 memo. "Where ten days was considered to be good time for the solution of an overlap at the beginning of the year, some are now recovered in three days." Individual messages could be solved much more quickly.

The 2468 code breakers operated on a twenty-four-hour basis. There were three shifts—day, swing, and graveyard—and women rotated between them. One report noted that the Japanese Army unit "probably handles the most enemy traffic for deciphering of any agency in the world." The unit had been fashioned along the principles of an American assembly line: routine, simplicity, flexibility. During training, instructors tried to root out people who were prone to hysteria or nervous breakdowns or who did not seem easily adjustable. "This is a business organization," said one memo, "and not a country club." Innovations were always being sought. A pneumatic tube was set up between Building A, where a machine unit was located, and Building B, where the code breaking happened, which sped up the delivery of traffic and eliminated a lot of courier work.

The assembly line also was designed by women. Alice Goodson set up a bank listing address preambles in alphabetical order, as well as stereotypes and other devices to help readers like Dot get a start. Helen O'Rourke designed the overlap unit. There was a ten-person unit that coordinated between the women working on 2468 and those assigned to other code systems. A color-coded system—Lavender, Orchid, Lilac—was devised to keep track of the time periods when the 2468 code or cipher books changed.

The women in Department K—Dot's unit—were a "very fine group," according to one Lieutenant Bradley, an Army officer who rotated into the unit late in 1943. He was there when a break into a new square led to "great glee over the entire place." He watched as dozens of new recruits joined and became adept. At first, readers needed overlaps that were ten or fifteen messages deep to recover code groups, but they soon needed fewer and fewer. "The pattern was the main thing," said Lieutenant Bradley. "There was no information service or cribbing section at that time. Each reader had to depend on what he could remember."

The women knew they were performing well. "The great value of the intelligence derived from the [2468] messages is a constant incentive to the department," a report noted, praising Dot's Department K, which was producing all kinds of shipping intelligence, foretelling what units were about to receive oil or gasoline; what ships were in a given harbor; what convoys were getting ready to sail and where they were headed.

That report also cited one consequence of all this foreknowledge. On May 3, 1944, Department K read a series of messages indicating the noon positions through May 8 of fifteen ships headed for New Guinea. Shortly thereafter, the U.S. Navy sank four of them. Another memo pointed out that the *New York Times* had reported on one of many successes. In September 1943, a *Times* piece had noted a Pacific engagement in which "our strongly escorted medium bombers attacked an enemy convoy of five cargo ships and two destroyers which arrived during the night with reinforcements and supplies for the enemy garrison. Coming in at masthead height, our bombers scored direct hits with 1,000 pound bombs on three freight transports, each of 7,000 tons, sinking them." The internal Arlington Hall memo noted that a *Times* reader probably thought it was "chance" that the bombing mission found the convoy. Not so. In fact, "a message had been intercepted and read" two weeks earlier.

November 1943, one month after Dot's arrival at Arlington Hall, marked the war's most devastating month for Japanese tonnage sunk. U.S. subs sank forty-three ships and damaged twenty-two. American sub captains received intelligence of seventy-six movements of enemy ships. In December, American subs sank or damaged about 350,000 tons, including thirty-two ships sunk and sixteen damaged.

Behind the success of the U.S. Navy were the code breakers. "The success of undersea warfare is to a certain extent due to the success with which Japanese code messages were translated," noted a naval report. An American naval commander pointed out in a postwar memo that sometimes a convoy might slip through, but only because U.S. submarines were kept so busy by information from decoded messages that they could not handle all the convoys they were alerted to. Over at the Naval Annex, the assembly line of WAVES identified the movements of marus supplying the Japanese Navy. Findings from both operations found their way to the submarine captains, who could hardly keep up with the bounty of intelligence.

After the war, a census would be taken of Japanese marus and their fates. The files take up boxes and boxes. Here is a single page from that immense archive:

On July 2, 1943, the Isuzu Maru was sunk by a submarine.

On December 2, 1944, the Hawaii Maru was sunk by a submarine.

On October 16, 1944, or thereabouts, the #23 Henshuu Maru was sunk by aircraft.

On August 31, 1944, the #20 Hinode Maru was sunk by submarine.

On October 16, 1944, the #16 Hoorai Maru was sunk by aircraft.

On January 27, 1945, the Hisn Yang Maru was sunk by a mine.

On January 2, 1944, the Isshin Maru was sunk by submarine.

On January 20, 1944, the Jintsuu Maru was sunk by aircraft.

On September 12, 1944, the Kachidoki Maru was sunk by submarine.

And so on.

The devastation of Japan's shipping had an enormous impact. Soldiers were deprived of food and medicine. Aircraft did not get spare parts and could not launch missions. Troops did not reach the places they were sent as reinforcements. On March 12, 1944, a broken 2468 message gave the route and schedule of the Twenty-First Wewak Transport convoy, sunk while leaving Wewak to return to Palau. When the Japanese Eighteenth Area Army made a "complete tabulation of shipping from Rabaul and Truk during January," in an attempt to convince Japanese Army headquarters that it was feasible to send them much-needed supplies, these messages laid out the shipping routes and sealed their doom. Only 50 percent of ships reached the destination; only 30 percent got home.

At the end of the war, a U.S. naval report found that "more than two-thirds of the entire Japanese merchant marine and numerous warships, including some of every category, were sunk. These sinkings resulted, by mid-1944, in isolation of Japan from her overseas sources of raw materials and petroleum, with far reaching effects on the capability of her war industry to produce and her armed forces to operate. Her outlying bases were weakened by lack of reinforcements and supplies and fell victim to our air, surface and amphibious assaults; heavy bombers moved into the captured bases." This report's author, C. A. Lockwood, commander of the submarine force of the U.S. Pacific Fleet, noted that his men got a "continuous flow of information on Japanese naval and merchant shipping, convoy routing and composition, damage sustained from submarine attacks, anti-submarine measures employed or to be employed, effectiveness of our torpedoes, and a wealth of other pertinent intelligence." Whenever code breaking was unavailable, he added, "its absence was keenly felt. The curve of enemy contacts and of consequent sinkings almost exactly paralleled the curve of volume of Communication Intelligence available."

He added: "There were many periods when every single U.S. sub in the Pacific was busy" responding.

In fact, he added, code-breaking intelligence made it seem to the Japanese that there were more American submarines in the Pacific than there really were. "In early 1945 it was learned from a Japanese prisoner of war that it was [a] common saying in Singapore that you could walk from that port to Japan on American periscopes. This feeling among the Japanese was undoubtedly created, not by the great number of submarines on patrol, but rather by the fact, thanks to communications intelligence, that submarines were always at the same place as Japanese ships."

The commander noted that dispatches that led to attacks usually had to be destroyed, so he took it upon himself to list just a few of the more notable achievements. They included the sinking of aircraft transport *Mogamigawa* by the submarine *Pogy* in August 1943; the sinking of escort carrier *Chuyo* by *Sailfish* in December 1943; and so on and so on, including "the contact and trailing of Yamato task force by *Threadfin* and

Hackleback in April 1945, which resulted in sinkings the following day by carrier air forces of the battleship *Yamato*, the cruiser *Yahagi*, and destroyers *Hamakaze, Isokaze, Asashimo* and *Kasumi*." He pointed out that code breaking was responsible for at least 50 percent of all marus sunk by subs. And "information concerning enemy minefields" enabled U.S. subs to avoid them, and forced Japanese ships "into relatively narrow sea lanes."

Arlington Hall worked closely with "Central Bureau Brisbane," its satellite unit in Australia, and with Australian and New Zealander code-breaking allies. CBB stayed in touch with the staff of General MacArthur, and a memo in 1944 noted how the breaking of all the Japanese Army codes—shipping, administrative, air force—contributed to the success of Operation Cartwheel, MacArthur's island-hopping campaign. Thanks to the penetration of most Japanese Army systems, it said, MacArthur knew about supplies, troop training, promotions, convoy sailings, reserves, reinforcements, and impending attacks. By May 1944, messages translated by Arlington Hall had alerted the U.S. Army to changes in the makeup of the Japanese Army, helping identify new armies, divisions, and brigades. The staff knew how many planes the Japanese Army air forces had and the condition of the railroads; they knew about shipping losses and were able to keep a running tally. Citing a few of the biggest achievements, the memo noted that "never has a commander gone into battle knowing so much about the enemy as did the Allied commander at Aitape" on July 10 and 11, 1944.

This is the kind of code-breaking intelligence Pacific commanders received: medical reports, incidence of disease, number wounded, percentage of total strength, casualties, losses incurred through submarine and plane attack, convoys delayed, marus sunk, brigades shipwrecked, and tools, arms, machinery, and codebooks lost. As the Americans planned to retake the Philippines, code breakers fed them information on reinforcements, plans to hamper U.S. air activity, units engaged in battle, army supplies, and reinforcement problems.

The code breakers also responded to requests from American military intelligence. "When they were planning some major moves against the Japanese—either against some of the islands or the last big move that they were planning was, of course, the invasion of the Japanese mainland—They would come and ask us, if possible, to concentrate on messages from one or two certain places," remembered Solomon Kullback. As a result, MacArthur "wasn't going in blind into a lot of these areas he invaded."

Code breaking also proved instrumental in reducing American casualties. George C. Kenney, MacArthur's air corps commander, was able to prevail in the air and shorten the ground war. Code breaking enabled the destruction of Japanese aircraft in Wewak in August 1943, and in Hollandia in March and April 1944, making possible MacArthur's "greatest leapfrog operation," along the northern New Guinea coast. In November 1944, Arlington Hall decoded messages saying that two convoys contained troops to reinforce the Philippines. The U.S. Navy sank at least six of the ships and disabled one, and one caught fire.

Joe Richard, the young officer working in the Australian unit who spotted the telltale digit pattern that led to the break in 2468, told later of how the recovery of a codebook on Okinawa, in June 1945, gave them all the translations for that period and "led to reading about the Japanese army's preparations to fight against any landing on their home islands. These were so extensive, involving every Japanese, that the Allied general staff estimated (based on experience at Iwo Jima and Okinawa) that 1 million casualties might be expected by our allied forces, which I think induced Truman to use the atom bomb and to moderate FDR's unconditional surrender ultimatum and accept Japanese surrender keeping the Emperor."

In the summer of 1944, the U.S. military retook Guam. The Americans got an intercept station up and running again, and Dot Braden, sitting at her wooden worktable, began to get a lot of intercepts from Guam. Never having been anywhere near the Pacific Ocean, she always visualized rather fancifully that Guam was a tiny little island with a single palm tree on which a lone American GI sat, sending intercepted messages over teletype, which ended up in her hands.

* * *

Even from her distant vantage point, Dot could tell that things were going much better for the Allies in the Pacific, thanks in part to the efforts of women like her and Crow and even Miriam. "Now we're getting somewhere," she would think as she ran between the table where she worked and the console where Miriam put together the overlaps. She and the other women knew that ship sinkings were the logical and desired consequence of their concerted efforts. They did not feel remorse. America was at war with Japan; Japan had started the war; the lives of American men were at stake, not to mention America itself. It really was that simple. Sometimes Frank Lewis would walk through the room and the women would take notice; they knew he was one of the people who had broken the 2468 code. She also recognized William Friedman, one of the men not in military uniform. Dot would hear high-ranking military officers talking, saying that things were going well. She had never thought the United States would be defeated. Now she could feel, almost viscerally, that the progress of the war was on the upswing.

So she did the best she could, giving her all on every shift she worked, using the stereotypes to get started, doing mental math, differencing up and down columns, going to the filing cabinets to retrieve cross-dupes, running over to Miriam and tolerating her contempt for southerners. Dot's unit was completely run by women, and she took pride in the work they did. She liked it much better than teaching school.

"It was like a puzzle," she remembered later. "We were getting somewhere. I was proud I was doing it."

During the course of 1943 and 1944, while Dot Braden scurried to and fro between her table and the overlapper's console, nearly the entire Japanese merchant fleet was wiped out. Beginning in 1943, starvation became the common lot of the Japanese soldier. Officials later estimated that two-thirds of Japanese military deaths were the result of starvation or lack of medical supplies.

Broken messages revealed the extent of the devastation. One message described how a group of Japanese soldiers were making a ten-day supply of rice last for twenty-five days. "By resorting to chewing it raw instead of cooking it," the message said, "the period of consumption had been prolonged somewhat."

At Arlington Hall, even the instructional materials revealed the profound impact that breaking 2468 and other codes was having on the Pacific War. One document provided a list of cribs and stereotypes for a code system called JEB. It noted that one Japanese transmitting station reported frequently on the arrival of personnel, as well as their nonarrival.

"If the latter," the document noted, "the question 'What has become of them?' may appear next."

PART III

The Tide Turns

CHAPTER ELEVEN

Sugar Camp

1943–1944

They boarded the train at midnight, leaving Washington under sealed orders. The women knew only that they were headed "west." The train departing the capital was an ordinary troop train—dirty, crowded—with no sleeping berths. The women slept sitting up, if they slept at all, and three who were unable to find seats took turns resting in a cubbyhole used by the brakeman. Some cherished the hope that they were being sent to California. But when the train arrived at their destination, it emerged that "west" didn't mean quite what they thought. They were in another Union Station, this one in Dayton, Ohio.

The women gathered their things and made their way into the chill morning air, where they mustered for roll call. Although this was a secret mission, a photographer stood waiting to greet them, so the women lined up and smiled for group photos, smart and polished despite the grime and fatigue of the all-night train ride. They wore their naval uniforms, of course: six-gored skirts and chic fitted jackets, white-and-blue caps, white gloves, blue belted greatcoats, stockings and pumps, smart pocketbooks strapped catty-corner over their shoulders. Each woman held a little hard-backed suitcase that contained everything else she needed.

A bus waited in the parking lot. They climbed on and it carried them away from downtown Dayton and into the nearby countryside, where after a short drive the bus swung into a driveway marked by stone gate-posts. The women entered a grassy compound: elevated, bucolic, peaceful. If they didn't know better they'd think the U.S. Navy was taking them to a Girl Scout camp. There were maple trees and rustic cabins clustered around a central clearing.

The women disembarked and mustered, once again, and a Navy color guard greeted them. The American flag was raised. They felt tired, but by now they were used to feeling tired. They lined up to receive linens and pillows and dispersed to find their cabins. As fresh Navy recruits, their lives for the past two months had been a series of unfamiliar lodgings, first at boot camp and then in Washington, D.C., where they spent several weeks sitting in the hard upright benches of the chapel in the Naval Communications Annex compound, taking tests, listening to security lectures, and waiting while their backgrounds were investigated. Nobody told them the nature of the work they had been brought to Dayton to perform.

The cabins were small but pleasant. The women pushed open the wooden shutters to let in sunlight. Each cabin was divided into two bedrooms, with two beds per bedroom. Each bedroom had two closets, and there was a small writing desk built into the wall beside each bed, with a gooseneck lamp over each desk. There were no screens for the latticed windows, but there didn't seem to be any bugs in Ohio. Nearly everything was made of wood, including the desks, the beds, the floors, the walls, and the ceilings. Between the bedrooms was a bathroom with a toilet, a shower, and two sinks. The cabins were unheated.

Four women were assigned to each cabin, except that soon—as more recruits began arriving, throughout April 1943 and into May—cots sometimes had to be squeezed in, to make room for an extra person. At first the camp offered no food service, so—as the women began training for their mysterious new occupation—they took some of their meals in Dayton. The city boasted several fine-dining establishments, including the Biltmore Hotel's Kitty Hawk Room, named after the place in North Carolina where the Wright brothers—Dayton natives—made the first airplane flight. On weekends, the women could ride the elevator to lunch on the fifth floor of Rike's, a department store, which was a treat because of the elegant presentation. In those early weeks the women sometimes had dinner in the basement of the ward room at camp, where they found that tomato soup and grilled cheese sandwiches made a perfectly satisfactory supper.

The women were billeted at a place called Sugar Camp, a thirty-one-acre property named for the magnificent maple trees on the grounds. For much of its history the state of Ohio was agrarian, and the trees at one time had been tapped for maple syrup. Dayton had been transformed during the War of 1812, when it served as a mobilization point for American attacks on Canada and British troops in the northwest U.S., bringing banks, businesses, and factories. This continued during the Civil War, when it served as a supplier for the Union Army. The city nurtured more than its fair share of inventors and entrepreneurs, including not only Orville and Wilbur Wright, the aviation pioneers, who had their bicycle shop here, but also Charles Kettering, who invented the automobile ignition system. Kettering's self-starter was the reason Dayton had a General Motors plant. It also had Dayton Electric; the headquarters of Frigidaire; and Wright Field, used for aviation and testing.

But the mainstay of the economy was the National Cash Register Company, which owned Sugar Camp. NCR made the machines that kept American commerce running—accounting machines, adding machines, and of course cash registers, which were big and bright and gleaming, as splendid and ornate as fairground calliopes. NCR sold its machines all over the world; it had set up its first overseas sales office before the twentieth century even started. Over the years, NCR's founder, John Patterson, had acquired huge swaths of land, including the Sugar Camp property. Patterson was a pioneer of modern sales culture, and before the war, the camp had served as a summer retreat for NCR salesmen, who spent intensive weeks taking classes, listening to motivational speeches, competing for cash prizes, and learning about things like annual quotas and regional sales territories and the stages of selling.

But gleaming cash registers required materials that were needed by the military, and NCR's ninety-acre industrial campus had been converted to producing the machinery of war. Around the country, major companies like Ford, IBM, Kodak, Bethlehem Steel, Martin Aircraft, and General Motors were cooperating with the war effort, producing weapons and war materiel and helping develop systems. So were universities like Harvard and MIT. NCR was fully committed: One hundred percent of its operations were war-related work.

Seeing as how there was nothing, just now, for the salesmen to sell, Sugar Camp had been turned over to the Navy women, six hundred in all. Though their project was secret, their presence in the town was not. NCR was not above seeking publicity for its contribution to the war effort, and comely uniformed women were a good way to do this. NCR had sent the photographer to the train station and continued to document the women's daily lives and NCR's warm hospitality toward them. The photos, which found their way into the NCR company newsletter, captured images of the WAVES marching, lounging, swimming, singing, eating—everything, that is, but doing their work.

The women labored seven days a week, twenty-four hours a day, in shifts. Three times a day, more than a hundred of them would muster at Sugar Camp and march four abreast into Dayton, up hills and down, in snow and rain and sunshine, passing a house where a girl they called Little Julie would come to her window and wave at them. Before long, people in Dayton were saying you could set your clock by the sight of the WAVES marching. Their destination was the NCR main campus, located about a mile from Sugar Camp. A cover story was devised to explain their presence. "The WAVES will take courses in the operation of special accounting machines," announced the NCR newsletter. It struck some of the women that the people of Dayton must think they were remarkably stupid, to take a whole summer to learn to use an accounting machine.

The NCR complex dominated downtown Dayton, occupying an area the equivalent of eleven city blocks. The facilities were so extensive that in addition to its yellow brick buildings and the offices and factories they contained, NCR had its own water wells, its own electrical power plant, its own movie theater. The women worked in Building 26, a modest structure tucked away from the rest that formerly housed a night school offering classes to NCR employees. Behind Building 26 stretched a spur of the Baltimore and Ohio Railroad. Armed Marine guards now lived in Building 26 and patrolled it, to make sure no unauthorized persons gained access. The women were locked inside the rooms where they worked. They were permitted to put on cotton smocks and to take off their white gloves and work barehanded.

They sat at big tables that had been installed in the former classrooms, the same seating arrangement every day. The rooms each held about a dozen people. Hanging from a cord that plugged into the ceiling, or nestled in a little dish on the worktable before each of them, was a soldering iron. On the table before each woman was a wheel made of Bakelite, brass, and copper. Shortly after they arrived, their supervising officer—female—taught the women how to use a soldering iron to fashion a little interlacing basket of wires that attached to each wheel. The wires were short and of varying colors. The women fashioned the wires according to diagrams, wrapping each wire around a prong and putting a dab of solder where the tip of the wire connected to the contact point of the wheel. Each woman would take the hot soldering iron and melt the solder, and when it cooled, she would tug the wire to make sure the seal held. It was exacting work, and their supervisors warned them there was no room for mistakes.

It took a while to become adept—some never did and were given easier jobs involving circuitry—but most of the women were good with their hands and mastered work that was no more difficult than making lace.

Many of the women already were familiar with machinery. Quite a few had worked as telephone operators before the war. Ronnie Mackey was one. She had grown up in Wilmington, Delaware, where after high school she worked in a fabric showroom, then moved to switchboard work because it paid better. Another, Millie Weatherly, a North Carolinian, had been working alone on the Sunday of the Pearl Harbor attack.

Her switchboard lit up as soldiers from a nearby base called home, some of them crying, to tell their parents they would not be back for Christmas. Millie plugged their calls as fast as she could. About a year later her mother remarked, "You know, the Navy is welcoming women of good character and high school education." And so here Millie was, in Ohio.

Jimmie Lee Hutchison was another. She was a tiny person, just nineteen when she was working at Southwestern Bell in McAlester, Oklahoma. Like every American town and city, her community had been transformed by war, with the arrival of a naval ammunition depot and an internment camp for Japanese American citizens. Jimmie Lee had four brothers in the service, and her fiancé, Robert Powers, was a pilot with the Army Air Force. The Navy sent a recruiter to the Southwestern Bell office where Jimmie Lee and her friend Beatrice Hughart were working. After perusing the materials, the two friends visited the recruiting station. They did not intend to join up then and there. But they liked the idea of helping bring the boys home. By the end of the day both had enlisted, Jimmie Lee by lying about her age.

At Hunter College, where Jimmie Lee took her naval aptitude test, she was surprised to learn she had a knack for reading blueprints. She had never seen one before, but something about the diagrams made intuitive sense. Switchboard work required operators to follow complex wiring patterns. Her good friend Bea also got orders to Dayton, so here they were, working together, once again.

Even those women who had not worked switchboards often came from households where they had developed an easy competence with minor industrial chores like replacing a worn-out cord on an iron. Some were independent to the point of being hard-bitten. There were two farm girls at Sugar Camp who liked to talk about how they went after their brothers with hoes during family arguments, and the way they talked about how they "meant business" with the hoes was unnerving to the other women.

Like their counterparts in Arlington and Washington, the women at

Sugar Camp had a variety of reasons for seeking government service. Iris Flaspoller, from New Orleans, was escaping a hasty marriage. She and her husband, August, had married in January 1942, just after Pearl Harbor, when so many couples were tying the knot. They had not yet graduated from high school, and it quickly became clear the marriage was a mistake. They agreed to divorce as readily as they'd agreed to marry. To obtain a no-contest divorce took a year and a day, and Iris figured the U.S. Navy would be as good a place as any to sit out that period. So now here was Iris, whom her coworkers called "Flash," literally sitting.

During rest periods, the women would put their heads down on the worktable, and the officer in charge, a former schoolteacher named Dot Firor, would read aloud from *The Bobbsey Twins* or *Little Women* or some other comforting storybook narrative, and give them twenty minutes to let their minds drift. On the graveyard shift, some would sing to stay awake as they soldered. There was a woman from an Irish family, Pat Rose, who would sing "It's the Same Old Shillelagh" in the most beautiful lilting soprano.

The women were not told what the wheels they were wiring would be used for. They figured the wheels would be attached to some kind of machine—that seemed obvious—but what that machine did, they had no idea. There were men working one floor above them, constructing a machine the likes of which had never been seen, but the women didn't know that. They did know this: Whatever the wheels did, it must be important. Circulating among them was a man named Joseph Desch, a Dayton inventor who struck the women as brilliant and seemed deeply involved in the secret project. Once the Sugar Camp cafeteria was up and running, Desch often would visit with them at mealtimes. With Desch over eggs, bacon, and hash browns would be his amiable wife, Dorothy, who was tall and dark-haired and slim and wore the most elegant hats. The pair always were accompanied by Lieutenant Commander Ralph I. Meader, a wiry Navy officer who lived with the Desches in their modest two-bedroom brick Cape Cod and seemed to occupy the unspoken role of designated Navy minder to Joe Desch.

Desch was friendly to the women, and so was his wife, but Commander Meader was friendly in a different way. He flirted and had the hearty, slightly false air of a politician. He was a women's man and the women developed ways of coping with his attentions. Some would avoid him. Some liked to rub lipstick on their fingers and grab his cheeks and say "Izzy bizzy boo," which delighted him.

But for much of the time, the women were left to their own devices. They worked hard, but—not having chores to do or houses to keep—enjoyed free time to read, write letters, and use the Olympic-size swimming pool on the grounds of Sugar Camp. In the early morning after finishing the graveyard shift, they would stroll back to camp, savoring the meadowlarks and the fresh smell of clover. One of the women, Betty Bemis, was a champion swimmer who had won several national titles, and men—Joe Desch, Ralph Meader, even Orville Wright—would come down to the pool to watch Betty practice.

The women were warned that when they went into Dayton, they should travel in pairs. Soldiers and airmen lived at nearby Wright and Patterson airfields, and it was fine for the women to date them, but they were not to say anything about their work. Parts of the city were out of bounds. Many German Americans had settled in the southwestern part of Ohio, and while most were loyal citizens, there were lingering remnants of the Bund, the organized movement of Nazi sympathizers. The women were told they could be kidnapped and that German spies would very much like to know what was taking shape in Building 26.

The women did what they were told. They traveled in pairs, and they did not ask questions. They tried not to speculate on what they were making. Even so, it was impossible, during the long hours with their diagrams and soldering irons, not to notice that there were twenty-six wires and that the wheels had twenty-six numbers on them. The numbers went from zero to twenty-five, but it didn't take much education to figure out this added up to twenty-six.

Twenty-six, of course, was the number of letters in the alphabet.

* * *

For the Allies, 1942 had marked the low point in the Battle of the Atlantic. In the last six months of that year, German U-boats sank nearly five hundred Allied ships as they tried to make the crossing between North America and England, destroying 2.6 million tons of shipping. Now 1943 was shaping up to be worse. March 1943 had been the most terrible month of the whole war, with ninety-five Allied merchant ships sunk by Nazi submarines. In one major convoy attack, the U-boats sank dozens of ships in just three days.

The American war machine, as mighty as it was, could not produce enough ships to make up for such heavy losses. The U-boat situation had always been a crisis, but now the crisis was acute. England needed wheat and other food supplies. Joseph Stalin needed weapons to drive the Germans out of the Russian heartland. And the Allies needed to sweep the Atlantic Ocean clear of the U-boat menace once and for all, if they were ever to make the waters safe for the vast lines of convoys that would be needed to transport troops and tanks and weapons in sufficient numbers to mount—at long last—an Allied invasion of France.

The British still had charge of code-breaking efforts in the Atlantic theater. But the Americans were becoming more than just a junior partner. It had taken a while for the two services to begin working smoothly together. Agnes Driscoll's early rebuff of the British had set the Allied relationship back in 1941, but beyond the recalcitrance of a single woman, the British Navy had been appalled by the amateurish nature of the U.S. Navy's overall intelligence operation. The British would share reports that were never responded to, possibly never received by the people who needed to see them. The United States, in return, felt their English friends were withholding details about the Enigma project. Both services were right: U.S. naval intelligence had been dysfunctional at the outset, and the British were indeed holding back. Part of their secretiveness sprang from a desire to ensure that the "special intelligence" produced by the Enigma code-breaking project was used only defensively, to reroute Allied convoys into safe waters. They feared if the more aggressive Americans used Enigma offensively, to sink U-boats, this would tip off the Germans that the cipher had been broken.

But an accord was reached over time, and secure lines set up between American and British naval intelligence, as both services worked hard to track the U-boats and predict their movements. The effort at no time was easy, but it became excruciatingly difficult after February 1942, when the German subs started using the four-rotor naval Enigma, and Allied efforts against it proved mostly futile. During this dark period the Allies struggled to predict U-boat movements using methods such as high-frequency direction finding, or HF/DF, which was a way of locating subs using their radio signals. The Allied tracking rooms availed themselves of every form of intelligence—HF/DF, news of sinkings and sightings—to locate the submarines and route the convoys to dodge them. But the impenetrability of the four-rotor Enigma kept them at a major disadvantage until, in late October 1942, four British destroyers patrolling the eastern Mediterranean targeted and attacked a U-boat, the U-559. The submarine, which had surfaced, began to sink, and a group of British sailors tore off their clothes, dove into the water, and swam over to it, to retrieve papers and equipment. Two of the sailors, an officer and a seaman, were coming up a ladder when a rush of incoming water overcame them, and they went down with the sub. The others were able to scramble into a whaler. The men had retrieved two books, one of them a weather cipher book giving current key settings, which found its way to Bletchley Park and helped the code breakers get into the four-rotor Enigma cipher. They broke Shark for the first time using the bombe, to find a message showing the position of fifteen U-boats. They were back in.

Even with this assistance, though, the Allied ability to break the Enigma ciphers was spotty. Cipher books changed, and often the Allied code breakers had to try to come up with the key using hand methods. The British bombe machines could not help. Because the extra, fourth, rotor created twenty-six times more ways to encipher each letter, the older British bombes would have to run twenty-six times faster, or there would have to be twenty-six times more of them, to test every possibility.

What was needed, to attack the naval Enigma, was a bombe much faster than the ones being run in England. Allied officials decided that America—which had more factories, more raw material, more engineers and mathematicians—would build scores, maybe even hundreds, of high-speed machines capable of handling the four-rotor Enigma cipher. The Navy engaged Joseph Desch, the Dayton inventor, to design an American bombe. Desch was an inspired choice. A graduate of the University of Dayton, he did not boast an Ivy League education or even a PhD. What he had was an engineer's genius, the ability to work with his hands, and real-life experience in how a factory floor functioned.

Raised and schooled in Dayton—an excellent proving ground for a kid with inventive instincts—Desch as a boy had been fascinated by electronics and ordered vacuum tubes from a mail-order company that sent him the supplies he wanted, assuming he was an adult. Early in his career he worked at Telecom Laboratories, founded by Charles Kettering, and then at Frigidaire. Moving to National Cash Register, he and his staff patented the first electronic accounting machine. Desch also designed a new type of vacuum tube he made by hand. The renowned MIT engineer Vannevar Bush, doing war-related work through his National Defense Research Committee, admired Desch and brought him to the attention of the right people. A plan was drawn up: Desch would design a high-speed bombe, working with Navy engineers and mathematicians. The machine would be built by Navy mechanics and NCR employees who had the requisite clearances. The Navy women—though they did not know it—had been brought to Dayton to wire thousands of "commutator wheels" to go on the front of the American bombe machine, fast-spinning wheels to test possible settings.

In short: What the Americans were going to produce was the very roomful of high-powered machines that the Germans thought could never be built.

* * *

Officials at the code-breaking compound in Washington were tasked with coordinating the bombe construction project, and what this meant, for top naval officers, was much running back and forth between Washington and Ohio. In Dayton, Building 26 was modified and renamed the U.S. Naval Computing Machine Laboratory. Meanwhile, the Naval Annex at the former Mount Vernon Seminary set up OP-20-G-M, a top secret "research" group of engineers and mathematicians, to work with Desch. First, though, they decided that Agnes Driscoll had to be neutralized. On January 31, 1943, a unit diary included a brief and rather merciless notation that on that day "Mrs. Driscoll, Mrs. Clark, Mrs. Talley, Mrs. Hamilton transferred to Japanese N.A.T. Project." (That was the naval attaché machine, which Frank Raven ultimately broke.) Lieutenant Commander Howard Engstrom, a mathematics professor from Yale, took over the Enigma project, aided by a formidable group of intellects that included John Howard, an engineer from MIT; Donald Menzel, the Harvard astronomer; Marshall Hall, a mathematician from Yale; and others.

Including women. A number of the research department's mathematical staff were female. Like everybody else, the Navy was eager to locate women capable of doing higher math—the very field that women long had been discouraged from entering—and the Naval Annex put out the word to boot-camp evaluators, asking them to be on the lookout for enlisted women who scored high on the math aptitude test. These women had not enjoyed anything like the same educational opportunities the men had, nor the chance to embark on distinguished careers as engineers and academic mathematicians. They did, however, have the aptitude, the desire, and the ability. Many found, in the bombe project, the kind of statistical and probability work that they had been looking for all their lives.

One such woman was Louise Pearsall, a twenty-two-year-old from Elgin, Illinois, an industrial suburb about forty miles outside of Chicago.

Before the war, her mother had been a member of the America First movement—an isolationist, along with the rest of her suburban bridge club—but all card-table chatter about isolationism ceased on December 7, 1941, when two Elgin boys went down on the USS *Arizona*. Louise, the oldest of four, attended high school on scholarship at the private Elgin Academy. She aspired to become an actuary, hoping to find work with an insurance company, making statistical projections about risks. It was a modest enough ambition, were it not for the fact that actuarial math—as she often was reminded by deans and potential employers—was a man's field.

Even so, Louise went to the University of Iowa, where she was the only woman in many of her math classes—the chalk-covered little calculus professor used to stare at her as he paced the room, disconcerted by her presence—and performed well. But she left after two years because her father struggled to afford the tuition and didn't think it would pay off with a job. Louise enlisted in the WAVES, expecting to be made an officer, but the first class of female officers filled so quickly that she and some other educated women—"real intelligent gals," as she recalled them—went in as ordinary seamen, so as to get into war service as soon as they could.

At specialized training camp in Madison, Wisconsin, Louise took classes in physics, Morse code, and radio operation, where she was held up from graduating because of an audio dyslexia that made it hard for her to master receiving Morse. The timing was perfect. Instead of becoming a radio operator, she was rerouted to the Enigma project. In March 1943 she got orders to travel to Washington, where, at the Naval Annex, she underwent more tests and interviews and found herself assigned to work for MIT's John Howard. As the design for the new high-speed bombes was refined, it was Louise's job to sit at a desk and do what the bombe would ultimately do faster: test Enigma key settings. She worked on permutations, figuring out, if X became M, and T became P, what was the mathematical formula that would take a letter through the right sequence. It was intense work, utterly absorbing.

"That," she told her daughter later, "was where I should have been all my life."

To come up with an Enigma key setting, the naval team had to understand more than math. They had to understand the nature of the messages that German U-boat commanders sent. And the commanders sent a great many messages, much as the Japanese did, and for the same reasons. German Admiral Dönitz was a micromanager who insisted upon total command and control of his U-boat fleet. Submarines were obliged to constantly communicate with headquarters, providing updates that enabled Dönitz to make tactical decisions and issue orders. Since the boats were often thousands of miles from Germany, doing so entailed sending messages over high-frequency circuits that could cover long distances. This—as with the Japanese in the Pacific—opened them up to enemy interception. The wolf-pack tactic, in which many U-boats would be summoned once a convoy was sighted, also meant the subs had to abandon the radio silence normally observed during a group naval operation.

So the Allied mathematicians went about learning German naval greetings, the names of U-boats and commanders, and how German messages tended to be phrased. They knew a short message from Dönitz might include an order to "report your position" or to head for a port on the French coast. Subs often reported location and fuel capacity. All of this helped the code breakers come up with cribs. If they suspected that a line of cipher such as

RWIVTYRESXBFOGKUHQBAISE

represented the German phrase "weather forecast Biscay":

WETTERVORHERSAGEBISKAYA

they would line up the two lines of type, write numbers over or under them,

1 2 3 4 5 6 7 and so on

and look for "loops," places where one letter turned into another, and then that second letter turned into another. In the above example, they would see that *E* paired with *T* at position 5, *T* with *V* at 4, *V* with *R* at 7, *R* with *W* at 1, and *W* with *E* at 2, closing the loop. When the bombes were up and running, they would be able to program these into the bombe machine, which sought a setting where all these loops would happen. Since they didn't yet have machines, Louise herself was the bombe, sitting in a small room with a tiny team of colleagues, "working in an office on figures," as she later put it. "We had no equipment. We didn't have anything, really, to do anything big with. We were just getting started."

The Enigma had a few weaknesses that helped them. No letter could be enciphered to itself—that is, *B* would never become *B*. The machine had a reciprocal quality, meaning that if *D* became *B* on a certain key setting, then *B* became *D*. These factors limited the encipherments, but only somewhat; the possibilities still ran into the billions. Sometimes the team would get a key setting for a couple of days; sometimes they could not get a setting at all. There would be moments of clarity and long periods of darkness. The inconsistency of their work—and the helplessness when they could not recover a key setting—was dreadful. Ann White, from Wellesley, was working in the unit that translated broken Enigma messages from German to English. She always remembered one terrible night when a high-ranking naval officer came in, gave them a message, and begged, "Can't you give us any clues?"

As the summer of 1943 approached, John Howard told Louise Pearsall she needed to learn how to shoot a pistol. Some members of the mathematical research unit were relocating to Dayton, and she was one of those picked to go. She began doing target practice with a .38. In early May she and four other women, with about the same number of men, got their guns, boarded the train, and headed west to Dayton. Soon they too found themselves billeted at Sugar Camp. Unlike their bunkmates, these women knew why they were there. They were going to help make the bombes work.

* * *

Joe Desch was working to perfect the first two experimental prototypes, called Adam and Eve. He was under terrible pressure. Commander Ralph Meader was a taskmaster who used guilt as a motivator, and he was always telling Desch to hurry up, hurry up, that he'd be responsible for the deaths of countless more boys, more sailors and merchant seamen, if he didn't come through with a high-speed bombe—and soon. Meader would tell the bombe designers that the U.S. Navy would have lost the Battle of the Coral Sea if they'd been the ones on that code-breaking team.

It wasn't just a matter of getting the math to work; it was a matter of getting the machinery to run. The bulky bombes, which stood seven feet tall and ten feet long and weighed more than two tons, had hundreds of moving parts, and even a bit of copper dust could foul the works. "The design of the Bombe eventually required material and components from some 12,000 different suppliers," noted one memo. Some components did not exist on the commercial market and had to be designed and made. The designers needed diodes, miniature gas tubes, high-speed commutators, and the carbon brushes the wheels would come into contact with. As the women in Building 26 wired commutators, the staff in NCR's Electrical Research unit swelled from seventeen in August 1942 to eight hundred in May 1943, building the machines and perfecting the design.

In the Navy, a newly launched and commissioned warship makes a "shakedown cruise" to work out the kinks and get ship and crew running smoothly. The bombe's shakedown cruise commenced in May 1943, around the time Louise Pearsall arrived. It was her team's job to troubleshoot, together with Desch and Howard and some of the men who built the bombe. "The first two experimental bombes were under preliminary tests," noted a daily log for May 3, 1943, showing the many things that could—and did—go wrong: "Encountered some incorrect wiring and shorting of the wheel segments by small copper particles" was one of innumerable entries recording snafus.

It soon turned hot and humid in Dayton, a midwestern river town, but fortunately Building 26 was air-conditioned. Even so, the unit worked in a fever. Louise Pearsall's team, which worked apart from the WAVES wiring the wheels, would make up a menu, set the commutators, then start the bombe going, the wheels spinning to see whether the menu produced a "hit," meaning that the permutation they'd entered into the machine could plausibly represent that day's key setting. They would test the hit on an M-9, a small machine that replicated an Enigma. The M-9 had four rotating wheels, just like Enigma. The team would feed a message in and see if it produced German. If so, they had hit the jackpot: Their permutation represented the key setting. "It was fun," Louise later said. "Because I was working for all these engineers and mathematicians."

Louise got one of the first jackpots. She came up with a menu that produced a hit, and when they sent the results to Washington, a colleague there called back and congratulated her. "You just cracked one," they told her. Her break provided evidence that the bombes could do what they were supposed to.

By June, the bombes were working, but fitfully. More machines were brought online and had to be shaken down. The daily log shows the problems the team had to cope with. On June 29, both Adam and Eve needed repair. There was "one bad red commutator"; Adam needed oil; "a short was present at the end of the run but disappeared during the test." On July 1, "Adam blew relay"; "Eve has become temperamental"; "now we are completely shut down while maintenance is finishing both machines." An hour later Eve was back in operation. Two hours later "Adam finally fixed." Forty-five minutes later "Eve is out again." And so on. On July 13 Eve was "out due to broken brushes." Two days later, brushes on the timers "had become so soaked with oil that they were continually causing shorts." Diodes and relays gave problems. So did things called pigtails. Another day, "Eve's troubles were conclusively found by Mr. Howard to be tied up with her rewinding trouble." Meanwhile, the log noted, "Girls are getting tired and are making errors again, causing reruns."

Louise worked herself into exhaustion. On July 6 she was able to take a weeklong leave and go home to Elgin, where her father "pumped the hell" out of her, as she put it, trying to get her to reveal what she was doing. She didn't crack. He lived the rest of his life without knowing that Louise had put his two years of college tuition payments to better use than he ever could have imagined.

* * *

By September the team had put the finishing touches on the first generation of high-speed bombes. Over the summer the Navy had constructed a "laboratory building" in the Mount Vernon Seminary compound in Washington, a big multistory structure with sturdy floors of reinforced concrete. In Dayton, flatbed cars pulled up in the dark of night on the railroad spur behind Building 26, and crates were loaded aboard. Other bombes would follow, more than a hundred in all, in the weeks that followed. NCR also manufactured M-9s and shipped those as well. Louise Pearsall rode with one of the first shipments. The train was late leaving. Louise was sitting in her seat, wondering why they were delayed, when her boss, John Howard, sat down and confided that men had been detained who seemed to be suspicious and possibly were going to sabotage the train.

She and Howard were the only ones who knew the reason for the delay. Louise Pearsall spent the long overnight trip back to Washington sitting bolt upright in her seat.

* * *

When they arrived, the male officers went to the main Navy headquarters to check in. Louise and the other enlisted WAVES had to go to a central facility at the Navy Yard, however, where the women's formal transfer back from Dayton was processed. The sailors at the processing station started making snide comments. "You're from Dayton!" the men exclaimed, as if this fact was something to be ashamed of. The women were taken aback.

Louise, tired and impatient, told the sailors she and the other women needed to be processed through quickly. "We have an assignment to go back to."

"That's what they all said," sneered the sailor.

Louise had no idea why she and her colleagues were being jeered. What the women could not know was that during the summer, Commander Meader had ejected a number of WAVES from Dayton, returning them to Washington for perceived misbehavior—in the Navy's view—that identified them as security risks. (Some of the other women in Dayton, who knew what was going on, referred to this as "pruning the group.") Records show that on August 20, 1943, an enlisted member of the WAVES had been transferred out of Dayton and sent back to central processing in D.C.; she had shown up at the dispensary in Dayton with heavy menstrual bleeding and an examination had shown an "incomplete abortion at six weeks before entering the navy." That same day, another yeoman was sent down who had shown up for menstrual bleeding and abdominal cramps; her past medical history showed "induced abortion before entering the service." Back in 1942, when the WAVES were formed and regulations drawn up, some officials worried that the no-pregnancy rule would prompt women to seek an abortion in order to join. It would appear they had been correct.

A few others had been sent down for violations that the memos coyly declined to spell out. On August 14, 1943, one WAVES enlistee was sent back from Dayton along with a memo saying "because of condition X she is considered unsuited for duty." The memo didn't say what "condition X" was, but it seems likely the WAVES enlistee was pregnant. On July 30, another Dayton WAVES member had been sent back. "Fails to meet qualification X," her memo said. Another was sent back for "malingering," another for being "undisciplined." Precisely because security was so tight, WAVES were even more likely to be expelled from Dayton than from other postings.

In short, women from Dayton had acquired a bad reputation in the processing center—by the sexist standards of the day—and the men assumed that this new bunch had done something reprehensible. Louise had no idea about any of this. She protested the disrespectful treatment. "If you don't believe me, call Lt. Howard," she told them. "We came back with him."

But the sailors wouldn't call her boss. Smirking, they ordered the women to start washing windows.

Finally a female officer from Navy headquarters came looking for them. She could not believe what she saw: her best picked group of female mathematicians, pride of the bombe project, scrubbing windows. "My God, they're our top girls!" she told the men.

When the women made it to the Naval Annex to help get the first shipment of bombes installed, Louise learned that John Howard had been desperately trying to find them, asking, "What the heck's happened to my girls?"

* * *

More bombes arrived from Dayton, and the crew in Washington worked overtime to get them up and running. The Annex created a crib watch, a decryption watch, a traffic preparation watch. Several hundred of the women at Sugar Camp made the return trip east from Dayton to Washington, to live in Barracks D and run the bombe machines, though they still did not know the machines' true purpose.

Louise Pearsall, who did, continued troubleshooting, assigned to evaluate the printouts the machines produced. Though the D.C. bombe crew would soon consist of nearly seven hundred women, it was a smaller group at first, and she worked seven days a week, twelve hours a day. Everybody socialized together—men, women, enlisted, officers, liaison workers from MIT and IBM—and they socialized intensely. The lone male sailor on the bombe crew threw a party at which Louise was introduced to Southern Comfort. It took her a day to recover from her hangover, furiously drinking Coke, since she didn't drink coffee.

Washington had always been a bibulous town, and wartime was no exception. The code breakers mastered the region's alcohol-related regulations, learning that it was necessary to travel to Maryland or Virginia to buy alcohol by the bottle and that Virginia's state-operated liquor stores closed early. Drinking was one way to relieve "the tension, the pressure, the trauma," as Louise put it. It wasn't just the code breakers who were stressed and overwhelmed. Everybody in Washington felt the pace of war. Louise knew a woman in charge of military train schedules, and she was tearing her hair out.

The bombe was a "high, high, high priority project," as Louise later put it, and everybody on it was important. Once, when a sloppy (or tired) operator threw a printout into a burn bag, John Howard had to stand on a chair and impress upon them the life-and-death gravity of their work.

Just how important they were became clear when Louise's brother Burt, a hotshot Marine pilot, tried to get into the Naval Annex compound to visit her. Both of Louise's brothers were in the military, and both were big deals at home in Elgin. Not here. Here, Louise was the big deal. When Burt approached the first set of Marine guards at the Naval Annex, accompanied by a couple of pilot buddies, he informed the guards— fellow Marines—that they were going inside to see his sister. "No, you're not," the guards replied, barring their way. Louise had to come outside after her shift. They hired a cab and asked the driver to take them slowly down Constitution Avenue, and Louise showed her younger brother the Washington sights.

That visit was a rare respite from a workday that was all-consuming. The Enigma project took its toll on everybody connected with it. In the fall of 1943, Joseph Wenger, one of the top officials in the Annex, had a nervous breakdown so severe that he had to spend six months in Florida. Joe Desch broke down as well. In 1944, weighed down by Commander Ralph Meader's tongue-lashings, Desch stormed out of NCR and spent several weeks on a friend's farm outside Dayton, felling trees and chopping wood. The bombe design project completed, he'd been assigned to develop a machine for Japanese codes. Desch had four nephews serving in the Pacific, one of whom died during this time. "He had nightmares for years about men dying," said his daughter, Deborah Anderson.

The people working on the Enigma project were mathematicians and engineers. They were precise, conscientious people who liked to solve problems and build beautiful things, not kill people. The work was hard on the women as well, particularly those like Louise who knew the stakes. One of the women in charge of maintaining the commutators, Charlotte McLeod, from Buffalo, New York, would time her visits home to coincide with a big predicted snowstorm coming in off the lake, so that she would be snowed in and would get a few extra days to recover. A daily log on February 25, 1944, noted that a WAVES member named Olson had been sent to her barracks, "a general wreck of jitters—unable to work." Another was reprimanded for coming in "slightly intoxicated," according to the log book. "She wasn't bad, but she obviously has been drinking." Another log entry spoke of "terrific pressure."

However they chose to relieve their stress, the women were loath to abandon their duties for long. Louise Pearsall was annoyed when, toward the end of 1943, she was told she could not remain in the ranks of the enlisted. Given her detailed knowledge of one of the war's most top secret projects, the Navy insisted she become an officer. In January 1944 she was sent to Smith, where she reencountered some of the enlisted women she had met at boot camp, also now being promoted. They were old salts and teased the young college women just coming in, giving them "little side instructions" about how to comport themselves as naval officers. Occasionally they would tell the new women the wrong thing, for fun. Louise was antsy to get back to work. For something to do, she played the snare drum in the band, and when they paraded during their graduation as officers, it was raining and the drum she was holding sank lower and lower as it filled with rainwater.

A newly minted naval officer is rarely sent back to the same place where he or she worked as an enlisted person, but Louise Pearsall was a special case. After two months at Smith she returned to the Naval Annex during the week ending March 18, 1944, wearing her ensign bars, and found herself accorded more respect than the last time she had been processed through. A lieutenant working in personnel said John Howard had been driving them crazy, asking when Louise would get back. She was sitting in a routine orientation class when a lieutenant commander came in and told her to go on back to work. "Louise, would you get out of here right now?" he told her. "I'm tired of listening to your boss."

* * *

Everybody became an old salt pretty quickly. After just six months in the Navy, the young women who had wired the commutators—and now were running the bombes in Washington—found themselves supervising women even younger and greener than they were, many of them eighteen-year-olds fresh out of boot camp at Hunter College. Jimmie Lee Hutchison, the switchboard operator from Oklahoma, was in charge of a four-person bombe bay. Jimmie Lee's friend Beatrice worked a machine nearby. Their workplace took up the entire bottom floor of the laboratory building, which had been built in the old Mount Vernon Seminary compound near the Navy chapel. It was a hangar-like space with three rooms, each room divided into "bays" containing four machines. There were 120 bombe machines all told. The machines were noisy; on summer days the room got so hot that the women opened the windows to keep from passing out. When they did, the racket could be heard from outside on Nebraska Avenue.

As a bay supervisor, Jimmie Lee Hutchison had an assistant and four operators. When she came on duty, she signed a logbook that lay on top of a printer near the bombes. It was Jimmie Lee's job to keep the log, supervise the bay, and set up one of the machines according to a menu she was given. Doing so entailed moving commutator wheels and rotating them to the starting position. The wheels were heavy—weighing nearly two pounds—and had to be carefully placed so they wouldn't fly off and break somebody's leg. She would set the wheels, sit on a stool, and wait to see if she got a hit. The machines ran so fast that they couldn't stop immediately. They had a rudimentary memory, so the wheels would run for a few seconds, then stop, back up, and generate a printout recording the setting that produced the hit. Jimmie Lee would take the printout to a window where a gloved hand belonging to an unseen female officer would emerge and take it. It was tiring work that required energy and concentration. The women hated the long "hoppities," when they were testing a possible wheel turnover and had to get up and down, up and down, changing the wheels several times on the same run.

Jimmie Lee by now had married her high school sweetheart, Bob Powers. By a lucky coincidence, Bob had been assigned to ferry planes into the airfields at Dayton, bringing planes up from North Carolina and from Bowman Field in Kentucky. They married at Bowman Field on June 18, 1943. Back in Dayton, Jimmie Lee's WAVES friends threw them a party. After that, the newlyweds had spent time together whenever Bob Powers flew into Dayton. Sometimes he would get in before she finished her shift, so he'd wait in the hotel room until she got cleaned up, and they'd go out to dinner. One time Jimmie Lee lost her engagement ring down the sink in her cabin and was frantic until the custodian kindly removed the trap and fished it out. Neither Jimmie Lee nor Bob had ever heard of trick-or-treating, which was not something people did in Oklahoma, so their first Halloween, in October 1943, they felt like kids.

After Jimmie Lee was sent to Washington, though, their visits were few and far between.

As Jimmie Lee and the other women settled into their duties, they became part of an Enigma code-breaking chain that was virtually all female. When a message arrived at the Annex, it would first go to the cribbing station. The cribbers had one of the hardest jobs, sifting through intelligence from the war theater, including ship sinkings, U-boat sightings, weather messages, and battle outcomes. Scanning intercepts, they had to select a message that was not too long—a long message might involve more than one setting—and guess what it likely said. Juxtaposing crib and message, they had to devise a menu. Louise Pearsall did this; so did Fran Steen, the biology major from Goucher College. Fran had spent a year on the Japanese project, then moved to the German ciphers. Promoted to watch officer, Fran had access to a secure line that connected her to a counterpart in England. Her code name was "Pretty Weather," and her British contact—male; she never met him—went by "Virgin Sturgeon."

From there, the menu would be passed to Jimmie Lee or another bombe deck operator. If there was a hit, it would go to somebody like Margaret Gilman, from Bryn Mawr, who would run it through the M-9 to see if the "hit" produced coherent German. Once they got a key setting, subsequent messages for that day could be run through the M-9 and translated, without having to use the bombes.

Soon, the operation was so smooth that most keys were obtained in hours and most messages decrypted immediately. The effect of the U.S. bombes on solving the Atlantic U-boat cipher "exceeded all expectations," noted one internal Navy memo. "Since 13 September 1943, every message in that cipher has been read and since 1 April 1944 the average delay in 'breaking' the daily key has been about twelve hours. This means that for the last half of each day, we can read messages to and from Atlantic and Indian Ocean U-boats simultaneously with the enemy. In fact, during these hours the translation of every message sent by a U-boat is at hand about twenty minutes after it was originally transmitted. At present, approximately 15 percent of these keys are solved by the British and the remainder by OP-20-G."

Once they were broken, the messages would pass to somebody like Janice Martin, a Latin major from the class of 1943 at Goucher, now working in the submarine tracking room, which was located one floor above the bombe deck. Janice, a lawyer's daughter from Baltimore, was stationed in a room set up so that anyone who opened the door would see, standing in the hallway, a blank wall. Inside, she and her colleagues could see a huge map of the North Atlantic. The broken messages from the M-9 were sent up to her office, translated, typed up by a secretary, handed to the senior watch officer—a man—and then passed to the junior officer, who would be either Janice; her childhood friend Jane Thornton, who had grown up in Baltimore near her and gone through Goucher with her; another Goucher classmate; or a woman from Radcliffe. The U-boats had to report whether they sank an Allied ship or whether any U-boats had been sunk, and the women used these Enigma messages—along with files on individual U-boats and their commanders—to track, with pins, every U-boat and convoy whose location was known. At another desk, several other Goucher women, including Jacqueline Jenkins (later the mother of Bill Nye, aka Bill Nye the Science Guy), tracked "neutral shipping" based on daily position reports. Neutral shipping mattered because if those ships deviated from their assigned sea lanes, it might mean they were surreptitiously supplying U-boats.

In addition to tracking the ships, researchers in Janice's room would compile a preliminary intelligence report overnight. Between seven thirty and eight a.m., there would be a knock on the door, and Janice's team would hand over an envelope containing the night's messages and the report. The messenger would put it in a locked pouch and take it to the Main Navy building downtown. There, a commander named Kenneth Knowles, working with a counterpart tracking room in England, would make decisions about whether to use the intelligence defensively to reroute convoys or offensively to sink the U-boats. The downtown tracking room at first was staffed by enlisted men, but as the war proceeded, WAVES took over there as well. One male officer said the WAVES did a better job because they had had to meet stiffer selection requirements. When the intelligence they generated was passed to the fleet for action, the source of the information was never revealed. By the end of 1943, Janice Martin recalled, "the British turned the whole operation over to us."

* * *

After the carnage of 1942 and early 1943, the Allies had seen a stunning turnaround in the Atlantic. By September 1943, most U-boats had been swept from the Atlantic waters. This was thanks not only to the new high-speed bombes but also to a host of other Allied war measures: advances in radar, sonar, and high-frequency direction finding; more aircraft carriers and long-range aircraft; better convoy systems. The Allies changed their convoy cipher, and Dönitz could no longer read it. The tables turned. During the summer, American hunter-killer units used code breaking along with other intelligence to find and sink big German submarines that were sent out to refuel U-boats. These refuelers were known as milch cows, and between June and August, American carrier planes sank five. In October, they finished off all but one. The refuelers were critical to the U-boats' ability to stay so far away from their home base, and as the milch cows went down, the U-boats began to drift homeward.

There was always the chance, however, that the U-boats could come back. And they did try. In October 1943, the U-boats reappeared. But now the costs were punishingly high. For every Allied merchant vessel sunk, seven U-boats were lost. Now Dönitz was the one who could not build boats fast enough to replace those he was losing. In November, thirty U-boats ventured into the North Atlantic and sank nothing. The U-boats began lurking elsewhere, clustering around the coast of Britain, hoping to intercept materiel brought in for an anticipated invasion of France. Dönitz was always trying to innovate the U-boats, adding a *Schnorchel* that enabled them to remain submerged longer. He was willing to sacrifice his boats, and his men, and kept the U-boats in the water even as a way to tie up Allied resources.

But it was a losing battle. In May 1944, the Allies sank half the U-boats in operation—more than the Germans could replace. More than three-quarters of the U-boat crews were killed, suffering terrible watery deaths. The women in the tracking room were privy to the full immensity and horror.

By now the British had indeed handed over the four-rotor bombe operations to the Americans. After the war, a U.S. Navy file was made of messages from grateful—and gracious—British colleagues. "Congratulations from Hut six on colossal...week," said one missive from Bletchley. An internal British memo acknowledged that "by half way through 1944" the Americans "had taken complete control of Shark and undoubtedly knew far more about the key than we did."

* * *

The Germans found other uses for the Enigma, which achieved less fame but were extremely dramatic. From June 1943 to the summer of 1944, the Nazis attempted to use submarines to run Allied blockades and travel between Europe and Japan, to load up on supplies to feed the war effort. Special Enigma keys were devised for use by both the Germans and the Japanese, and other ciphers also were employed.

These were long and desperate sea journeys. On April 16, 1944, a sub the Japanese code-named the *Matsu* departed Lorient, France, bound for Japan, carrying four German technicians and thirteen Japanese personnel, as well as German antisubmarine countermeasure equipment; torpedoes; radar apparatus; plans for high-submerged-speed subs; and the influenza virus. The Allies tracked its journey via messages sent and received at stages. One message gave the sub's route through the Balintang Channel, and on July 26, the USS *Sawfish* sent its own message back to U.S. naval headquarters, saying, "He did not pass...put three fish into Nip sub which disintegrated in a cloud of smoke and fire." After this hit, the Allied code breakers read messages from Tokyo to Berlin noting that the sub had been headed to Japan to bring supplies, but "our inability to utilize them owing to the loss of the ill-fated ship is truly unfortunate and will have a great effect throughout the Imperial Army and Navy."

Similarly, on February 7, 1944, a message from Tokyo confirmed that a sub code-named *Momi* would soon depart Kure, Japan, for Germany. The sub made a four-month journey, arriving in Europe during the D-Day invasion. Berlin sent a message to the sub noting that "Anglo-American forces have landed on French coast between Le Havre and Cherbourg, but your destination is still Lorient." The cargo included 80 tons of rubber; 2 tons of gold bullion; and 228 tons of tin, molybdenum, and tungsten; as well as opium and quinine. This sub was lost, causing Berlin to message Tokyo: "The disaster which has befallen these liaison submarines one after another, at a time when they were playing such an important role in transportation between Japan and Germany, is indeed an extremely regrettable loss to both countries."

* * *

As 1943 gave way to 1944, the American bombes ran on a twenty-four-hour basis—and their mandate expanded. With the U-boats under control, Bletchley Park asked the Americans to help break the daily keys of the three-rotor Enigmas used by the German Army and Air Force. Louise Pearsall moved to the Luftwaffe effort. The pace remained relentless as the European war thundered on. The women would spend the morning working naval U-boat ciphers and the rest of the day breaking the others. "The attack against the U-boat cipher has been so successful that only about 40 percent of Op-20-G's 'bombe' capacity is utilized for it each day," said a U.S. Navy memo. "To further the common good, the remainder of the 'bombe' time is used to run successful attacks on German Army and Air Force ciphers under the direction of the British." The bombe machines would be called into heavy use just prior to—and during—D-Day; as the memo noted, "this resulted in a considerable gain in intelligence during a very critical phase of the invasion of France."

In Dayton, meanwhile, a cadre of WAVES remained at their work-tables, wiring wheels to be used as replacement parts. The women felt almost guilty enjoying the war in such lovely surroundings. They played the piano; sang; went into town. The Beverly Hills Country Club, a distance away but worth the trip, had shrimp cocktail for $1, Russian caviar for $2.75, and two floor shows nightly. When the weather got cold, the women moved into a heated barracks. In the mornings Esther Hotten-stein, the lieutenant in charge, would come skipping stark naked out of the shower, singing "Oh, What a Beautiful Mornin'" from *Oklahoma!* The women were pretty much on their own in Dayton, and they liked it.

These women were a select group, and they got along beautifully. There were, to be sure, occasional dramas, and had been all along. One night a commotion developed when it emerged that Hottenstein had inadvertently come upon two women having sex with each other. The others gathered, somebody used the word "queer," and Ronnie Mackey, one of the youngest, said, "What do you mean, they're queer?" Charlotte McLeod took her aside and explained. Another WAVES member began wearing her raincoat everywhere. Nobody paid much attention until she was taken to the hospital and gave birth to a boy. The women put the newborn in the Sugar Camp sick bay, using a supermarket cart for a crib. They called him "Scuttlebutt," the old Navy term for a water cask around which sailors cluster to gossip, and loved him dearly until both baby and mother were obliged, like the same-sex couple, to depart.

Other women at Sugar Camp were in stages of romantic commitment that might have seemed bizarre in peacetime but felt normal now. A WAVES enlistee in Betty Bemis's cabin was writing letters to her boyfriend, and that boyfriend had a buddy named Ed "Shorty" Robarts with no immediate family to write him. All the girls in the cabin started writing letters to Shorty. Gradually the other women dropped out, but soon Betty, the champion swimmer, and Shorty, the unseen soldier, were engaging in a serious correspondence.

This was not uncommon. The quasi-divorced Iris Flaspoller also was writing a sailor, stationed on Tinian Island, she had never laid eyes on. She and her correspondent, one Rupert Trumble (his nickname was, of course, Trouble), wrote each other every day and developed private jokes. One of the jokes was that they were married and had children. Trumble would write and ask how the children were doing and Iris would make up something amusing to report. Other times he would write to say that he was dreaming of her. One time he sent her an envelope that contained only ashes, to demonstrate the burning love he felt.

CHAPTER TWELVE

"All My Love, Jim"

May 1944

The letters accumulated each day in the mailbox that was attached to Apartment 632A at 609 Walter Reed Drive in Arlington. Dot Braden would take them out of the box when she came home after her shift or find them waiting on the table when she got home, if Crow or Louise had beaten her home and picked them up first. There was usually a big sheaf. All the women in the apartment were getting and sending letters. Dot corresponded often with both of her brothers, Teedy and Bubba. Curtis Paris's letters had dropped off, and she lost track of him. She and George Rush kept writing. Dot wanted to dump George but couldn't quite bring herself to do it. Morale and all that. Over time, though, the letters from Jim Bruce began to take precedence. Jim Bruce was a diligent and faithful letter writer, his handwriting small and precise, his *D*'s very loopy when he wrote "Dear Dot." His letters were written on feather-thin airmail stationery, neatly folded in sixths, and addressed to "Miss Dorothy Braden." Like the other military men, he sent them through the Army Post Office system, which disguised GIs' overseas locations by using, as a return address, the APO address of the American processing station. The envelopes from Jim always had the distinctive airmail edging, striped like a barber's pole. The return address would say Lieutenant James T. Bruce Jr., with a U.S. Army Postal Service postmark, usually in Miami or New York.

While some soldiers were serving in secret locations, Jim wasn't. He was working as a meteorologist for the Army's strategic air command in Ghana. It was not life-threatening work, but it was an important assignment. His job was to safeguard the lives of airmen, making forecasts that would protect pilots from dust storms and other mishaps.

Dot always experienced a little skip of her heart when she came home to find a Jim Bruce letter. After he went overseas toward the end of 1943, he initiated their correspondence by sending her an affectionate letter, whose tone was in keeping with the serious interest he'd shown before he left, including that harebrained, last-minute trip up to Arlington to see her. The interesting thing about Jim was that he was laconic, almost taciturn, in person, but chatty and sentimental in his letters. He used her first name a lot, which conveyed that he was really thinking about her.

Both Jim and Dot had to be careful not to reveal war-related details—government censors read each overseas letter—so their writing tended to swing from the emotional to the flatly day-to-day, without much in between. In the spring of 1944, Jim mentioned that his unit was forbidden to go in to the nearest town, Accra, because of an epidemic of trench mouth, or gum disease, on the base. "I don't have anything exciting to write you, Dot," he continued.

> I went to the beach a while this morning. Last night we had the premier or first showing of "Going My Way." Bing Crosby played the part of a Catholic Priest. Can you picture that. It was a very good movie. It opens in New York next week so maybe you will get to see it before long. I am glad I don't have to wait until all the movies get old before I see them like you have to do in Washington.

For his part, Jim was always eager to hear from Dot and loved getting her witty, literate letters. "Dot, you be good and write to me real often. Your letters are good for my moral[e]," he told her. She had sent him a photo. "How about sending me some more recent shots," he wrote her. "I know that one can't be beat but I want some more anyway."

Sometimes—often—Jim complained when he didn't hear from Dot for any length of time. Letters to soldiers overseas tended to get held up in transit, and the upshot was that letters sometimes arrived in jumbled bunches. During the long waiting period, Jim's letters would get somewhat pitiful and Dot would smile when she read them. "I get treated worse than anyone I know," he wrote her on April 30. "It has been fifteen days now since I got a letter from you." He added, "Dot, if you wait as long to write as you did before, well, you can guess what I am thinking."

But he was an even-tempered person and didn't hold it against her. In the next paragraph, he said he had gone to a dance in Accra given by some British officers and civilians. "It was a very wet party," he confided. "First it rained and then there was lots of whiskey. I didn't get intoxicated, Dot, but just had a few drinks to be sociable." He said that the "girls" at the dance were mostly nurses, and wives and daughters of the British "gents in town." "I had a very interesting time talking to this old maid who was in her forties I would say."

The next month, after receiving a batch of letters from Dot, Jim was in a much-improved mood. "Dearest Dot," he now was greeting her. He wrote with exhilaration to say that he had been invited onto a military plane to Brazil, to provide midflight forecasting. It was a fourteen-hour trip each way, and he had just gotten back. He couldn't name the places they visited but said the "most exciting part" was when they flew in a small plane from one town to another and he was able to act as copilot. At one point the pilot handed over the controls and told him to take over. Jim had never flown a plane. "The plane went every way but the right way at first but soon I was able to keep it on the right course. It was very easy to handle once I knew what each gadget was used for," he wrote her.

I don't think I have ever seen anything in nature as pretty as low clouds when viewed from 10,000 ft over the ocean. The tops of them are all shapes and sizes and look a lot like snow-capped mountain peaks. I wish I had

some pictures of them to send you, Dot. Better than that though when this
war is over and I get my airplane I will take you out over the ocean and
show them to you.

It was a rather fanciful explosion of sentiment—the desire to take her up in a private plane. But the letter suggests that the two of them were still figuring out their relationship, which, like so many during the war, was evolving purely through correspondence, with no phone calls, no in-person visits. Given the time and distance, developments were halting and tentative. During Jim's brief stay in Brazil, where there wasn't as much rationing—he'd been able to eat ice cream and drink a Coke—he debated getting Dot a present but wasn't sure they had progressed to the point where he could send her intimate things. "Dot, I started to send you some silk stockings from Brazil but didn't know how you would like the idea of me buying you clothes." He had sent his sisters each a pair and told her "if you would like some just send me your size" and he would ask a friend stationed in Brazil to get some for her.

In July 1944, Dot sent Jim a photo of herself with a friend at Beverly Beach. Her own letters must have given him the idea that she was leading a carefree, almost dilettantish existence. "Note the Coney Island background," she wrote on the back. She sent another of her sunbathing with a different female friend, noting, "The man behind us in all of these fixed the camera every time it got stuck." She also sent a photo of her with five Arlington Hall colleagues, including Crow, on an excursion to Colonial Beach. Dot was half-hidden behind a pole, but the group of code breakers looked very glamorous and carefree, wearing summer dresses or shorts and blouses. Crow was perched on a ledge on the far left-hand side, wearing shorts and smiling. "All together!" Dot wrote on the back. "Carolyn is the first on the left. It's a pity the pole wasn't larger! P.S. Carolyn isn't really that pumpkin-headed!"

In August 1944 she sent one of herself and Crow lying on their stomachs on towels on the sand. It said, "Aren't we cute? Carolyn is the other of this twosome. P.S. We did get burned."

Her letters were lighthearted, but the stress of wartime did manifest itself, as they tried to navigate spats and uncertainties. Jim continued to complain if time went by and he didn't get any letters. One night Dot drafted an angry response, writing late at night after a long day. She wrote: "I was just sitting here in bed, waiting for Carolyn to turn out the light, and re-reading the last epistle received from you. I thought that I was a past master of hammer-on-the-head technique, but, J. Bruce, you have me beat all hollow! In some spots that letter blesses me out twice in the same sentence!... The truth of the matter is that I really envy you. Did you think that was just soft soap when I said it before? I'll trade places with you just any day you say." But the letter seemed too strong, so in the rational light of morning, she didn't send it.

The good thing about the pace of their correspondence was that it gave them time to cool off. In August, Jim was pleased that he had received two letters from her, each of which had taken two weeks to reach him. "I enjoy very much reading your letters, Dot," he said. Casting around for things he could safely tell her, he reported that the enlisted men were playing a "hot game of bridge"; that he had finished his meteorological charts for the day; that the officers didn't seem much interested in poker, which was good since he played some recently and it was "hard on the nerves." But there was a Ping-Pong table and "a victrola with some good records that the Navy left us," as well as a piano.

The war went on and on. As early as the beginning of 1944, Jim had floated the idea of marriage—on top of his hinting around and taking her to meet his sisters before he left—and Dot struggled to decide whether she wanted to settle down, and if so, when, and with whom. She went back and forth. In October 1944, Jim wrote Dot that he was being sent to another posting and complained that he hadn't heard from her. It had been nearly a year since his hasty trip to Washington to see her. "I guess you are still having a good time with your friends in Lynchburg," he said, a little plaintively. He said that he could only take sixty-five pounds of personal items to his new station, so was having to pick and choose what to put in his suitcase, and figured he could get two pounds of random items in his pocket. "Dot, don't you worry your pretty self about me because I think my new assignment will be better than the one here."

By November, Jim Bruce was stationed in Iran. "The weather situations are rather interesting here," he wrote. "I noticed a while ago that we have a few clouds that I didn't forecast and can't explain." He sent her a candlestick and pen, and she wrote saying that she liked them. He still did not seem certain of her affections. "Dot, you should stop worrying about your brothers and me if you really do worry about me. I am safer here than I would be in the States where there are so many ways to be hurt. There are no bad girls here to harm me either."

He continued, seriously: "Dot, what worries me is this war is going to be longer than we thought and it is going to be a long time before I get home to see you." In December 1944 he had not heard from her "since I wrote last," and reflected with some anxiety that he now must be six thousand miles away from her. "That is a long ways, in fact it is six thousand miles further than I would like to be." He concluded: "Be good and write to me often, Dot. I enjoy reading about all the things you do."

Even as she vacillated, Dot was writing letters that were affectionate, which he took note of. "In a letter from you that I received yesterday, you wanted to know if I still had a dimple in my chin so the purpose of the enclosed photograph is to show that I do have a dimple and that I can raise a mustache," he wrote her. "I promise to cut off the mustache before I come home though."

Dot, I have been receiving your letters in about ten days, much to my happiness, and I enjoy them all very much. You wrote that if I wanted to know how much you love me that you would tell me. Please tell me because I do want to know.

In that letter, however, Jim was worried about her old friend Bill Randolph, who had received a diplomatic posting. "Dot, you are lucky to be such a good friend of a vice-consul. I hope he stays in South America until after I get back home because I am afraid that you will decide to be more than brother and sister." He was planning to throw a party with Scotch and fruitcake. "Dot, it will be close to Christmas when you receive this so I hope you have a very merry one and a happy new year. I sincerely hope that I can be with you next Christmas."

In January he was elated to have two letters from her and a Christmas card. "I love you and am looking forward with great anxiety to the day when I get home and see you." In a subsequent letter he was still worried, though, that he hadn't heard from her. He admitted that he was not going to church much and urged her to write him both at the APO address she had been using and another APO address that some of the other officers were using, "and number them consecutively" to see which Army Post Office got the letters to him faster.

He was pleased with a recent forecast he had made.

It stopped raining and cleared up when I predicted, the fog came in this morning and then it cleared up again as my forecast said it would. When I went to the mess hall this morning some of the pilots congratulated me which is very unusual. Usually they give us H! if the forecast is bad and forget it if it turns out OK.

He noted that he had been overseas for almost thirteen months and figured that in four or five more he could start hoping to be transferred. "I will be working all the angles I know of trying to get home to see you."

Be a good girl Dot and write to me. Some day you can tell me all that you are thinking and not have to write it.

All my love, Jim

Jim continued to broach getting married. They corresponded over this in the fall of 1944. Enjoying her independence, Dot put him off; then gave her assent; then wasn't sure. Around February 1945 she suggested they write a letter every day. But at the end of that month, Jim wrote that "the letter that I received from you today is the one in which you turned down my request to become engaged now."

He was clearly disappointed. "Dot, I know that you did the right thing but I must admit that I was surprised," he wrote. "When we were corresponding over the question last fall you stated that you could think of nothing better than becoming engaged to me." He speculated that "maybe you didn't really mean it last fall" and worried that he had been bragging about his family and that had put her off. "I didn't mention my good family thinking that it would have any effect on our marriage," he wrote apologetically. "I just thought that you might be interested in knowing something about them. I was asking you to marry me and not my family."

But here was the good thing about Jim Bruce: He didn't browbeat Dot or try to force her. He told her he respected her decision. And neither did he give up. "Well, I guess we had best drop the subject for a while," he wrote mildly. She had mentioned that Crow and Louise were going home to Mississippi for a visit, and he wished them well.

Dot, I must go to work now and help to keep 'em flying.

All my love, Jim

CHAPTER THIRTEEN

"Enemy Landing at the Mouth of the Seine"

June 1944

By the middle of 1944, as a result of his many conquests, German chancellor Adolf Hitler had the impossible task of defending the entire northwestern coast of Europe from enemy invasion. The Nazis knew an Allied invasion of occupied Europe was likely—Japanese diplomats stationed around Europe had chattered about an invasion in the Purple traffic since at least 1943—but they did not know when or where it would happen. For quite some time the Allies did not know, either.

Code breaking—of many kinds—helped cement the Allied decision. In November 1943, the Purple machine at Arlington Hall rattled out one of the most valuable contributions that Hiroshi Oshima, the Japanese ambassador to Nazi Germany, made to the American intelligence effort. It was a wordy, effusive, somewhat emotional, meticulous description of German fortifications along the northwestern coast of France, from Brittany to Belgium and everything in between.

Actually, it was a series of messages. The first was a missive in which Oshima bragged to colleagues in Tokyo about a trip he had taken to see Nazi operations in France. He and some colleagues had traveled by train from Berlin to Brest to inspect defenses there and along the French coast—what the Germans called their Atlantic Wall.

Oshima, who admired the Nazis, was full of details about who in the German Army was who—he described being "feted by Marshal Rundstedt"—and what an honor it had been to meet them. He reported that he and his companions had observed German defenses around Lorient, the Brittany seaport. They had seen night maneuvers; inspected encampments and harbors; spent the night in Paris; and gone on to Bordeaux, where they watched blockade runners carry out maneuvers. They headed to Poitiers and Nantes, meeting with German commanders. "When we were wined and dined we got the chance to talk with the right people everywhere."

This dispatch was followed by a longer message, which was somewhat like reading the writings of a military tour guide. The gist of it was: As formidable as Hitler's Atlantic Wall might be, it had sections that were less well protected than others.

"All the German fortifications on the French coast are very close to the shore and it is quite clear that the Germans plan to smash any enemy attempt to land as close to the water edge as possible," Oshima reported with satisfaction. He said that "individual machine gun nests are, without stint, strengthened with ferro-concrete"; that in the event of an Allied landing, "naturally it cannot necessarily be expected that they could be stopped everywhere along the line; but even if some men did succeed in getting ashore, it would not be easy for them to smash the counter-attack of the powerful German Reserves, who can rally with lightning speed." He commended the "morale and military spirit" of the Nazi soldiers, who treated their weapons with "love and affection and also confidence" and went about their work cheerfully. "Everywhere I engaged in easy-going chats with the soldiers, and their respect and affection for Chancellor HITLER—the depth of it—I actually observed many times."

He could not say enough about the soldiers:

The intimate unity—the complete unanimity of the German soldiers, high and low—the honest seriousness with which they apply themselves to their work is not only due to the general nature of the German people, but cannot but be regarded as also due to the Nazi education that they have received. This spirit permeates the very last soldier; I could see that and I breathed a sigh of relief and my heart was at rest.

Despite his infatuation with the Nazi troops and their training—or because of it—Oshima provided the kind of granular on-the-ground intelligence the Allies needed to defeat them. He noted that "the Straits area is given first place in the German Army's fortification scheme and troop dispositions." By this he meant the Strait of Dover, which is the narrowest part of the English Channel—the point where the crossing is shortest—connecting Dover in England to Calais, the French port. "Normandy and the Brittany peninsula come next in importance," his message continued. "The other parts are regarded as only secondary fronts." He then detailed where German troops were located, explaining that the army occupying the Netherlands stretched to the mouth of the Rhine; another army extended from there to west of Le Havre; and so forth. He provided a tally of divisions—infantry, armored, mechanized, airborne—and their strengths and numbers.

All of this was run through the embassy's Purple machine in Berlin, transmitted to Tokyo, plucked out of the air by WACs working at the Vint Hill intercept facility, deciphered by the Purple unit at Arlington Hall—mostly young civilian women sitting side by side at a table in Building A—and rendered into English by linguists in the translating division. If Genevieve Grotjan had tied it up with a ribbon, she could not have made a prettier present to Allied military commanders. The messages were supplemented by others sent through the Japanese naval attaché machine being read by Frank Raven and his team at the Naval Annex. As it happened, Japan's military attaché in Berlin, Katsuo Abe, was the polar opposite of Oshima; he hated and distrusted the Nazis and built his own spy network throughout France. So much the better for the Allies reading his dispatches. Raven and his crew called him Honest Abe, and what he learned about German coastal defenses also found its way to the D-Day planning operation. So did German Army Enigma messages read with the help of the bombe machines at the Naval Annex. At Bletchley, code breakers broke a long message from German field marshal Erwin Rommel, describing defenses along the Normandy beaches.

Together, the intelligence from these code-breaking efforts helped Allied commanders decide that they would concentrate their forces away from Calais, and make their D-Day landing in Normandy.

* * *

But for the attack to succeed, the Allies needed to ensure that the Germans were taken by surprise. The success of a cross-channel invasion depended on making sure that Allied soldiers did not encounter a full complement of German coastal defenders when they landed. The Allies also needed to be sure the Germans did not figure out what was happening in time to pull mobile reserve troops out of other areas and send them to help defend Normandy. One way to achieve surprise was by creating what Winston Churchill called a "bodyguard of lies" to protect the kernel of truth about the exact time and place of the landings.

And so, in the months running up to the invasion, the Allies created a brilliant deception program—the aptly named Operation Bodyguard—designed to sow confusion about when and where an Allied attack would happen. The goal was to persuade the Germans that Allied forces were bigger and more spread out than they actually were and that an invasion of Europe would occur in several places near simultaneously. They wanted the Germans to believe that the central attack, when it came, would come in the Pas de Calais, the region around Calais. To do this, the Allies created a "phantom army," a fictitious force to throw the Germans off the scent. They engaged double agents—spies for Germany in England who were turned by the British and made false radio reports back to Germany, spreading the word that the fake army was real and was massing for an attack. But for a fake army to seem truly convincing, it needed something else, something invisible and yet powerful: fake communications.

* * *

When Allied commanders were planning the multitude of logistics around D-Day, they had to think about many things, and communications was one of them. How would incoming GIs lay down phone lines and erect radio stations as they landed on the beaches and proceeded to race across France and into Belgium? This is the central mission of the Army Signal Corps.

The Allied commanders also had to assume that—even before the landing took place, as troops were assembling in England—Germans were busily doing to the Allies what the Allies were busy doing to them: monitoring radio traffic to construct the Allied order of battle and figure out who was moving where. The Germans controlled so much of the coast that they could monitor even low-frequency Allied traffic. There was no way the American, British, and Canadian units could communicate and be sure that the Germans did not pick up their transmissions. Even if the Germans didn't break their coded conversations, the enemy could learn a lot from traffic analysis. "So great were the chances of all the traffic that we transmitted above a certain power being received somewhere in occupied territory, that it had to be assumed that all of it *was* read," said a postwar Signal Corps memo. The Allies' solution? Fill the airwaves with fake traffic.

Actually, the Allies created two phantom armies. One existed to persuade the Nazis that an Allied force was massing in Scotland to mount an invasion of Norway. The goal of that deception, known as Operation Fortitude North, was to persuade the Germans to keep the troops that were stationed in Norway where they were, and not to divert them to France once the landing began.

The other phantom army was known as the First U.S. Army Group, or FUSAG, and it was supposedly led by George Patton, the American general whom Rommel most respected and feared. This was known as Operation Fortitude South. Patton's fictitious FUSAG was supposed to be massing in Kent and Sussex, to make a cross-channel attack on the Pas de Calais. Patton was, in fact, in England; the First Army, however, was not, nor was it anywhere.

It was crucial that the Germans not only believe in Patton's fictional First Army Group, but also continue to believe in it even after the D-Day landing. They must continue to think that the Normandy invasion was a diversionary attack to distract attention from the big one coming in the Pas de Calais. Believing that, the Germans would keep the bulk of their defensive forces in the Pas de Calais, giving the Allies time to establish a Normandy beachhead and begin the liberation march to Paris.

For a fictitious army to be believed, it had to send the exact sort of radio traffic that a real army would send. The radio traffic had to come into existence well before the attack was launched, and it had to stay in place for weeks after, even as the same transmitting stations were being used for real military communications. Creating and directing this traffic was a complex and highly important job; there could be no mistakes, nothing strange or untoward to attract notice. Much of it would be done at Arlington Hall by the women who had fallen for that inviting pamphlet and joined the WACs.

* * *

The ploughman homeward plods his weary way. The ploughman plods homeward his weary way. The ploughman plods his weary way homeward. The ploughman his weary way homeward plods. The ploughman his weary way plods homeward. His weary way the ploughman homeward plods. His weary way homeward the ploughman plods. His weary way the ploughman plods homeward. Homeward the ploughman plods his weary way. Homeward his weary way the ploughman plods.

Everywhere in the cryptographic unit of Arlington Hall, posters on the walls reminded staffers sending out coded messages to vary the order of the texts. "There's always another way to say it," exhorted one poster, demonstrating how lines from Thomas Gray's "Elegy Written in a Country Churchyard" could be rearranged. At Arlington, the staff of eight thousand did more than break enemy messages. They encoded American traffic and monitored that traffic to make sure it was secure. They were obsessively reminded to avoid the sorts of stereotypes and predictable repetitions that had given Americans an entering wedge into Japanese and German codes. "Parallel texts lost a battle," the posters pointed out, reminding the encoders that in World War I, a battle had been lost because a single message was sent both in cipher and in the clear. "Shifting position of words and substituting synonyms and using passive voice of verb" are all ways to vary the order of a sentence, the posters reminded them.

A whole section of Arlington Hall was devoted to "protective security"—and this section, like others, was staffed mostly by women, many of them WACs. The women operated the SIGABA machines, which were America's version of the Enigma. The SIGABA was initially conceived by William Friedman to encipher U.S. Army traffic, its design then improved by Frank Rowlett; the reason it was never as famous as the German Enigma was because unlike the German Enigma, it was never cracked. The simple reason it was never cracked was because Rowlett designed it so well (also it was heavier than Enigma, not quite so portable, and did not get overused as Enigma did). But it was also thanks to the care and competence of the people who used it. The WACs at Arlington Hall produced manuals instructing field soldiers how to use the machine; they maintained the Arlington Hall SIGABAs and tested them to make recommendations for operational and mechanical changes. They monitored their use to see if operators in the field were violating security. (The United States was rather notorious for poor radio security, from which the Germans learned a lot, so it was a ceaseless job.) They closely studied American message traffic, looking for "cryptographic error." They cryptanalyzed American traffic, to see how easy or hard it was to break. They looked for code-room errors and insecure practices.

Today all of this would be known as "communications security," and it went far beyond sending encrypted messages. One unit of women bird-dogged the American military units to make sure their radio traffic did not reveal too much about their whereabouts. The women intently studied the flow of U.S. military traffic to make sure that the Allies were not revealing the kinds of things that the enemy was revealing to them. They made charts and graphs to study American communications in specific regions, at specific times, during specific conflicts and events, to see what—if anything—might have been disclosed to the enemy. And they studied the characteristics of certain circuits.

These same skills enabled the WACs to create dummy traffic: fake radio traffic that so exactly resembled real American traffic that it persuaded the enemy that the fictitious units existed and were on the move. Such transmissions were useful for concealing military as well as political movements, and not just during the Normandy invasion. Exactly as Churchill described, the effect was to create a bodyguard, or protective area, around troops or leaders. In the Pacific theater, when a real attack was planned against Guam and Saipan, Arlington Hall created fake traffic to divert Japanese attention to Alaska. They created fake traffic to enable the deployment of the Fifth Infantry Division from the Iceland Base Command to the United Kingdom. They used fake traffic to disguise movements to and from the Yalta Conference.

The women analyzed Allied traffic, so as to be able to convincingly re-create a fake version of it. The calculations unit worked to determine "the various circuit characteristics, such as group-count frequency distribution, percentage in each precedence and security classification, filing time distributions, address combination, and cryptonets employed for each station of which the traffic is to be manipulated," as one report put it.

In order to create fake traffic, the women had to understand every last thing about the real traffic that went out, and the circuits and stations it traveled through. Once created, fake traffic had to be routed. The women had to create a plausible schedule; release prearranged dummy traffic to be transmitted during times when such traffic might be expected; maintain circuit flow; and monitor what went out. They had to understand the circuits, the call signs, the frequency, the peak volume times: everything.

Meanwhile, the women also were conducting analysis of German radio traffic. "They were really pushing things very hard at that time, all the way around," recalled Ann Brown, a WAC working in traffic analysis at Arlington during the Normandy invasion.

The bogus force included a dummy landing-craft tank, a fake headquarters, and two assault forces with associated ships and craft. The Allies began sending out dummy traffic months before the D-Day landing, as Patton's fictitious Army traveled around England and began to gather. Meanwhile, the double agents were hard at work communicating with Germany, ably encouraging the notion that FUSAG was poised to assault the Pas de Calais.

The Purple machine delivered the happy news that the deception plan had worked. On June 1, 1944, the tireless and always obliging Baron Oshima crafted a message to Tokyo and sent it over the Purple circuit. The encrypted message revealed that Hitler, anticipating an Allied invasion, expected that diversionary landings would take place in Norway and Denmark and on the French Mediterranean coast.

Oshima added—and this was crucial; this was exactly what the Allies had hoped—that the Führer expected the real Allied attack, when it came, to come sailing through the Strait of Dover, toward the Pas de Calais.

* * *

"No invasion tonight," thought Wellesley's Ann White as she walked along the peaceful streets of northwest Washington late in the evening of June 5, 1944. It was warm in the capital by this time of year, and rosebuds were nearing bloom in the manicured neighborhood where the code-breaking facility was located. It was nearing midnight as Ann, wearing her uniform, turned off Nebraska Avenue and entered the Naval

Communications Annex compound. She passed the now familiar double line of Marine guards, showed her badge, saluted the men as she went past, and made her way into the rooms that held the German Enigma section, where she relieved the officer heading the evening watch. Ann was a watch officer, now—a lieutenant—and supervised the midnight watch of women reading Enigma traffic. She had trained on the pistol range and knew how to shoot the gun that was always kept lying on the watch officer's desk. The room was so secure that buzzers controlled the entry, and people wore badges displaying the level of confidential material they could access.

The women working in the Enigma unit knew an invasion of France was in the offing, though those elsewhere in the Annex did not. The Enigma team had known this for several days and had been told to keep their mouths shut. The bombes were pounding full force to break the German traffic, and an air of suspense and tension hung heavy as the women waited, wondering when the landing would occur. It could happen anytime. Tonight, however, Ann had looked up and taken note of the gorgeous full moon as she walked into the unit, which was known as OP-20-GY-A-1.

Surely, she thought, the night would be too bright for a surprise landing across the English Channel.

But then, not quite two hours into her shift, her team received an intercepted message that suggested differently. "At 0130, messages on coastal circuits were translated which gave news to those U-boats of the invasion of France," noted the log for the unit. There was a flurry of rapt activity as the women ran the message through the M-9, to get the text. Ann, knowing German, was able to read the words even before the Enigma message was taken upstairs and translated into English. It was a terse message, sent by central command to all U-boats on the circuit. "Enemy landing at the mouth of the Seine." All up and down the coast of France, the Enigma machines were chattering, spitting out the same warning. "Enemy landing at the mouth of the Seine." And here in northwest Washington, the women read the words as quickly as the U-boat crews themselves.

D-Day was under way. After a weather-driven delay of more than a day, the night of June 6 had cleared enough to enable—barely—a nighttime crossing of the English Channel. The full moon gave the Allies the tide they wanted, and the storm that had been whipping up the channel abated for just long enough to launch ships that had been poised for days. During the night, nearly twenty-five thousand airborne troops dropped by parachute and glider into the fields above the beaches. The Normandy beach landings—the largest seaborne invasion in human history—were commencing. It was finally happening. The Allies were invading France.

Now the traffic really started coming. The next message described thousands of Allied ships—nearly seven thousand warships, minesweepers, landing craft, and support vessels—filling the English Channel off the coast of Normandy. The message enumerated the destroyers, the cruisers, the tankers, the supply ships. More bulletins followed. Far away—thousands of miles from where the women were frantically working, full of hope and dread and curiosity—the ships' hulking silhouettes loomed against the morning sky, bringing more than 160,000 American, British, and Canadian soldiers and the weapons and supplies they needed to assault the shores of occupied France.

And now the Americans were taking Utah Beach, the amphibious tanks making their way through the waves exactly as they had been designed to do, powered through the water by little propellers on their bottoms. The Army Rangers were scaling Pointe du Hoc under terrible fire; the frogmen were swimming in; the sappers were blowing paths that would enable the soldiers to leave the beaches and continue on toward the hedgerows and into northern France. The Royal Winnipeg Rifles and the Queen's Own Rifles of Canada were emerging onto Juno, taking very heavy casualties. The tanks at windswept Omaha were sinking, terribly, in the wind and the unforgiving swells, as Germans rained devastation from the cliffs above. Seasick, undaunted, the Allied soldiers kept coming, wading ashore, facing mortars, flamethrowers, machine guns, long-range artillery fire.

The women at the Naval Annex followed the D-Day message traffic all night, reading the invasion from the vantage point of the opposing Germans. The Allied soldiers were braving the beach, scaling the cliffs. The women read it all, into the morning and through the rest of that day. At 1:40 in the morning they were warned to "make no reference of any kind to the fact that we know about the invasion, even after the news comes out officially," according to the watch officer log. All night the women bolted up and down the stairs of the laboratory building, taking the messages to the translators one floor above—even though Ann could read them herself. They felt excitement, relief, horror. They well knew how momentous the event was. But how was it evolving? They could not be sure. The messages told them some things, but there was so much more they wanted to know. The women wondered how many men were dying; whether the Nazis were counterattacking; how the assault was unfolding.

The women worked as hard as they could. In the twelve-hour period between 0730 and 1930 on June 6, the crews scored eleven jackpots on the bombe machines, eight of those coming in one watch, as the Germans shared news about the invasion, beyond what was taking place on the beaches. The women learned that the French Resistance had acted swiftly to cut German communications. So much resistance, so many brave men working to defend the free world, had come together. "Even seated at our desks," Ann White would later write, "we felt the power of our country."

The Normandy landings came as a complete surprise to the Germans—a surprise that saved an estimated 16,500 Allied lives. The Allies over the next weeks were able to establish a true beachhead, linking Omaha, Utah, Juno, Gold, and Sword, and then to break out past the hedgerows and begin the liberation drive toward Paris.

At 0800 on the day of the invasion—afternoon now, in France—Ann White finished her all-night shift and walked out of the Naval Annex, feeling bleary and spiritually unsettled. There was a bus line that ran from nearby Ward Circle to a stop near National Cathedral, the neo-Gothic landmark on Wisconsin Avenue just north of Georgetown. Beside the cathedral was St. Alban's Church, which was smaller but exquisitely beautiful, and open around the clock. During the war St. Alban's had never once been empty. At least, it had never been empty while Ann had been there. She and some of the other women code breakers slipped inside and found places in the pews.

D-Day was a great achievement. They knew that. But somehow it did not seem cause for celebration, or not exactly, or not yet. Going to church was the only way they could think of to honor the tragedy and loss, which they sensed though they did not yet know the full extent: the Allied soldiers bobbing facedown in the water, drowned under their packs; the Rangers shot down as they dug handholds with their knives to scramble up the cliffs; the bodies on the beaches; the pilots who crashed in the fog and the smoke; the parachutists who drowned in marshes. It was the only way they could think of to honor the men who had made this sacrifice.

Ann would remember the Normandy invasion as one of the great moments of the war, and she would remember her wartime code-breaking service as the great moment of her life.

But for now, all she could do was pray for the souls of the dead.

* * *

Up and down, up and down. Set the wheels on the spindles, program the machine, sit down, wait. Then up again. Set the wheels on the spindles. Program the machine. Sit down, wait. Days after the Normandy invasion, the WAVES worked the bombe machines as the messages kept pouring in. "The invasion has caused a great increase in the amount of our traffic," read the daily log for OP-20-GY-A-1. "A great quantity of administrative traffic, representing several circuits, has appeared." Jimmie Lee Hutchison Powers worked her bombe bay all during the Normandy landing and its aftermath, and so did her hometown friend

Bea Hughart. The two former Oklahoma switchboard operators jumped up and down, changing wheels, inputting menus, day after day, as the Allied soldiers began to fight their way toward Paris. These women's experience of the glorious liberation of France was the heat and noise and urgency of the bombes—and the dread knowledge that the men they loved were taking part in the action overseas.

Jimmie Lee's husband, Bob Powers, had piloted one of the gliders supporting the Normandy landing. Bob had gone in on the first wave of aircraft, taking off from England. Gliders were engineless aircraft towed by planes, then released over fields and forests, sailing through the skies carrying troops as well as weapons and even vehicles such as jeeps, which would be waiting for the paratroopers and the men coming up from the beaches. The gliders were known as "flying coffins" because of their flimsy structure and the danger of the work. The pilots, going in, knew the odds.

By now, every American knew that there was a particular look to a telegram that arrived announcing a military death. It would come on a Sunday morning. Telegrams had little clear window boxes bearing the address, and if the soldier in question was dead, there would be blue stars around the address. A few days after the invasion, Jimmie Lee got something that was not quite that. Her high school sweetheart and husband of one year, Bob Powers, had been downed over Sainte-Mère-Église, a French town near one of the glider landing areas. The fog and smoke had been terrible, as had the antiaircraft fire. Her telegram said he was missing in action. There was a period of terrible uncertainty and then in September, she got the worst news: Her young husband was indeed dead. Bea Hughart's fiancé had been killed at D-Day as well. The two women had joined the Navy to try to save the lives of American men, especially the ones they knew and loved. Even while succeeding at the larger mission, they had failed at the intimate and personal one.

It was only now that Jimmie Lee understood the import of the work she was doing. When she asked for leave to attend her husband's funeral in Oklahoma, her request was denied. There were other bombe operators getting the same telegrams, and they could not all be allowed to leave. Jimmie Lee stayed at her post. Her father died not long after. She was never able to go home and unburden herself, never able to talk to her father about how much she missed her husband. Nor had she been able to tell her own father good-bye.

There was so much loss even amid the victories. Ten months after D-Day, in April 1945, President Franklin Roosevelt died. The women cried like babies. There was a funeral procession when the president's body was brought to Washington from Warm Springs, Georgia, and much public mourning, and the WAVES marched in a parade to honor him. People wondered whether the new president, Harry Truman, would be up to the job. Some of the women made individual trips to the White House, standing at night in the eerie quiet of Lafayette Square, where the sound of water dripping from the trees was the only noise, along with the changing of the guard.

* * *

In many ways the waning days of the war were the bloodiest and worst. In both war theaters, Axis leaders resolved to make any Allied victory as costly as possible in terms of lives lost. Japan hoped that if it could create sufficiently terrible casualties among the sailors, Marines, airmen, and soldiers launching attacks on the occupied islands, the United States might seek an early termination and Japan might gain a negotiated peace. The kamikaze air attacks began in the Pacific, and attacks by suicide boats as well. "There were many signs the enemy was disintegrating," recalled Elizabeth Bigelow, who was breaking Japanese codes in the Naval Annex, working in a unit called "keys." Even as they could see the Allies were winning, the women lived in fear of messages bearing the worst possible news.

The code breakers did what they could to monitor the well-being of loved ones, and sometimes succeeded. At the Naval Annex, Georgia O'Connor was working in the library unit with her friends. Thanks to the U.S. Navy messages coming in over the ECM—the machine that transmitted internal American messages—she was able to follow the USS *Marcus Island*, the escort carrier on which her brother was serving. She tracked the carrier through all of the war's final Pacific battles: the invasion of the Philippines, the Battle of Okinawa, the Battle of Leyte Gulf. The *Marcus Island* suffered kamikazes and near misses as it passed through the Surigao Strait and Lingayen Gulf and along the Luzon coast. Georgia's brother was in the radio room of the ship. She did not communicate with him personally, but she was able to tell her family he was safe, though she could not tell them how she knew this. "We always knew what was going on in the South Pacific," she said later.

Others were not so lucky. Elizabeth Bigelow, recruited out of Vassar, had two brothers serving in the Pacific. One was in the Marines. Her other brother, Jack, a Princeton graduate, was serving on the *Suwannee*, an escort carrier that belonged to the carrier group known as Taffy 1. Jack was the oldest son in the family, a golden boy whom everybody, Elizabeth most especially, adored. He was small but well-built and had been a gymnast at Princeton. Early photos of him in a Navy uniform showed him to be "a young man with a hat that looked too big for his slight form," she later remembered. Quiet and good-natured, he had started at Princeton in 1938 and studied electrical engineering. After Pearl Harbor, he enlisted in the naval reserve, and that Christmas he gathered with friends who were also newly enlisted, and Elizabeth was struck by "the excitement and hope in their voices." In 1942 he was inducted into the Navy, where he became a radar officer. The *Suwannee* saw action in the Gilbert Islands, Kwajalein, and Palau, and by early 1944 a shipmate had sent her a photo that showed "Jack totally exhausted and looking decades older."

In late October 1944, the campaign to retake the Philippines began. The Taffy 1 group participated in the Battle of Leyte Gulf, the decisive battle to win back the Philippines. Leyte Gulf was the largest naval battle of the war, possibly the largest naval battle in history. The battle itself involved successful code breaking but at one point it also was nearly disastrous for the United States. It was the first battle that saw organized attacks by kamikaze airmen. During the days-long engagement, Admiral William Halsey and his Third Fleet were lured out to sea; the Japanese attacked the Seventh Fleet, which the *Suwannee* was part of, and the escort carriers found themselves the first line of defense. A kamikaze tore a hole in the flight deck of the *Suwannee*. The plane's bomb pierced the deck and exploded between the flight and hangar decks, setting off a terrible fire fed by aviation fuel dripping down from burning planes on the deck above. Many men in that part of the ship were incinerated. They jumped overboard if they could.

Elizabeth's family got the awful telegram. She was given leave to go home for two days. Her father's hair turned gray, it seemed to her, overnight. None of them ever recovered from the loss. The same shipmate who sent the photo told her that Jack's body was unharmed, but later she was able to see his Navy file and learned the full truth. Her big brother Jack had been one of the men who burned to death.

Some of the women broke messages warning about attacks before they happened but were helpless to avert them. Goucher's Fran Steen—a lieutenant now—was working her shift as watch officer when a message came in saying that the destroyer captained by her brother, Egil, now serving in the Pacific, was targeted for a kamikaze raid. Her team alerted the Navy, but there was no way to prevent the attack. Fran kept working, knowing that the only thing she could do was her job. The kamikaze struck and her brother's ship was sunk. At the time she thought Egil Steen was dead. It later emerged that he was one of only a few to survive, thanks to the spot in the ward room where he had been standing.

Donna Doe Southall was one of two hundred WAVES officers staffing the code room where the Annex received U.S. Navy dispatches with news about ship sinkings. Though she was responsible for Pacific messages, she was looking through the Atlantic ship-sinking traffic when she saw one saying that her brother's ship had been sunk. She didn't know it at the time, but a third of the crew survived, and the British destroyer *Zanzibar* picked her brother out of the ocean. He was taken to England, where he was treated for pneumonia and given clothing by the Red Cross.

For years, Donna's mother sent packages to the woman who donated those clothes. One of the packages contained a blue dress Donna had worn as a bridesmaid, which any number of English girls got married in. But her brother was never the same. When he came home he developed schizophrenia, was made a ward of the Veterans Administration, and died suddenly at age fifty-nine. For her part, Donna married a naval officer who was on a ship off the coast of Okinawa that was hit by a kamikaze. He was blown out of his shoes, put in a pile with the dead, regained consciousness before he was disposed of, and lived to have a family with her.

CHAPTER FOURTEEN

Teedy

December 1944

Teedy Braden was five years younger than Dot. The Braden siblings were close, and Teedy and Dot both had an excellent sense of humor. They loved to needle each other. When the siblings were growing up, Teedy and Dot's other brother, Bubba, liked to size up Dot's boyfriends. The two brothers would hang around the front yard of 511 Federal Street, passing judgment on whoever was going in and out visiting their older sister. The brothers also liked to clamber onto a crowded streetcar, at the other end from where Dot was standing, and loudly say things like, "Who would ever want to go out with that little girl with the permanent wave in her hair?" Dot likewise enjoyed teasing her brothers about their romantic lives. One summer a photo was taken of their family during a swimming outing, and Teedy was the only person looking off to the side while everybody else looked at the camera.

"What do you think is holding Teedy's gaze so intently?" Dot wrote teasingly on the back of the photo. "Could it be the lady lifeguard?"

Teedy Braden finished high school on a Friday in June 1943 and by Monday he was in the U.S. Army. He started basic training at Camp Fannin in Texas. Within a year Eisenhower needed men to reinforce the troops who landed at Normandy and were now fighting their way through the fields and forests of France and Belgium. So the Army sent

Teedy to Camp Breckinridge, Kentucky, for training. At first it wasn't clear what they were going to do with him. In July 1944, a month after D-Day, Teedy was still at Camp Breckinridge and was able to get a two-week furlough to go home to Lynchburg, leaving July 3, 1944, and returning to camp on July 16.

"I sure do hope that you won't [be] too busy to run down as I sho would like to see you," he wrote to Dot, but Dot was unable to get time off. He wrote her again after he got back to training camp. "How's everything, gal?...I sure would've like to have come up and stayed with you for a while but I reckon it would've upset my schedule a little." He said it was tough readjusting after leave. "Man, man, it sure is hard to get back into harness after being out of it a little while. By gum, they had all sorts of details waiting for me. Everything from KP to latrine orderly. That latrine orderly is a pip!...whew."

Back at camp, he sensed something was up. "There's a lot of rumors going around since we're having 'shakedown' inspections and equipment check-ups."

On July 31 Teedy wrote Dot to say that he might be coming to Fort Meade, Maryland, in about two weeks. "If I do go it'll mean that it's the first step toward taking a boat ride which we've all been expecting soon." "Boat ride" was a euphemism for sailing on a troop ship across the Atlantic and into the heavy European fighting. "We'll all go as riflemen," he told her. His unit was practicing nighttime river crossings. He had sent a photo of Dot to "Gus and Johnny," some mutual friends, but had not heard from either. "There's a chance that they've gone on a boat ride as they were expecting it."

Teedy did make the trip to Fort Meade. He would soon be part of the 112th Infantry Regiment of the Twenty-Eighth Infantry Division, a Pennsylvania unit nicknamed the "Bloody Bucket" because of its bucket-shaped insignia and its eventful battle history. He was one of thousands of very young men shipped over to replace the men lost in the fighting during and after Normandy. These new men were at a disadvantage in every way: Not only were they hastily trained and not hardened to battle, but some veterans resented them for replacing fallen comrades and shunned them as green and inexperienced. They tended to "become casualties very fast," as one officer put it. This was the situation Teedy Braden was thrust into. Dot's mother came up before he left, and she and Dot both went to see him off. With the Atlantic Ocean swept clear of U-boats, Teedy Braden made the ten-day Atlantic crossing to England, not yet twenty years old, sleeping in the bottom of the ship in a hammock slung from pipes, eating a diet abundant in canned apricots.

When he alighted, Teedy entered some of the worst fighting American troops would endure in the European war theater. The Western Allies were pursuing the German Army, but the Nazi soldiers were putting up fierce resistance, and Allied units and supply lines were getting strung out in the rush to push through France. Hitler was seeking one last, smashing win, a victory that would thwart the Allied offensive and deplete their resources. At the beginning of November Teedy's unit found itself caught in the Battle of Hürtgen Forest, a terrible engagement along the border between Belgium and Germany, where Nazi troops laid mines and booby traps, strung barbed wire, and built bunkers amid the trees. Hürtgen was a dense and dark pine forest with steep slopes pillowed by deep ravines. Teedy's unit, the 112th Infantry, suffered extremely high casualties; at one point they were down to three hundred men from more than two thousand. Even the Germans later said that the fighting in Hürtgen Forest was worse than that of the First World War; one officer called it a "death mill."

And it was just a warm-up. Barely two weeks later, the Germans attacked in what became known as the Battle of the Bulge. It was Hitler's last big roll of the dice, the biggest, bloodiest battle the United States fought in Europe. And it was one of the war's worst intelligence failures. Allied code breakers had noticed a radio silence suggesting the Germans were planning an attack, but the military did not pay sufficient attention and the soldiers were taken by surprise.

The Americans by now had become strung out in a ragged line, and the Germans decided to try to push through with the hope of breaking out and reaching Antwerp. The exhausted Twenty-Eighth, sent for a rest in the southern Ardennes—a quiet paradise resembling Switzerland— was surprised at a time when it was short of both men and weapons. What was left of Teedy's unit sustained enormous casualties in several days of pitched fighting as the Germans attempted to break through Allied lines. Shattered, the Twenty-Eighth fought and fought. As Dot heard it later, her mother, Virginia, was visiting with friends in Lynchburg when she received the terrible message saying Teedy was missing in action. People in Lynchburg visited her to express their condolences.

Anguished and distraught in her grief, Virginia Braden did not tell Dot, who remained blissfully unaware. At Arlington Hall, the code breakers worked hard throughout the battle, even though the cafeteria had served tainted ham and there was an epidemic of vomiting and sickness. Over at the Navy facility the bombes whirred away, and the women at Sugar Camp also knew the Battle of the Bulge was unfolding. "I used to feel guilty because of enjoying the war years in such a beautiful, comfortable station while the slaughter was going on in Europe and the high seas," Lieutenant Esther Hottenstein would later write. "I remember especially the winter of 1944 (December) the Battle of the Bulge where we worked overtime."

When the fog of war cleared, however, it emerged that Teedy Braden had survived. Speaking years later, from the vantage point of decades, Teedy was able to relate how he managed to do that. He recalled how the battle lines were jagged and shifting, and at one point he and some other GIs found themselves behind enemy lines. "I was on an armored car, holding on to it for dear life," Teedy said. "As we come out the other side I see this German come out of the ditch carrying a Panzerfaust," which was a handheld rocket launcher. The German pumped it into the side of the armored car; Teedy was blown across the road and into a tree, where he was knocked out. When he came to, he saw burning tanks, burning ambulances, green flares as German tanks fired into the sides of

Allied vehicles. Nazi soldiers were running up and down shooting American men. Somehow, his rural instincts saved him. Unarmed, he managed to nip around a tree and spotted some fellow GIs moving cautiously through the forest. He joined them and they made their way through the woods, hitting the ground whenever the Germans opened fire, then standing up and scampering forward.

"Then all of a sudden a .50-caliber machine gun in front of us opened up and we knew we had hit the Eighty-Second Airborne," Teedy Braden remembered. They had made it to safety. The Eighty-Second Airborne told the exhausted Americans to head down the road until they saw a palace. Teedy started off in the direction they indicated, but he must have been concussed, and he passed out again. An American tank scooped him out of the road and carried him to a château packed with weary men. He found a spot to sleep on the floor of a tiled bathroom, wedged between a toilet and a wall. In the morning he got coffee and stood on the once-manicured château lawn, watching American bombers flying into Germany. It took a long time for survivors to get sorted and reequipped, having straggled in from many devastated units.

Teedy did not attempt to convey any of this to the people back home. "I suppose that you've been kinda worried since I haven't had a chance to write you for some time," Teedy wrote Dot in January 1945. He apologized that the letter was written in red ink and said it was the only pen he'd been able to find. "I've only been able to write mom a couple of times," he told Dot. In understated fashion, he explained that Christmas, for him, was a "kinda hectic one."

Being Teedy, he was still able to joke. He had taken French in high school because his big sister Dot had, and it was proving useful in Europe. He'd been able to eat a meal in a fine Belgian restaurant. "I can now snap my fingers and yell 'garcon' with the best of them," he told her. He reported that the Belgian women were "quite an eyeful but of course I was too interested in the old architecture and the city's history to pay much attention."

"Well, Dot, I just wanted to let you know that I am still percolating," he finished. He enclosed five Belgian francs as a souvenir and told her it was worth about twelve cents. "It sure is fancy money for .12, isn't it?" he said. "Well, I hope that all of you have a pretty fruity list of New Year's resolutions now. So long! Love, Teedy."

When he arrived safely home from Europe, Dot called their mother, and Virginia Braden got on a bus and raced up to Washington so that she could see for herself that Teedy was alive.

* * *

By the spring of 1945, with tens of thousands of American men sacrificed to the final war effort—the Battle of the Bulge alone cost nearly twenty thousand American lives—the Allies had managed to regain momentum. They fought off the German counterpush and crossed the Rhine and into Germany, which was being subjected to a heavy bombing campaign that destroyed factories and munitions and, infamously, the city of Dresden. On the eastern front, Russian soldiers routed the German invaders and pushed toward Berlin, taking massive casualties. As the Russians drew nearer, Adolf Hitler committed suicide in his bunker on April 30. It was now a matter of time. In Italy—one of the toughest, longest campaigns for the Western Allies—German soldiers who might have reinforced their comrades were pinned down and the fascists overthrown. Benito Mussolini and his mistress, Clara Petacci, were killed on April 28 and then strung up by partisans. On May 7, 1945, Germany surrendered. The Third Reich was no more.

The Allies had won the Battle of the Atlantic—and the European war. Admiral Dönitz—the new head of state in Germany—ordered his U-boats to stand down. The Enigma unit at the Naval Annex read a message from Dönitz to his surviving captains, which told them, "You have fought like lions...unbroken and unashamed you are laying down your arms after a heroic battle." As GIs liberated concentration camps, the world would learn the full horror that had unfolded in Dachau and Buchenwald, a permanent stain on human history. Many of the women, and their families, also would never recover from the losses of sons and brothers and loved ones. But the boys in Europe—those who were left—were coming home. In Washington, conga lines pranced along the streets. Some of the WAVES went to the roofs of hotels to watch the celebrations. One WAVES member played a celebratory game of Ping-Pong at an officers' club and later married her opponent. A number of code breakers would recall the magical experience of watching the nighttime lights, long dimmed for the war, come back on in the nation's capital. Dot, Crow, and Louise absorbed the happy news at the apartment on Walter Reed Drive, though their workload did not abate.

To the contrary. The code breakers in both D.C. facilities—the Naval Annex and Arlington Hall—were reminded that the Pacific War was ongoing and the need for secrecy was as great as ever. "I have been informed and understand that the termination of the war in Europe does not affect the necessity for continuing to maintain secrecy concerning the classified activities and operation of the Signal Security Agency," read a form that Dot signed. (The Signal Intelligence Service had been renamed the Signal Security Agency, with several iterations in between.) "And that existing security standards must be maintained for the remainder of the war and after the war is terminated."

* * *

One month later, in June 1945, Virginia Braden wrote a fond letter to Teedy at his Army camp in Mississippi. "My Dear Teedy," she wrote, "Hope everything is going all right with you Son. We got back from Va Beach Thursday night, we went Sat and I had a very nice time. You should have seen me riding a float in on a wave believe it or not. I got a real good tan and enjoyed just relaxing on the sand."

She also let Teedy know what was going on with his big sister. Dot had written George Rush to tell him that she was not going to marry him, and George Rush had written her back to say that she was the only woman he would ever love. But now George had shown up in Lynchburg with a different Dorothy altogether, and this different Dorothy was his new wife!

> George Rush has at last gotten married and is here with his bride. I haven't seen her yet, but her name is Dorothy. He wrote to Dot around Xmas that she was the only girl he would ever love but she didn't answer it. How fickle.

The real news, she added, was that Dot had told George Rush she had decided to marry Jim Bruce. "Ha! I don't think she is really settled on anyone," Virginia Braden wrote. "More power to her!"

CHAPTER FIFTEEN

The Surrender Message

August 1945

At Yalta on February 11, 1945, and again at Potsdam in July and August, Allied leaders insisted on unconditional surrender in which the Japanese would concede defeat and the emperor would step down. The Japanese government refused. On August 6, the *Enola Gay*, a B-29 Superfortress, passed over the Japanese city of Hiroshima and dropped the first atomic bomb in history to be used in battle.

Not long after, Alethea Chamberlain came to her station, sat down, and put on her headphones. She was a WAC intercept operator at Two Rock Ranch, a listening station the Signal Corps maintained near Petaluma, California, in a beautiful agricultural area north of San Francisco. It was a nice posting; the intercept operators could hitchhike into San Francisco.

Chamberlain began fiddling with her dial, trying to pick up the Hiroshima station she received. Hiroshima sent out a very good signal. Now all she got was dead air. There was nothing at all. She could not figure out why this was or what had gone wrong, why there was no signal at all coming from Hiroshima.

On August 9, another atomic bomb was dropped, this one over Nagasaki. The Japanese said they would consider surrender but insisted that the emperor be allowed to remain. Many women code breakers had brothers serving in Pacific units preparing to invade the Japanese home-land, an attack that, if it happened, would take place in November 1945. Ruth "Crow" Weston's youngest brother was among them. Such an assault would entail the invasion of the southern island of Kyushu and could cost as many as a million American lives. During the first half of August, kamikaze attacks continued against American warships and aircraft. At Arlington Hall, Japanese Army message traffic told of the number of soldiers waiting to repel an invasion. The diplomatic traffic, however, was saying something slightly different.

* * *

The minute Ann Caracristi set foot in Arlington Hall she knew something important was up. It was around two o'clock in the afternoon on August 14, 1945, and Ann, the inventor's daughter from Bronxville, was arriving for her shift. There was a palpable euphoria boiling down the halls of the building, moving ineluctably in her direction. And there was no question where the tsunami of excitement was coming from: the language unit.

The language unit at Arlington was an exalted group of individuals who knew Japanese and could translate messages into English. This was an unusual skill for an American in the 1940s to have. Some translators were young officers—the j-boys—who had been sent by the Army to language training in Berkeley, California, and Boulder, Colorado. Others were scholars such as Edwin Reischauer. Others were missionaries who had lived in Japan. Many in the last two categories had learned the Japanese language out of interest in, and love for, the land and culture. Most felt an emotional attachment to the country and knew people who lived and worked there. They were now working to defeat the nation they once had proselytized and studied.

The translators were crucial to Arlington Hall. There were never quite enough. They helped with book-breaking—recovering the meanings of code groups—and Ann Caracristi worked closely with them.

Their other main job was to translate diplomatic messages from Romaji into English. The diplomatic messages were subtle and complex—they were diplomatic, after all—and getting nuances right was all-important. The language unit read every diplomatic message that came into Arlington Hall, and so far there had been half a million. They served as the linchpin between the code-breaking staff and the military intelligence unit that put together reports for the Pentagon and the Navy. The language unit had their finger on the pulse of the war, privy to the most high-level communications.

For the past six months those communications had been intense and wretched, as Japanese diplomats living in Europe reacted viscerally to what was happening in their homeland. The translators followed their wretchedness and even grew to feel attached to some of them.

The diplomatic ciphers coming into Arlington Hall included not only Purple messages, but those in other systems as well. Not every diplomatic missive was enciphered by the high-level Purple system. Other systems carried traffic about heavy industry, financial dealings, espionage, air raids, and commodities. There was one called JBB, used in the occupied islands and territories, and another called JAH, which was employed around the world and carried the largest volume of diplomatic traffic. JAH was an all-purpose code with origins that went back decades. The Japanese had used a version of it in Herbert Yardley's time, when the Americans called it LA, and it had been one of the first codes Friedman's acolytes tackled during their training, thanks to the fact that Yardley had stashed some old intercepts in his filing cabinets. The Japanese had updated it and continued to use it.

The odd thing about JAH was that, despite being a general utility-type workhorse—carrying news about pay and vacation—it also carried what one report called "a wealth of first-rate material." The report noted with bemusement that "in spite of their security consciousness, the Japanese frequently transmit messages of real significance in JAH while at the same time including extremely inconsequential information in their most highly classified systems." JAH "theoretically was restricted to low grade traffic of an informative and administrative nature," but some messages gave U.S. intelligence officers insight into problems and personalities. JAH also contained "documentary material"—papers, speeches, orders, memos, publications—that could be used as cribs. It also gave economic and political data on occupied territories and was used to disseminate propaganda.

While many of Arlington Hall's language units were headed up by j-boys, JAH was handled by a woman named Virginia Dare Aderholdt. According to a memo, Aderholdt graduated from Bethany College in West Virginia—Wilma Berryman's alma mater—which was a four-year college founded by the Christian Church (Disciples of Christ), and offered a first-rate language department and a commitment to good causes. Many graduates did missionary work abroad. Virginia Aderholdt had spent four years in Japan, and the JAH code now was her baby. She owned that code. She knew it backward and forward. She could scan and decode and translate almost simultaneously as it was coming through the machine built to receive it.

The Arlington Hall translators had experienced the war from a unique vantage point. They followed, intently, the conversations of Japanese officials. They followed what was taking place in captured territories whose Japanese-run governments were starting to weaken. Beginning in January 1945, the puppet government in the Philippine capital of Manila was having a hard time of it. When the titular head fled, the translators followed his flight into Taiwan by reading JBB. Later that month translators found themselves reading dispatches sent by the foreign office in Tokyo, reporting on the raids over the city by U.S. military aircraft. They knew when Japan began evacuating Japanese nationals from southern China.

And they were able to track the vigorous efforts of Naotake Sato, the Japanese ambassador to Moscow. Sato was assigned to entreat the Soviets to broker a peace deal on Japan's behalf. In the spring and summer of 1945, Sato was incessantly busy. In April, it emerged that Russia was no longer willing to observe a neutrality pact with Japan and was massing troops to attack. Sato was doing everything he could to manage the situation and turn it to Japan's advantage: begging for an audience with the Soviet foreign minister, sending messages, running back and forth. Arlington Hall translators followed his travels and travails. Moscow traffic became frenzied; the translators worked as hard as Sato did. When his efforts failed and Sato offered his resignation, feeling he had not lived up to his obligations to the emperor, the translators felt oddly moved on his behalf.

Sato's resignation message was "received with great consternation by the Diplomatic translators here who had begun to really love that man," read a postwar history. Tokyo responded by asking him to stay in his post as long as he could.

The translators also followed the movements of Ambassador Oshima, that stalwart Nazi sympathizer, as he left Berlin for Bad Gastein and was captured in May. And they could sense the endgame approaching. "In May began the bombardment, by Sato in Moscow, Kase in Bern, and Okamoto in Stockholm, of the Japanese foreign office: message after message contained the advice that Japan had best think about getting out of the struggle. Air raids on Japan became intense, reducing the diplomats abroad to misery when they thought of their homeland," the history related.

So destructive were the air raids that the Tokyo foreign office lost power on May 24 and could not use its Angooki Taipu B, though it soon was back up and running. By midsummer events began to move very fast. Sato was having global conversations about ending the conflict; in Bern, Switzerland, a Japanese banking official was engaged in undercover conversations with Allen Dulles of the OSS. "Traffic came into the Diplomatic section and went out of it on greased skids," the history noted. When one message revealed Japanese efforts to cut a deal with the Soviets, the message was flown to President Harry Truman at Potsdam. The diplomatic messages helped the United States monitor side conversations the Soviets were having with the Japanese. When Truman was informed about one of these by Churchill, he already knew about it, thanks to the quick work of Arlington Hall.

In early August, Arlington Hall translators started seeing traffic suggesting that the Japanese—who could not communicate with the United States directly, because lines were literally cut and there was no real way to do so—were planning to send a message via the neutral Swiss, who often acted as go-betweens, announcing their intent to surrender. The translators—and U.S. military intelligence—awaited the all-important message. It would be sent from Tokyo to the Japanese ambassador in Bern, whose job it was to take it to the Swiss foreign office. The U.S. military set up a special intercept net to snatch it. They knew from prior messages that it was going to arrive not in Purple, but in lowly, overworked, undervalued JAH.

On August 14, the whole translating unit was on pins and needles. People were afraid to go to lunch. Finally a garbled message came through in JAH, announcing the impending arrival of more messages. All these premessage messages were driving the translators crazy. "Shortly thereafter, the two texts arrived—the Japanese text first, the English text immediately thereafter," the history recounts. "Excitement mounted as the messages were being decoded, shot sky-high when it became clear that this was it."

Virginia Aderholdt worked as fast as lightning to get that surrender message decoded. She "had worked on that code and loved to work on it and she had memorized the code and we put her at a table right next to the teletype and when the message came in this young lady looked at it and wrote down the plaintext just as almost in the real time it was being typed out," Frank Rowlett later remembered, with a kind of awe. (He does not name her, but does describe her as being from West Virginia, and it's hard to see who else it could have been.) They telephoned the translation to military intelligence, where a stenographer rapidly typed it up.

According to Rowlett, who was a bit of a raconteur, he had to save Virginia Aderholdt from being trampled by the translators wanting to crowd around the message and behold it. "Well now every God damn translator in B4 began hunkering and hovering over that little girl who was doing the best she could to decode this message and I walked down...when I saw this gang coming down there I just put on my Colonel's bars or whatever you wear when you're a Colonel and I told them to get the hell out of there, they were going to break the floor through and leave her alone."

The message had to pass through two transmitting stations to get to Switzerland. The Americans snatched it from the first and worked so fast that Arlington Hall had it decoded before the Japanese received it on the other end. Word got to the president as soon as they could get him a clean copy. Arlington Hall was also reading the neutral Swiss, so they were able to double-check for garbles when the Swiss sent their own message through.

At Arlington Hall, the rule was that translators must keep the contents of all messages to themselves, and up to then, they had. But this time, staying quiet was not humanly possible, not with a message of this magnitude. The Second World War was over. Or imminently would be. The hubbub was what Ann Caracristi sensed when she stepped into the building to start her shift.

* * *

At first, Frank Rowlett, Solomon Kullback, and a few others went around telling people to put a lid on it. Soon, though, the news had spread, and everyone in Arlington Hall was gathered and asked to raise their right hands and take a vow of silence. Dot Braden was among these. She was working her shift, sitting at the big wooden table. Like everybody else, Dot eagerly left her workstation, assembled with the others, and raised her right hand. The code breakers were told that Japan had surrendered and that they were not to divulge this news until the president announced it later that day. Dot felt excited and glad, but not surprised. The gravity of the knowledge scared her—World War II was over and she was among the few in the world who knew. The giddy truth surged inside the place, bubbling to come out. But they kept it in.

At seven p.m., President Truman announced the Japanese surrender to a weary but euphoric nation. Dot Braden, Ann Caracristi, Virginia Aderholdt, and the rest of the code breakers poured out of Arlington Hall.

So did the women working at the Naval Annex. "The city exploded," recalled Elizabeth Bigelow. Lyn Ramsdell, one of the friends from the Navy's library unit, was sitting in a movie theater when a bulletin flashed across the screen. "Everybody just got up and left the movie, they were so excited, and the streets were just mobbed," she remembered. Outside, traffic was terrible: Cars were gridlocked, the buses were all full, people were shouting and dancing and singing. Trying to make her way back to the Naval Annex, Lyn Ramsdell ended up riding on the top of somebody's car. Other people perched atop trolleys, while above them, hotel residents flung toilet tissue out the windows. Groups ran arm in arm singing "Happy Days Are Here Again!" A crowd tried to break into the White House grounds, crying, "Give us Harry!"

From Arlington Hall, thousands of people crossed the river from Virginia into Washington. One of the Arlington code breakers, Jeuel Bannister, met one of the j-boys as they were all linking their arms and singing. She had never seen him before but sensed—correctly—that she had met her future husband. After that, she always referred to V-J Day as "Victory for Jeuel" Day. The next day, August 15, Truman declared a two-day holiday to celebrate the surrender of Japan. They had done it. The Allies had won. The world war was over.

* * *

The Japanese messages dried up. Delia Taylor Sinkov, the Sweet Briar graduate who had risen to direct the research unit for all Japanese Army codes, would later tell her son that there was nothing to do, now, at Arlington Hall, except sit around working crossword puzzles. On August 18, 1945, Brigadier General Preston Corderman gathered Arlington Hall employees into a grassy clearing. It was his intention to "speak to all Arlington Hall personnel regarding the past activities of the Signal Security Agency and the conversion of the Signal Security Agency to a peacetime status," as a postwar memo put it. Ann Caracristi and her friends called it his "Here's your hat, what's your hurry" speech.

The gist was: Thanks very much, everybody. Time to go. Job well done. The Arlington Hall code breakers were thanked for their service and told that it was their patriotic duty to get off the government payroll.

This seemed fair to Ann Caracristi. She loved the work, but she could see that the government no longer needed her service. She returned home to Bronxville, where a family friend helped her get a job in the subscription office of the *New York Daily News*. It was her job to sift through data and pinpoint who the paper's subscribers were. Ann's bosses wanted her to prove that their subscribers were more highbrow than most people thought. It was not bad work, but it was not as much fun as breaking codes. So she was delighted when she got a call from her good friends Wilma Berryman and Gertrude Kirtland, both still at Arlington Hall. Gert was high in the personnel office and had clout. So did Wilma. Together they formed something of an old-girls' network, and they wanted Ann Caracristi back at Arlington and were prepared to make it happen. Ann said yes in a hot minute, and before she knew it she was headed back to Washington, where she would live for the rest of her life. It turned out that Arlington Hall was not being shuttered.

In fact, it was just getting started.

CHAPTER SIXTEEN

Good-Bye to Crow

December 1945

Jim Bruce returned from overseas in September 1945. After nearly two years of a purely epistolary relationship—during which they had agreed to marry—the couple's feelings were confirmed when they saw each other in person. "We've got to get a ring," Jim told Dot. She had just about had her fill of rings, but even so, she "knew Jim was the one." Crow had started crying when she heard Jim Bruce was coming back. It wasn't that she was jealous, because Crow wasn't that kind of friend. It was just that she was going to miss their easy and companionable friendship, their excursions, their breakfasts, their shared jokes.

In December, Dot's mother wrote her a newsy letter talking about Christmas preparations. "I went down town yesterday to do some shopping but things are so picked over here, even now, that we didn't do much but go from one store to another and wear ourselves out." She mentioned the upcoming wedding—ceremonies remained small and often casual affairs—saying the seamstress had a cold and that parts of Dot's outfit were delayed. All would be ready in time for the wedding, however, and Virginia Braden had gotten the buttons for Dot's negligee. She joked about a woman they knew who was getting married in her sixties "and the old man is seventysome." She traded a bit of gossip, saying that "he gave her a $250 diamond ring which she selected with her daughter's help. (I still have hopes! Ha!)"

She told Dot how to cook a chicken, something she would need to know. "All you have to do to cook a chicken is—after boiling it until tender put it in a pan with a little of the broth sprinkle it with flour put in oven, and every now and then baste it with the broth, to help it from being so dry until it is brown."

Dot Braden and Jim Bruce were married on December 29, 1945, a few days after Christmas. It was a small wedding at Court Street Methodist Church in Lynchburg. The bride wore a gray silk suit and a daring fuchsia feathered hat that she—citified now—felt was the height of fashion. "I thought I was Miss Style," Dot would later say with a laugh. She and Jim took a short honeymoon in the North Carolina mountains, but from there Jim had to travel back to Oklahoma, where he would be mustered out. Leaving the Army took a while. Time seemed to be operating in slow motion.

And so Jim put Dot on the train back to Arlington after their honeymoon ended, and set out to drive to Oklahoma. On the train ride back, Dot noticed some faint red spots on her face. When she got back to the Arlington apartment, Crow and Louise looked at her with alarm and helped her get to the dispensary. She was coming down with the measles. "Is this something that happens to you when you get married?" Dot asked Crow and Louise. It was a real question. She meant it genuinely. Neither of them knew the answer. It seemed plausible that marriage would do something like that to you, give you measles, create some kind of physical reaction.

Dot had not heard the "Here's your hat, what's your hurry" speech, or if she did, she ignored it. On December 14, Dot had taken a test evaluating her written French and did so well that she was put in the French decoding section, which was up and running in Section B-III of Arlington Hall, clearing up messages left over from the German occupation. Promoted from CAF-4 to SP-6, she was making $2,320 a year, about

$700 more than when she arrived, and far more than she had made as a teacher. Her evaluations all had been positive; in quarterly reports she had received "outstanding" on qualities including skill, attention to pertinent detail, and resourcefulness.

While Jim was in Oklahoma, they both felt it made sense for Dot to keep her job. Housing was hard to come by, and the money would come in handy. But separation was hard on the newlyweds, and Jim's letters from Oklahoma vacillated between pragmatism and romantic impatience. They were unfailingly affectionate. "My dear wife," they began. One letter noted that he had just received two letters from her. "I think I enjoyed them more than any letters I received from you before. I am sure that I am really in love with you," he said. "I am glad that you enjoyed our honeymoon, Dot. I enjoyed it very much too. Since I received those letters this afternoon I can't stop thinking about you. It is bad in a way because I bought something at the PX and left my wallet on the counter." He then went into a long description of getting the wallet back.

He was torn over whether Dot should come to Oklahoma to join him. "The boys who brought their wives are out having a terrible time finding a place to live," he related. "I think I could find a room some place for us to sleep but I don't know how good it would be." He knew she liked her work and that it might not be practical for her to come. "We can't always be practical, though," he reflected. And so it went, back and forth for several letters. "Dot, I do know that I love you very much and want to see you again as soon as possible." He confessed that he felt "very lonesome for you."

On January 31, 1946, Dot Braden Bruce resigned from Arlington Hall and prepared to travel to Oklahoma to join her husband. She sold her share of the furniture to Crow, who was keeping the one-bedroom apartment with her sister Louise. Louise was working as an astronomer at the Naval Observatory, and Crow was still working at Arlington Hall. Their little sister Kitty was finishing college, and would later embark on a career in computer programming. None of the Weston girls would ever go back to live in Bourbon. Nor would Dot live in Lynchburg again.

Union Station—so much more familiar than when she first arrived—was chaos. The ticket office told Dot she could get a ticket as far as Cincinnati and then would have to take her chances. Crow came to see her off, and so did Dot's mother, Virginia Braden. Dot's memory of Crow as she left was of her best friend crying and calling out, "If you think I'm going to stay here all my life with Sister, you've got another think coming!" Dot would always remember her mother turning to comfort Crow, saying, "You'll find a man. There are plenty of men coming back after the war."

Dot traveled on. Soldiers had not seen women in months. They all wanted to talk to her. She never paid for a meal. One GI asked if she would get off the train and have dinner with his family. She told him she was married and he said they wouldn't care; his folks would just be happy to meet her. Another sat down next to her, feigned sleep, and snuck his arm around her; a sailor intervened, calling him a "bastard," and made him move away. When she got to Cincinnati, helpful soldiers lifted her up and bodily put her on the train to Oklahoma. Jim was waiting. He managed to get discharged over the course of a few weeks, and they moved to Richmond, where DuPont had held his job.

In February 1946, Dot received a letter from what was now called the Army Security Agency, the new name for Arlington Hall. It thanked her for her wartime service. "Because you were found to be a person of excellent character and unquestioned loyalty to the United States," it also said, "you were entrusted with information which should not, under any circumstance, be revealed to unauthorized persons." Information she was privy to "should not now, or at any future time, be revealed."

It was a nightmare finding housing after the war. Everybody in America needed a place to live. In Richmond, the newlyweds set up in what Dot came to think of as the apartment from hell: The walls were thin and there was a couple next door who fought, arguing and throwing dishes. Not all postwar marriages were working out as well as theirs was. The government trucked in prefabricated houses for veterans. Jim stood in line and they got one. The new neighborhood was a nice place to live. The women drank coffee and talked about their new babies. Nobody asked Dot what she had done during the war.

Dot got pregnant, but Jimmy was born prematurely, and that was another fresh hell. When they got Jimmy home, he would cry and cry. The doctor said whatever she did, not to overfeed him, telling her to give him some kind of formula that mostly seemed to be water, and the hunger made him cry more. Dot had never had a baby before and thought all babies cried that way. So she obeyed the doctor's bad advice and had to listen to Jimmy cry. Crow came down from Washington and brought her flowers and got her through that awful period. Jimmy would grow up just fine and Dot and Jim had two more children, both daughters, everybody healthy and happy.

Crow stayed at Arlington Hall, but she couldn't tell Dot what top secret project she was working on. Several years later, Virginia Braden's prediction came true and on a blind date Crow met a man, Bill Cable, who worked for the Veterans Administration and had the good sense not to pump his date for information about what she did. Dot always said Bill Cable couldn't have fit Crow better than a glove. "She couldn't have gotten a better husband." He was calm and good-natured and every bit as slow and deliberate as Crow was. Crow would not marry him until Dot and Jim came up to meet him and give their stamp of approval. Crow continued working at Arlington Hall, as a mathematician in a unit with some of the elite wartime code breakers. Her personnel record showed that she received excellent ratings, and in 1948 her unit of mathematicians received a commendation. But then she and her husband started their family. On December 31, 1952, Ruth Weston, pregnant with their first child, submitted a handwritten note to her government bosses saying, "I wish to resign my position as Mathematician," explaining that "My time is needed at home to care for my baby."

Women now were expected to quit work when they started having babies. The postwar U.S. government made this clear. There was no more state-sponsored child care. In a postwar, Cold War America, child care was viewed with suspicion, as the kind of thing communists used to raise their children collectively. The U.S. government began doing the opposite of its wartime recruiting; it made propaganda-type films telling women it was important to leave their jobs, return home, and tend their households. The films pointed out that it was unnatural for women to be breadwinners, taking jobs from men. Quitting one's job became a matter of patriotism. And so, many of the wartime women workers did leave their jobs when they had children. Among them was Crow, even though she very much liked what she did.

When Ruth "Crow" Weston Cable's daughters were growing up, they were under the impression that they were somehow related to Dot's son, Jimmy Bruce. The two former code breakers remained so close that their children assumed they were family. Ruth Weston Cable would call up Dot Braden Bruce and say, "Dot! This is Crow!" and Dot would bring up the mattress escapade and they would laugh and laugh. Dot always kept the gold earrings Crow gave her. Both women grew bored at home and went back to work when their children were old enough. Dot became a real estate agent. Crow, still living in Arlington, took a job as a cartographer with a transportation consulting agency. Crow loved maps, and she loved her work. She also worked the polls every Election Day, honoring the patriotism and sense of civic duty that her father, back in Bourbon, Mississippi, had instilled in her. That day of commitment to democracy remained sacred to her.

_____

The Mitten

January 2016

On a bright day in January 2016—sunny and chilly, but not too punishingly cold—a discreet line of sedans, SUVs, and pickup trucks files into a cemetery in the northern Virginia foothills. It's not an undisclosed location, or not exactly. Just an obscure one, about seventy miles outside of Washington. The cars pull into a grassy parking area and people get out: men wearing dark suits and overcoats, ties fluttering in the wind, younger people helping elderly ones. The bundled-up mourners are wearing hats and gloves, picking their way across the soft ground to sit in folding chairs set up in rows under an awning. The group consists of family, neighbors, well-wishers, and eminent members of the U.S. intelligence and national security community. They have come to pay their respects to the woman who rose to become the first female deputy director of the National Security Agency, the federal entity the wartime code breakers begat.

In the crowd are men who worked for Ann Caracristi—many found her intimidating; she had this mesmerizing thing she could do, flipping a pencil between her fingers and never dropping it—and younger women who found her example inspiring. Ann, who died at ninety-four, maintained her razor-sharp intellect till the end, reading *The New Yorker* and watching the news on CNN, surrounded by stacks of books on subjects from Shakespeare to Kierkegaard. And she kept her sense of humor. When her tall caregiver would enter Ann's tiny Georgetown home, Ann, confined in her final days to a cot in her sunny first-floor kitchen, would laugh every time the caregiver had to stoop to get through an interior doorway.

Ann Caracristi, once the twenty-three-year-old head of the Japanese Army address research section at Arlington Hall, did more than just return to Arlington Hall after the war. She worked on some of the toughest code-breaking challenges the Cold War era brought with it. After her brief foray into the newspaper business, Ann was assigned to the "Soviet problem," a multifaceted effort. Her first assignment was working ciphers about Soviet weapon systems. The project was terminated after its existence was revealed to the Russians by a spy. She would move on to the East German problem. It was difficult and serious work at a difficult and serious time.

During her career Ann Caracristi rose through the ranks to become a senior member of the NSA brain trust. She received honors including the National Security Medal and the Distinguished Civilian Service Award, the Department of Defense's highest civilian honor. When she went to the White House to receive an award from President Ronald Reagan, she asked her friend Gert's nephew to accompany her, without once revealing how important the award was. Her public recognition was unusual: Most of the women who served during and after the war received no recognition, at least no public one. True, of course, of most of the men as well.

During a visiting session at a Georgetown funeral home, snapshots on display included one of Ann Caracristi looking wonderfully glamorous, wearing a white evening gown and a fur stole. She had come a long way from the bobby-soxer who washed her hair with laundry soap. During her career she was seen as fair and smart and tough. She was what one man who knew her described as a "magnificent bureaucrat," meaning that she knew how to move a federal bureaucracy and get results. She worked with top military men and was respected by all of them. She did not like mistakes but she knew that mistakes happened. In the workplace, she did not display the easy humor she did at home. "She's not a smiley person," said Jo Palumbo Fannon, the young high school graduate who swore in Arlington Hall newcomers and made a career in the personnel operation afterward, and who liked her. Hugh Erskine, a younger relative of the wartime Erskines who moved en masse from Ohio, spent a summer between high school and college sorting messages, long after the war was over—the same job Ann had held when she started. He found her "a little scary."

You would never know any of this to talk to her in her home. During five interviews I conducted with her before her death, Ann Caracristi was good-humored and never condescending, as she explained her long-ago work on the Japanese address codes. "It sounds so dumb," she said, self-deprecatingly: the idea that sussing out such a thing as addresses could be helpful.

Ann spent most of her postwar life in the same little red house on a side street in Georgetown, so small it looked like a hobbit must live there. She dwelled in quiet companionship with Gertrude Kirtland, who left government work to become a published author of children's books.

At the cemetery, the minister begins to talk about Virginia's Blue Ridge Mountains, which Ann—northerner though she was—came to love and know well. These are the same mountains that Ann Caracristi, Wilma Berryman, and Gertrude Kirtland used to visit for those rare respites during the war, driving Wilma's car, for restorative breaks. Ann and Gert bought a weekend house not far from Mount Weather, the bunkered facility where the government would relocate in the event of a national disaster such as a nuclear attack. For NSA coworkers, to be invited to spend a quiet weekend in the hills with Ann and Gert was a treat and an honor. People always said yes.

Ann's grave has been dug parallel to the grave of Gertrude Kirtland. They will rest side by side. In the funeral notice placed in the *Washington Post* by Ann's family, Gert, who predeceased her, is described as Ann Caracristi's "longtime companion." This meaningful phrase came as a bit of a surprise to some in the national security community, who thought Ann and Gert were simply single ladies living together as dear friends. Her family never knew what the nature of the relationship was, and never felt they could ask: She was a peerless secret-keeper in every way. But it was a decades-long committed partnership. During the service, the minister talks about how Ann will be able to join Gert in eternity. Their relationship is worth remarking upon only because, for much of the postwar Cold War era, for a man working in intelligence or national security to be living in any kind of committed domestic relationship with a person of the same sex was a career deal breaker. In England, Alan Turing was persecuted until he poisoned himself. In the United States, NSA employees, like those at other federal agencies, for many years would be obliged to resign if they were found to be homosexual. There were purges.

In an interview before her death, I asked Ann Caracristi whether the postwar private lives of women working in the clandestine mail-reading business perhaps did not receive quite the same scrutiny men's did. She agreed that this was likely the case. Being allowed a bit more latitude in one's personal life seems to have been a rare instance in which being female was a career advantage after the war. It could be that women didn't matter quite as much as men did. It could be that people didn't care what women did in private. It could be that women working in intelligence were thought of as "honey traps," capable of using their sex to lure a man into betraying his country, rather than as complicated human beings with quiet but rich interior lives.

After the war, Ann and Gert stayed good friends with Elizebeth and William Friedman, whose own relationship with the NSA did not end well. William Friedman felt the postwar NSA was going too far with secrecy and overclassification, and the NSA felt he had brought some papers home that he should not have, and there was a bitter rupture. There is now an NSA auditorium named after him, however. All has been forgiven. The Friedmans devoted their own retirement to driving a stake through the heart of the theory that Francis Bacon wrote the plays of William Shakespeare.

* * *

A number of women code breakers who distinguished themselves during World War II also went on to high posts at the NSA, which is the federal agency responsible for monitoring enemy communications and protecting those of the United States. Ann Caracristi at one point worked for Carrie Berry, a former Texas schoolteacher who became the first woman sent to Cheltenham, England, to serve as NSA liaison with the British. Ann's good friend Wilma Berryman—later Wilma Davis—worked on the Chinese problem.

The tenor of the place was different after the war. The Army and Navy operations merged, and eventually the whole operation relocated to Maryland. But wartime affinities lingered, and so did wartime grievances. Polly Budenbach, a Smith College graduate who helped Frank Raven break the Japanese naval attaché cipher, stayed on, and after the war she had a funny run-in with Agnes Driscoll. She had never met the legendary Miss Aggie in person, but one day she saw the great lady approaching down a hallway. Even at her advanced age Agnes Driscoll was capable of taking names and holding grudges. Though they had never been introduced, she must have known that Budenbach's mentor was Raven, the very man who did Miss Aggie in. As they passed each other in the hallway, neither woman spoke, but Agnes Driscoll did let out an unmistakable hiss.

A few other women stayed on and rose high. Within the NSA, many of the early "supergrades"—the top civil service rating—were female. Among these was Bryn Mawr's Julia Ward, who built Arlington Hall's library and information service. Also prominent was Juanita Morris, the young woman who left her North Carolina college and found herself deposited in the eccentric German section. During the war, Morris—later Juanita Moody—had played a leading role in solving a difficult German diplomatic code considered so hopeless that she had been forbidden to work on it. After the war, she found herself manning the desk responsible for breaking Cuban messages. People considered Cuba a little bit of a backwater. That notion was disproved during the Cuban Missile Crisis.

For some of the wartime women who stayed on, friendship took the place of a nuclear family. In the very early days of 1943, an ex-schoolteacher named Gene Grabeel had been seated at a table and presented with some Soviet intercepts. Gene, a graduate of Mars Hill, a two-year college in North Carolina, and Farmville State Teachers College in Virginia, had been unhappily teaching home economics to eighth graders when she was recruited by Frank Rowlett, who knew her from childhood. At her little table, Gene Grabeel was given a jumble of Soviet messages and told to see what she could make of them. The Russians were allies and Arlington Hall was not supposed to be reading their mail. But the climate was such that the Russians were not communicating their intentions. Would Russia come into the war against Japan? Would they pull out of the war against Germany and negotiate a separate peace? The intent of the top-top secret effort was to quietly look for Soviet diplomatic messages that might tell what the Russians had in mind.

But what Arlington Hall had—Gene Grabeel perceived—were messages in a number of different systems. These included communications from the KGB and GRU—the foreign and military intelligence services—containing the names of Americans and other Allies who spied for Russia during the war. The messages were enciphered using a "one-time pad," a kind of additive book in which additives were rarely reused. Each page was to be used only once, so as to make it impossible for a code breaker to accumulate depth. But as Soviet factories were moved to the Ural Mountains during the German advance on Moscow, it became impossible to produce new pads and a few were reused. A few was all it took.

Sitting at her table, Gene Grabeel helped launch what became known as the Venona project. The name associated with Venona is that of Meredith Gardner, a linguist and book breaker who brilliantly was able to interpret messages and recover code groups, leading to several prosecutions of Soviet spies in the United States, and ruining the lives of others. But 90 percent of those working on Venona were women. Gene

Grabeel separated the traffic into systems; none other than Genevieve Grotjan Feinstein (she married a chemist, Hyman Feinstein) found one of the coincidences. A linguist named Marie Meyer found another. The Russians employed the cipher in question for only a few years, but Arlington Hall worked that system for decades, mining the old material and digging out names. It was a group of former teachers—Carrie Berry, Mildred Hayes, Gene Grabeel, others—who devoted their careers to this. Many never married. In the personal photos of the Venona code breakers, they are wearing shifts and carrying handbags and look like a gardening club. They were best friends, and sometimes lived together, sometimes alone. "Gene was just an independent person that didn't want the responsibility of a marriage," said Gene Grabeel's sister-in-law, Eleanor Grabeel. "She enjoyed her freedom." When the Venona project was rolled up in 1980, the former schoolteachers had been working on it for more than thirty years.

After them, though, there was an institutional falling off. By far the majority of women at Arlington Hall packed up and went home after the war. Even many of the top women, among them Delia Taylor Sinkov and Genevieve Grotjan Feinstein, at some point stopped working. Often this occurred when they began having children. Motherhood was the dividing line between brilliant women who stayed in the work and those who did not. For a woman with children, there were few resources to make a career feasible. The nation lost talent that the war had developed. The 1950s and 1960s would not bring another critical mass of women to succeed the wartime code breakers, and in the 1970s and 1980s, women at the NSA would have to fight a battle for parity and recognition all over again.

* * *

For the women who left the field but wanted to continue working or studying, postwar opportunities were mixed. Women who served in the U.S. Navy qualified for the GI Bill, at least in theory. For returning servicemen, the GI Bill was a life-changing benefit that put college within reach for middle-class men.

For the Navy women, however, their experience of GI benefits was hampered by the old idea that women are not suited for the highest levels of learning. Elizabeth Bigelow at age twenty had been recruited by the Navy from Vassar. In college, she shone in her mechanical drawing classes—as close to a drafting class as Vassar offered—and aspired to be an architect. Her professor was the legendary Grace Hopper, a computing pioneer who became a rear admiral for the Navy and helped develop the computer programming language COBOL. Elizabeth Bigelow always believed Hopper identified her for the code-breaking program. When Elizabeth got out of the Navy she applied to three leading schools of architecture. "In every case the response was the same," she recalled later. "We're sorry, but we are saving all our spaces for the men who have been in the armed services." Elizabeth wrote back protesting that she had been in the WAVES for two years. She could not say she had sunk a convoy, because that was top secret. "The answers all came back, 'We're sorry, but no.'" So she married and raised a family with her husband. When they moved to Cincinnati, Elizabeth ended up running the computer system at the University of Cincinnati. She taught herself how to do it.

Janice Martin Benario, the Goucher classics major who worked in the submarine tracking room, did use the GI Bill to get a PhD at Johns Hopkins. She met her husband there and spent a productive career teaching in the Classics Department at Georgia State University.

Dorothy Ramale, the aspiring math teacher who grew up in Cochran's Mills and longed to visit every continent on earth, got a master's degree using the GI Bill, which meant she earned a higher salary working as, yes, a math teacher, in Arlington County. She taught at the public middle school my own children later attended, where the kids in Miss Ramale's algebra classes doubtless had no idea that this sweet, good-natured woman broke codes that sank enemy ships. And she did visit every continent on earth, including Antarctica—twice.

Jimmie Lee Hutchison Powers, who lost her husband on D-Day, used the GI Bill to get a community college degree in cosmetology. She opened a salon back home in Oklahoma and supported herself and her widowed mother. After three years, she remarried. On their first anniversary, her new husband gave her a card. She opened it and thought the wording seemed familiar. Sometime afterward, her mother asked if she wanted her trunk of things from her wartime naval service. She opened it and found that Bob Powers had sent her the exact same anniversary card, right before he flew out and died.

Betty Bemis, the champion swimmer who worked at Sugar Camp, corresponded during the entire war with Ed Robarts, a bomber pilot who flew thirty-five missions. One day she was summoned to the phone. "Hi, Betty," he said. "I'm home." Other women by then had ended their casual correspondences with men they'd never met; Iris Flaspoller had written a Dear John letter to Rupert Trumble, which somewhat devastated him. But when Ed asked Betty to fly to have Easter dinner with him and his aunt and uncle in Miami, Betty hitched a ride on a military plane to see him. Three days after she arrived, he asked her to marry him. "We had a beautiful marriage," she told me in November 2015.

For some of the women, especially those who worked in top spots on the Enigma project, life after the war was harder. Louise Pearsall, who worked on Enigma, was tempted to stay on with the work after the war. Some engineers and mathematicians who worked as Navy code breakers formed a partnership to develop code-breaking computers for the Navy, and she considered joining them. But she was exhausted—mentally, emotionally, physically—and there was a boyfriend who had come back to Elgin, Illinois, so she got discharged and returned to Elgin, and they broke up. She went to work but quit when she suffered a nervous breakdown. She married a wealthy man, and her in-laws, who were uppity and pompous (her brother remembered) would not permit her to work. A year after the war she was living in a small apartment in Chicago, looking at an elementary school outside the window. Her daughter, Sarah, felt she suffered from depression much of her life, possibly brought on by having three children in quick succession, compounded by secrecy and the fact that she could not talk to anybody about her wartime experience. "She was a total wreck," said her brother William. "She was a total nervous wreck. You couldn't even look cross-eyed at her and she'd break down." She eventually would divorce her husband and take a job with IBM that she quite liked.

Betty Allen, one of the group of friends in the library unit at the Naval Annex, also had a hard time after the war. Most jobs went to men. For three years she drifted. Meanwhile her friends from OP-20-G-L were coping with their own postwar challenges. The women, mostly married, had new babies and were living isolated lives in small spaces. Housekeeping was hard. There were few new appliances, since American industry for years had been churning out tanks and planes and weapons. New mothers washed bedsheets in the bathtub and wrung out diapers by hand.

During the war the library unit had been a productive group of bright and active women, having meals prepared by Hot Shoppes and never having to think about what to fix for dinner. Now they were up to their elbows in domestic labor. So the former cryptanalytic librarians came up with a solution to their loneliness. The former enlisted women would write a round-robin letter.

Here is how the round-robin letter worked: One former code breaker would write a letter about what was going on in her life. She would send it to a second woman, who would write her own letter. That second woman would send both letters to a third woman, who would write her own and send all three to a fourth. The thickening sheaf would travel full circle until it came back to the first woman, who would remove her old letter, write a new one, insert it, and send the sheaf around the circle again.

The round-robin letter was a source of comfort to Betty Allen, who did find a job, and to the women shut up in tiny apartments with little babies. The round-robin kept going, all through the 1950s and the rearing of children; through the 1960s and the Vietnam War; through the 1970s and the feminist and civil rights movements; through the 1980s and the election of Ronald Reagan; through the terrible shock of the Twin Towers falling on September 11, 2001; through the Iraq War, and the election of Barack Obama, and the devastation of Hurricane Sandy, and the ascension of Donald J. Trump to the U.S. presidency. The women of the code-breaking library unit kept the round-robin letter going, and it is going still, as of the writing of this book.

One by one, of course, most of the women died. When Georgia O'Connor Ludington died, it was a very hard effort for her son, Bill Ludington, to insert a letter of his own into the round-robin letter, telling Georgia's wartime friends—whom he knew and loved—that his mother had died. But as of 2015, when I visited her, Ruth Schoen Mirsky was still writing to Lyn Ramsdell Stewart. They and one widowed husband were the only three left. When Lyn died, Ruth and the widower still kept in touch.

Ruth now lives in a second-floor apartment in the Rockaway neighborhood of Queens, New York. Harry Mirsky has died. Ruth was a tiny woman then and is a tiny woman now, still so proud of her service that her email address begins with "RuththeWAVE." She has scrapbooks documenting her courtship with Harry; photos taken from the rooftops of Washington hotels; and photos of her wedding, which her code-breaking friends attended, slipping away even if they didn't have leave.

She also has the ribbon commemorating the special unit citation that all the women at the Naval Annex received after the war. When I asked to see it, she reluctantly agreed. The women were instructed not to show it to anybody, and she still can't bring herself to let it be photographed.

The Navy women treasured that unit citation, but most never displayed it. Some didn't even purchase the naval ribbons to which they were entitled. "We hadn't won any battles and didn't feel it was appropriate," said Edith Reynolds White, a WAVES officer recruited from Vassar. Edith was working at the Naval Annex when a recovered codebook was brought in, dripping wet, having been fished out of a sinking submarine by an alert American naval officer. After the war Edith stayed in the Navy for a while and was transferred to a hospital in New York where men with tuberculosis were being treated. One day, she was told that she was to receive her unit citation. She was to appear under the flag at nine a.m. "in full uniform with the ribbons to which I was entitled." Since she had not purchased any, she had to find a male officer willing to lend her his. When she asked a young doctor, he said, "Only if I can come and look."

After the ceremony, the young doctor told her he was taking her to dinner. "It's traditional," he assured her. "Whoever you lend your medals to, you have to take to dinner." There was no such tradition, and, reader, she married him. Many years later, she was living in Norfolk and encountered the officer who fished out the codebook. She told him how valuable it had been. He was astonished. Nobody had told him. The officer later showed up on her doorstep in full gold braid with a box of chocolates, which was the first indication her son, Forrest, had that his mother had done something important during the war.

Fran Steen, the Goucher biology major who put aside her ambition to be a doctor, married a naval officer. She kept her pilot's license until she got pregnant. Her husband was killed in 1960, struck by lightning playing golf. She got remarried to a naval submariner, and they settled in Charleston, South Carolina. She worked as a census taker, an artist, and a fashion model. She was always reluctant to talk about her wartime service. But eventually she did tell her son, Jed, about being a watch officer when they got the message that her brother's ship was hit by a kamikaze, and about learning to shoot and bringing the bombes back from Dayton. They went to an air show of vintage planes and she mentioned offhandedly that one model was the one she had learned to fly at Washington National Airport.

Jed always sensed that his mother's mind worked differently than many other people's. Enigma cribbers had to start with the finished message and work backward to the likely key setting. Late in her life, she was a member of a women's group that called itself the Low Country Cocktail Club. Her son was driving her to a club meeting and asked her for directions. "Wait a minute," she said, thinking. To figure them out, she had to start with the address and work backward. "Her thought processes were highly analytic and different from what most people's were."

Rear Admiral (ret.) David Shimp met Fran Steen Suddeth Josephson at a different Charleston cocktail party and was astonished to encounter her in person. He had always heard her mentioned as one of the code breakers who labored over the messages that led to the shootdown of Yamamoto. He had met an old "cryppie" (the slang term for cryptanalyst) at a veterans' gathering, and Fran's name had come up. "She's the one that got that son of a bitch Yamamoto," the veteran recalled, saying that "only a damn woman could have figured out that blanking code," implying that something about it was irrational. It was just an over-a-beer kind of story, and Dave Shimp dismissed it until he met Fran herself. He arranged for her to be given a surprise award at a dinner for cryptologic veterans. Her son Jed got her there without telling her what the meeting was for. As Fran was listening to the speech, she began to realize they were talking about her. Jed, also a naval officer, stood and said, "As an American citizen, with all the freedoms we have, I thank you; as a fellow naval officer, I salute you; and as a son, I love you." Whenever Shimp tried to get her to share more details, though, she refused. When she did talk, she would dwell on the lives she hadn't been able to save, rather than the ones she had. "Those regrets were always foremost in her mind."

Over time public views changed about the war. One was not always well advised to mention what one had done. Jeuel Bannister Esmacher, the band director who worked at Arlington Hall, knew that a message she broke helped sink a convoy. She saw certain code words, hurried the message to the "big boys," and later heard over the radio that the ship had been sunk. At the time she felt proud. But when she started a family with the linguist she met during the V-J Day celebration, Harry Esmacher, she came to reflect on all the Japanese families who lost sons, and her feelings became more layered and complex. She felt more sorrow. "There were Japanese that went down with that ship that had mothers and sisters and wives," she reflected when I spoke with her. "You think about that also, at this point. I did not think of that back then."

When Elizabeth Bigelow Stewart mentioned to her own children the convoy she had helped sink, her daughter replied, "Mother, how dreadful! You killed all those Japanese sailors, and you were pleased about it!"

Elizabeth was dumbfounded. America quickly forgot what the war had felt like—how real the menace had been.

Jane Case Tuttle, the wealthy physicist's daughter, also got married after the war, and it was a disaster. Her husband had written her funny letters during a time when she was feeling lonely. She wanted a sense of normalcy and to have a family, and "I had always done everything I was told to do." She managed to extricate herself from the marriage and found that the memory of working during the war helped her retain her self-worth. Late in life she married a man who had been madly in love with her during the war. When I visited her, she was living at an assisted-living facility in Maine, an ardent supporter of the presidential candidacy of Bernie Sanders. Because she could no longer walk easily, she would sit in a recliner and throw clean, balled-up socks at the television when a politician she hated came on.

Ann White Kurtz also married during the war. In November 1944 she had to seek a discharge because her husband came back in bad shape—he was disoriented and had a tropical disease—and needed care. "Oh golly, did I miss it," she said. "I made the wrong decision." Some of her Wellesley classmates described their postwar lives as "disappearing into marriage." Her husband, she later put it, "needed a 'wife'" and "could not understand why I was so unruly." She got her PhD on the GI Bill, got a divorce, and became a professor. Late in life she joined the Peace Corps.

Anne Barus Seeley also married; she never pursued a career in international relations, but she did work in other capacities including running a weaving business and teaching. In her mid-nineties, she was still sailing and kayaking near her home on Cape Cod.

Many of the code-breaking women helped advance the feminist movement—through their postwar employment, but also, sometimes, their postwar dissatisfaction. One woman I interviewed, whose mother worked at Arlington Hall, always sensed something was missing in her mother's life, something she had had, once, and lost. This awareness, she said, "seeded feminism in our house." But other women felt left out by the feminist movement. Erma Hughes Kirkpatrick, the bricklayer's daughter, became a mother, housewife, and volunteer, and enjoyed it. She always felt feminism disrespected her contributions, even as her husband respected them. She started the first soup kitchen in Chapel Hill, North Carolina. "He and I were equals," she said.

Well, rather more than equals. Erma was a full naval lieutenant by the time she was discharged, and she stayed in the naval reserves. Her husband had been a Marine. Once, when they wanted to show their children Quantico Marine base, her husband couldn't produce an ID that would get them in, but Erma had her naval reservist identification. She showed it and they were waved through. The Marine guard saluted her. There was silence in the car for some time. "You don't do that to a Marine," she joked, later.

* * *

The assisted-living facility on the outskirts of Richmond, Virginia, is a nice one. There is a dining hall that serves good butternut squash soup and ham biscuits, and the place has lots of parties to keep residents busy. Dot Braden Bruce had to move here after she slipped in her garage and fell and hit her head. But in 2017 she is alive and recovered. At ninety-seven, she keeps her French skills sharp by chatting with caregivers from French-speaking West Africa. "A lot of people don't bother to learn their names, but I do," she says. Once a schoolteacher, always a schoolteacher.

Her life has come full circle, and once again she is living in a one-bedroom apartment. Crow died in 2012, Jim Bruce in 2007. Dot herself is still lively, still literate, still prone to reciting snatches of doggerel such as "Why does the lamb love Mary so? Because Mary loves the lamb, you know."

Photos of her family cover every piece of furniture. Jim Bruce in his later years was a dead ringer for James Stewart, tall and steady and agreeable-looking, and Dot jokes that she herself was a dead ringer for Elizabeth Taylor.

Toward the end of his life, Jim's memory started to fade, and he would ask Dot if they had a good marriage. "Did we get along?" he would ask.

She would assure him that they did. It was true. He was a good husband. "Long-suffering," Dot says with a laugh. He understood her lively independence. When she took work as a substitute teacher, he would watch their three children on weekends and give her time to grade papers. On Saturdays he liked making hot dogs for neighborhood kids, and real French fries from scratch. When she embarked on a real estate career, he would drive her to open houses. Whatever Dot wanted to do, it was okay with Jim. They never fought. Dot wouldn't have minded a bracing argument now and then, but Jim was peace-loving. "My husband was a very laid-back person," she says. "He had to live for sixty-three years with me. Not me with him."

The only gender-related dispute they had was when she tried to drag the trash cans out to the curb. He felt pulling the cans to the curb was a man's job.

The Bruce family did well. Jim had an excellent career in the postwar industrial economy, and so did Dot's brothers. Teedy, once declared missing in action, was living when this book was first published, then died peacefully in 2018. In Dot's family, there are so many Jims and Jamies and Jameses, sons and grandsons and great-grandsons, named after her husband, that I found it hard to keep them straight. There are Virginias as well, and a little boy named Braden. Every year, Dot treats the whole family—some twenty people—to a big holiday meal at Richmond's posh Jefferson Hotel. Every year they come, flying in from New Orleans, New York, California, bringing babies in snugglies, dressing little cousins in matching outfits. Many photos are taken. It is the happy family tableau she longed for as a girl. She complains about the cost, but she is not really complaining. During some of our many interviews, she had a file case containing her stock holdings under her chair. Her broker had died and she was shopping for a new one. She let several candidates take her to lunch.

Looking back, Dot wonders sometimes why she decided to marry Jim Bruce rather than George Rush. "My life could have turned out very differently," she reflects. Make no mistake; she felt she made the right choice. George Rush was a perfectly nice man. But she didn't want to move to California. She is so glad she took the train to Washington and embarked on her code-breaking service, together with her friend Crow Weston Cable. "I wouldn't take anything for it," she says. She thinks what tipped the balance in favor of Jim Bruce was the fact that he was steady and kind. And persistent. And he had a good sense of humor. Dot's favorite song has always been "Somewhere, My Love," and after they married Jim used to tease her when she played it. "Haven't you found your love yet?" he would say.

And, she reflects, "he wrote me all those darling letters."

After the war, Dot told nobody what she did. At some point, maybe fifty years after the war ended, she started giving hints. They did not believe her—her brother Bubba said it was "just a little job and I was trying to make it a big deal," she remembers now. But then they started to believe her, or sort of. It became a tenet among her grandchildren that Dot single-handedly broke the Japanese codes. And yet, nobody really took it seriously.

Memories come at odd times. Dot was reading aloud to one of her great-grandchildren a children's book called *The Mitten*, in which forest animals take refuge from a snowstorm by climbing, one by one, into a cast-off woolen mitten. So many animals climb in that one sneeze is enough to eject all of them. Reading it, she could not help but think of the Arlington apartment and all the girls who stayed in that one-bedroom place.

Her son Jim has always been intrigued by her wartime code-breaking service. As kids, he and his sisters used to go up in the attic and read the letters their dad wrote to their mother. His sentimental side was a revelation. But they never could get their mother to tell details about what she did. Now she has gotten the okay from none other than the NSA and has been assured that it's fine to tell her story: The long-ago ban was lifted several decades ago. The government would *like* her to tell her story. But she still has her doubts. She cannot quite believe it. Then again, what would they do to her? At her age? Put her in prison?

On a Wednesday afternoon in 2014, during the first interview for this book, Jim, her son, is sitting in an upholstered wing chair in her one-bedroom apartment. "Let it rip, Mom!" he urges. By now so many male code breakers have written their memoirs: Edwin Layton and Frank Rowlett and others, with book titles like *And I Was There* and *The Story of Magic*. Dot relaxes, a bit, about telling the part she played in this dramatic story, and Jim listens as his mother begins to talk. She mentions Miriam the overlapper—awful Miriam! with her yellow diamond!—and claps her hand to her mouth. Never has she uttered the word "overlap" outside the confines of Arlington Hall.

Even now, it has the feeling to her of something illicit, something forbidden, something dangerous and important, no matter how long ago this all occurred. It feels as if an enemy might still be at the window, listening in.

Afterword for the Paperback Edition

The messages began arriving even before this book was published and have continued steadily since. I print out each email, save every letter, and have a stack, now, of hundreds. "My favorite Aunt, Betsey Wynne, was a code girl!" Gwynne Gigon wrote. "She passed away just shy of her 91st birthday, and I miss her like crazy...she never mentioned the specifics of the work she did, and after her passing I found the Navy's letter of commendation."

"My mother never would tell me what she did during the war," wrote Karen Scott Johnson, daughter of Anna Mae Barrett. Karen did know that her mother enlisted in the U.S. Navy as a member of the WAVES, was a math whiz, and attained very high security clearances in her work. *Code Girls*, she wrote, offered her a window into what her mother's life entailed.

"I feel as though my life has been turned upside down," wrote Gail Simmons, provost of Hofstra University. Simmons knew that her mother, Barbara June Whitt, was in some fashion a code girl. "She served in the WAVES, went to boot camp at Hunter College, and would if pressed say that she did work with codes—but like the women you describe, she would not crack and give anyone in the family any details. It was clear that something about the experience had affected her deeply, but she opened up to no one." Reading the book, Gail wrote, "It all suddenly makes the snippets I knew about my mother add up, and it has taken my breath away. It has brought a feeling of closeness with my mother that I almost never had while she was living."

I knew, of course, that with more than ten thousand American women involved in the code-breaking programs detailed in the book, they would have hundreds of thousands of descendants. But nothing prepared me for the volume of letters and emails from readers—people young and old who feel a personal connection to the events of seventy-five years ago and to the women who served. How proud people are of their mother's or grandmother's or great-aunt's service; how eager they were to tell their siblings, their children, to impress upon the family the import of what this foremother did. Phil Cagney wrote that his daughter, Erin, a grad student at American University, had walked by the former Naval Communications Annex (now the Department of Homeland Security) every day, without realizing that her grandmother, Mary V. Lauer Cagney, had run the bombe machines that broke the German Enigma cipher from that very place. "She's literally walking in her grandmother's footsteps," he wrote.

Karen Veverka wrote to say that her mother, Ruth, was a code-breaking member of the WAVES who went on to work at the National Security Agency, as did her husband. "She would not talk about anything specific, and went to her grave with secrets," Karen wrote. "I lost her last year on October 10, 2016, at the age of 98. I am an only child, and was extremely close to her. She was a very strong and faithful woman, an inspiration to me her entire life."

Sarah Casseday wrote to say that her mother, Evelyn Boyette White, worked at Arlington Hall but would only describe her work as "secretarial stuff." Her mom had finished two years at Atlantic Christian College in Wilson, North Carolina, when she was recruited. "She told me that it was always hard for her to believe that my grandparents let her go to Washington, but how she loved those times."

"For so many of us, you have helped fill in the holes of our heritage," said Julie Mickler, whose aunt, Marjorie Mickler, worked at Arlington Hall; she had once showed her brother a medal she received for helping break the Japanese surrender message, and told him never to tell anyone. The family had never been certain whether to believe her.

John Day said that his aunt, Ruth G. Van Horn, had been an English professor in Michigan when she came to Washington. "She always said that this was the most exciting period of her life" but wouldn't talk about her work except for a single moment. "She boasted with great pride that she had seen the news of the Japanese surrender at the end of World War II even before President Truman had."

Some of the memories were especially poignant; a few readers wrote to say that their mothers had died young, and the book offered them a new way to feel close to them. For others, reading about the code breakers made them consider their mothers' lives in a new light.

Ginny Landes wrote to me about her mother, Mary, who had graduated from Montana State College with a degree in math and science before joining the WAVES. "Mom never mentioned her service until at least 50 years had passed," Ginny wrote. "She died in 1992, and we had not asked her any detailed questions. . . . It is difficult to realize that our mother, a brilliant and incredible woman, was resigned to a life of housewifery, post-WWII."

Other memories were uplifting and sometimes funny. Norman Torkelson wrote to me about his mother, Jean Theresa Pugh. He recalled that once he was sitting in his parents' den with his parents and siblings, watching a *60 Minutes* segment about World War II. The show mentioned the Battle of the Coral Sea and the code breaking that preceded that engagement. "I guess I can tell you what I did during the war now," his mother abruptly said. Torkelson's father began explaining that their mom had been a secretary, when she interrupted him. "No, Harold, I broke codes for the Navy and my girls worked on and broke that code." She stood up and high-fived everybody in the room. "Dad was speechless," Norman remembers.

Some readers are lucky to have their mothers still with them, and the book often led to conversations about what life was like back then, and forged a new bond. It made me think we should all interview our parents and grandparents more often. John Witt, son of Margaret Payne Witt, overnighted a copy of the book to his mother. "I know my mother feels tremendous pride in her contribution and now can express it openly for the first time."

Ada Nestor's son, John, gave her the book for Christmas, and, reading it at her home in Ohio, "she tuned us out and never put the book down."

John Elliott, an official with the State Department, has a Facebook page for his mom, Annette Dyar Sherman, who is still alive. Both his mother and his father worked in World War II cryptanalysis.

As I've traveled around the country giving talks about *Code Girls*, there sometimes have been real-live code girls in the audience, and audiences were always glad to get a chance to honor and thank them.

Before a talk in New Orleans, I received an email from a public servant named Val Spencer. "My 97 year old mother, Valora Spencer, lives with me," the message said. Valora the elder, who is deaf, had seen the talk advertised in the *New Orleans Advocate* and scrawled a note in the margin beside the ad, which she left on the kitchen table. "This is what I did during the war," it said. She then wrote another note on a notepad. "Honey Please try to get there and have her sign a book for me. I know I can't go but please go and get a signed book for me. Tell her I was a code girl. Take her picture on your phone for me." Then she wrote yet another note: "Valora D—I would like to have a copy of this book. Could you call and tell them I was a code girl and I'd like a copy." In the end both Vals were able to attend the talk. When the elder Val saw a projected image of Arlington Hall, she visibly trembled. The crowd gave her an ovation.

And I got emails (and Facebook messages) from code girls themselves, which was thrilling. "Thanks for all the research on subjects that I couldn't speak about, subject to the death penalty," wrote Jerry Scott. She described being recruited during her senior year at Smith and going to D.C. to work JN-25, living in a house with six other WAVES officers. "We worked watches and took a bus to the annex," she said. They had a rule that whoever cooked did not have to do dishes. "As I look back at those days, I sometimes feel guilty that we had such a good time during that brutal war."

"I was a code girl too, and knew most of the girls in your book—especially Ann White Kurtz," wrote Judy Parsons.

Relatives of code breakers have also approached me in person, including a grandnephew of none other than the great Agnes Driscoll, who came to a talk wearing his own naval uniform. I treasure every email and conversation, and the connection that this book has given me not only to the women I interviewed, but to hundreds of Americans for whom this book is part of their family story. I also heard from men who did this kind of work in Korea or Vietnam and who appreciated the pioneering work the women did. I met women who did the same work in Iraq in Operation Desert Storm. One man wrote to say he had been stationed at Arlington Hall decades after the war, and "I would like to think that I worked in the same room as Dot and Crow."

Wonderfully, some of the women I wrote about have been honored in their communities: Anne Seeley was profiled in her local paper, as was Ruth Mirsky; Jeuel Esmacher has given book talks to hundreds of people.

And then there's Dot! Dot has been profiled in any number of places, including her hometown paper in Lynchburg, and appeared with me on a panel at a literary festival in Charlottesville, where she brought down the house. She has attended book parties, and the after-parties of book parties. Most important to her, she has gotten her photo on the Wall of Honor at her assisted-living facility.

So many people in Arlington, Virginia, where I live, have written to tell me that they had Dorothy Ramale as a math teacher. "Mrs. Ramale (this salutation was the default back in the '70s) was a wonderful wonderful teacher. I loved math and it was a delight to be in her class (for this teen it was terrific to see a woman teaching this subject)," wrote Christine Payne.

Living in the Washington, D.C., area—where there are now seventeen intelligence agencies—I also came to understand how many officials in these agencies are descendants of code girls. These women begat much of our modern intelligence community, and many of our diplomats, it turns out, establishing a tradition of public service that echoed through the generations. I heard from Elizabeth Rood, a foreign service officer who is deputy chief of mission at the US embassy in Tbilisi, Georgia. "My mother, Mary Frances Munch Rood," was a code girl, she said; she graduated from high school at age sixteen and was working on a master's in classical languages at Yale when war broke out and she went to work at Arlington Hall. She went on to become a medical doctor and psychiatrist.

They also set a tradition, for women, of military service. I heard from Bill Nye the Science Guy, who believes that the wartime cryptologic background of his mother, Jacqueline Jenkins-Nye, is one reason he himself became a science educator (and a lover of swing music and the Lindy Hop). Bill said he often reflects on her funeral; as a naval veteran she was buried in Arlington National Cemetery, where three of the seven military "riflemen" firing her honorary salute were women. "My mom would have been absolutely delighted," he wrote; she was a staunch feminist.

Other readers were simply glad to know more about the times their own parents lived through. "It is the story of my mother's generation that I knew nothing about," wrote Deborah Rothman. People also understood that their own existence was often due to the fact that their mothers answered the call. I heard from Margaret Ellen Porter, whose mother, Margaret Jane Woods, met her future husband while she was in Washington breaking codes.

I think my favorite anecdote was shared at a book talk in Fredericksburg, Virginia. I am sorry that I did not catch the man's name. He recalled that both his parents served with the Navy during the war; his mother in Washington, his father in the Pacific. He knew they both were in code breaking but they would not say more than that, not even to each other. One day he took them to the cryptologic museum in Fort Meade, where there is a version of the Purple machine used by the Japanese diplomats. His mother was startled to see it; she had not realized, even now, that it is okay to acknowledge that it existed and that we broke it. "I worked on that machine!" she told them proudly. To which her husband replied, "*You* worked on Purple? *I* worked on Purple!" They belonged to the same code-breaking chain and did not even know it.

I invite you to read more personal accounts of code girls and their families—and to share your own—at the following website: www.codegirlsstories.com.

Liza Mundy
April 2018

Acknowledgments

My grateful thanks go, first, to the women who did this work during the war. Most took the secret to their graves, and it is too late, unfortunately, to thank them in person. I also am grateful to the women who consented to be interviewed for this book, many in circumstances that were not easy. Janice Martin Benario broke her wrist the night before our interview, so we conducted it in a hospital emergency room in Atlanta. Dot Braden Bruce took me to lunch, met with my own family, and always walked me to the door despite using a walker. Anne Barus Seeley invited my daughter and me to her Cape Cod home and drew columns on a piece of paper showing how she recovered additives. Margaret Gilman McKenna chatted over Skype. Ruth Schoen Mirsky brought out her scrapbooks. Viola Moore Blount shared recollections by email. Dorothy Ramale and Edith Reynolds White were confined to wheelchairs, but you'd never know it given how smartly they dressed. Suzanne Harpole Embree shared memories over Bloody Marys at the downtown D.C. Cosmos Club. When the Metro broke down, she walked several blocks and stood in line for the bus. Jo Fannon shared pamphlets she had saved for more than seventy years. Jane Case Tuttle wore the most awesome leopard-print bathrobe and gave me a gift bag of clean balled socks to throw at the television whenever a politician said something inane. It was easy to understand how women with this much spirit and fortitude helped the Allies win the war.

I also would like to thank the women in my family—my mother and grandmothers—who attended college. I still recall coming upon my grandmother Anna's old zoology notebooks from Hood College, and that kind of example makes an impression.

That this book exists is thanks to the efforts of many people who wished to see the women's story come to light. NSA historian Betsy Smoot, chief among these, shared advice, articles, contacts, links, and many patient explanations. Jennifer Wilcox, administrator at the National Cryptologic Museum, broke ground on this topic in several museum publications, and shared her files. At the Cryptologic Museum Library, Rene Stein, who knows where everything is and has digitized most of it, provided an instant response to my every question. In Dayton, Ohio, Deborah Anderson, the daughter of Joseph Desch, is one of the few Americans who sought to contact former code breakers, and perhaps the only person who orchestrated reunions to honor them. She has amassed a rich collection of photos and clippings and letters, which she shared over the course of several very fun days.

I am grateful to Mark Bradley, who was reading a declassified history of Venona, written by Robert L. "Lou" Benson, which mentioned the surprising number of female schoolteachers working on that project, and brought it to my attention. Many experts provided patient guidance, including Lou Benson himself. In addition, Robert Hanyok, Chris Christensen, and Jonathan Beard individually met with me to look over documents, give advice, explain (and re-explain), and share expertise. All were kind enough to read this book in manuscript form. Michael Warner was similarly encouraging, taking time from his own work to read the manuscript and offer insight. Julie Tate provided rigorous fact-checking and moral support. Any errors that have escaped their close and expert attention are mine alone.

Kristie Miller helped in many ways, including facilitating interviews with Ann Caracristi and sharing her knowledge of women's history. At New America, Brigid Schulte, Anne-Marie Slaughter, and the Better Life Lab provided crucial support without which this book could not have been written.

Several people did a service to the women by sharing notes from their own earlier projects. Mary Carpenter wrote a wonderful piece for the Wellesley alumnae magazine and, bless her, saved her notes, which she gave me. She has no idea how many times I reread them. Curt Dalton interviewed the women of Sugar Camp for his own excellent book, and generously provided the audiotapes. These are priceless records. Kerry Feduk at South Carolina ETV went to great lengths to locate and reproduce the full uncut interview with the late Frances Steen Suddeth Josephson, portions of which appear on their *South Carolina's Greatest Generation* DVD. Lieutenant Colonel Mike Bigelow from the U.S. Army Intelligence and Security Command History Office shared information on Arlington Hall as well as the oral histories conducted with WACs by Karen Kovach, who also met with me. Regina Akers, naval historian in the Histories and Archives Division at the Naval History and Heritage Command, shared her insights and guided me to those archives.

Any number of archivists helped bring records to light. Among these are Megan Harris, reference specialist with the Veterans History Project at the Library of Congress; Susanna Ola Lee, archivist at Winthrop University; Beth Ann Koelsch, curator of the Betty H. Carter Women Veterans Historical Project in the Hodges Special Collections and University Archives at the University of North Carolina at Greensboro; Tara Olivero, curator of special collections and archives at Goucher College; Nathaniel Patch, archivist in the reference section at the National Archives II facility at College Park; Paul Barron, director of Library and Archives at the George C. Marshall Foundation; Curt Dalton at Dayton History; Ellen Shea, head of research services at the Schlesinger Library at Radcliffe; Nanci Young, archivist at Smith College; Mary Yearl, archivist at Wellesley College; Jessica Smith, research services librarian at the Historical Society of Washington, D.C.; John Stanton, archivist at the Center for Local History at the Arlington County public library; Frances Webb and Ted Hostetler at Randolph College; Amy Hedrick at the Women Veterans Oral History Project at the University of North Texas; and Daniel A. Martinez, chief historian of the WWII Valor in the Pacific National Monument. Leila Kamgar at the U.S. State Department facilitated my tour of Arlington Hall, and Brandon Montgomery at the Department of Homeland Security showed me around the former naval code-breaking compound on Nebraska Avenue. David Sherman at

NSA did his best to get more wartime records declassified, though this remains an uphill battle.

I would also like to thank the family members who facilitated interviews and provided recollections. Chief among these is Jim Bruce, who put me in touch with his wonderful mother, Dot Braden Bruce, and was a cheerleader for this book all along. Others are Forrest White, who set up my interview with his mother, Edith Reynolds White; Cam Weber, who shared essays by her mother, Elizabeth Bigelow Stewart; Kitty Beller-McKenna, who helped me Skype with Margaret Gilman McKenna; Larry Gray, who wrote an essay about his late mother, Virginia Caroline Wiley; Sarah Jackson, who conducted an oral history with her late mother, (Miriam) Louise Pearsall Canby. Others who gave interviews were Barbara Dahlinger, Bill Cable, Carolyn Carter, and Kitty and Clyde Weston on behalf of Ruth Weston; Mike Sinkov for Delia Taylor Sinkov; Graham Cameron for Charlotte McLeod Cameron; Janice McKelvey for Sara Virginia Dalton; Laura Burke for Helen C. Masters; William Ludington for Georgia O'Connor Ludington; Linda Hund for Muriel Stewart; Eddie and Jonathan Horton, Virginia Cole, Eleanor Grabeel, and Daphne and Jerry Cole for Gene Grabeel; Pam Emmanuel for Martha Odum; Gerry Thompson for Nancy Abbott Thompson; Betty Dowse for the Wellesley code breakers; Jed Suddeth, Mary Isabel Randall Baker, Mabel Frowe, and Charlotte Anderson Stradford for Fran Steen Suddeth Josephson, as well as Rear Admiral (ret.) David K. Shimp, who did his best to honor and recognize her when she was alive.

Experts who shared their knowledge include Tom Johnson, David Hatch, Robert Lewand, William Wright, Suzanne Gould of AAUW, and Karen Kovach. Providing help and support along the way were Elizabeth Weingarten, Jaclyn Ostrowski, Christine Erskine, Rosalind Donald, Madonna Lebling, Nell Minow, Margaret Talbot, Ann Hulbert, Kate Julian, Denise Wills, Meagan Roper, Michael Dolan, Nancy Tipton, John Kirtland, Roy Caracristi, Allison Wood, and my family.

I am grateful to my longtime book agent, Todd Shuster of Aevitas Creative, who provided every kind of support and guided me into the office of Paul Whitlatch. Truly, this book could not have had a better editor. From our early meetings about form and content to the finishing touches, Paul has been a trusted wellspring of ideas and expertise. At Hachette Books, publisher Mauro DiPreta was supportive from our first meeting, as were marketing director Betsy Hulsebosch and associate publisher Michelle Aielli. Art director Amanda Kain came up with the perfect cover. I am grateful to publicity director Joanna Pinsker and production editor Carolyn Kurek, as well as Michael Gaudet, Jennifer Runty, Marisol Salaman, Odette Fleming, Carlos Esparza, Mark Harrington, and assistant editor Lauren Hummel, who kept everything running, and to Eileen Chetti for her expert copy editing. And to Chelsey Heller and Elias Altman at Aevitas.

I also would like to express my gratitude to those many authors whose books about aspects of war, code breaking, and twentieth-century history were so helpful in learning this terrain, as well as those whose writing about women and their achievements I found inspiring. This list includes, but by no means is limited to, Karen Abbott, David Alvarez, Christopher Andrew, Rick Atkinson, Julia Baird, Antony Beevor, Rosa Brooks, Stephen Budiansky, Elliot Carlson, Edward Drea, Glenn Frankel, David Garrow, Nathalia Holt, Ann Hulbert, Walter Isaacson, John Keegan, Denise Kiernan, Gayle Tzemach Lemmon, Jill Lepore, Candice Millard, Lynn Povich, John Prados, Gordon Prange, Stacy Schiff, Margot Lee Shetterly, Michael Smith, Dava Sobel, Margaret Talbot, and Katherine Zoepf. And I would like to especially thank David Kahn, who pioneered work in this field and has been a friend to so many in it. I am grateful for his generosity, insight, and of course the lunch at his club, to which he treated me, as he has done for so many authors, in the spirit of research and camaraderie.

Appendix 1
Glossary of Code-Breaking Terms

additive: A number added to a code group to make the message harder to crack.

cipher: A secret message system in which a single letter or number is replaced by another single letter or number.

code: A secret message system in which an entire word or phrase is replaced by another word, a series of letters, or a string of numbers known as a "code group."

crib: An educated guess, usually a word or phrase, about what a coded message says, used to help break the code or cipher. Cribs are often obtained when code breakers have access to a supplementary document, such as a public speech given by an official that might be quoted in the message, or when code breakers know a common expected word, like *maru*, the Japanese term for "supply ship," or *Wetter*, the German word for "weather."

cryptanalysis: The art and science of breaking codes and ciphers.

cryptography: The art and science of making codes and ciphers.

cryptology: Both making and breaking codes and ciphers.

enciphered (or superenciphered) system: A system in which words (or letters or phrases) are rendered as code groups consisting (usually) of digits, then further enciphered by adding a new digit to each digit. The Japanese Navy fleet code (JN-25) and the Japanese Army water transport code (2468) were both superenciphered systems.

indicator: A code group that specifies which part of an additive book has been used to further encipher a message.

key: This is a confusing word. Sometimes it's a synonym for "additive." Sometimes it means the daily "key" setting used to denote the order of rotors (and other mechanisms) for the Enigma and other ciphering machines, and sometimes it indicates a governing principle that controls how certain letters from enciphering tables are selected.

Appendix 2
World War II Timeline

World War II lasted for almost six years and involved thirty-eight countries on five continents. The timeline that follows outlines some of the key events of the war and the American code-breaking efforts.

1939
September 1

Nazi Germany invades Poland in a massive attack. Within weeks, Poland surrenders.

September 3

France and Great Britain declare war on Germany following the attack on Poland.

September 5

The United States declares that it will remain neutral.

1940
April 9

Nazi troops invade Denmark and Norway.

May 10

The Nazi invasion of Belgium, the Netherlands, Luxembourg, and France begins. Within a few weeks, all but France surrender.

May 26

The Battle of Dunkirk begins. British and French forces appear to be cut off, but over the next nine days more than three hundred thousand soldiers manage to avoid capture by the Nazis in an evacuation using fishing boats, yachts, lifeboats, and anything else that could float.

June 22

France officially surrenders to Germany.

July 10

Germany begins nightly bombings over Great Britain in what will come to be called the Battle of Britain.

September 20

Genevieve Grotjan finds the key to cracking the Japanese Purple machine, giving the United States access to all of Japan's diplomatic communications. The vital, top secret intelligence derived from these communications—in which Japanese diplomats in Europe reported back to Tokyo on what Hitler and other Axis leaders were planning and saying—was known as "Magic."

September 27

Germany, Italy, and Japan sign a pact to support one another and become the Axis powers.

1941
June 22

Germany invades its former ally, the Soviet Union. Within a few weeks, the Soviet Union joins Great Britain and becomes one of the Allied nations.

October

The U.S. Navy, sensing that it will need many more code breakers for America's entry into the war, makes a decision: It will recruit female college students. Navy liaison officers secretly reach out to the deans and presidents of women's colleges to identify and train top women from the senior classes. Training begins in secret and continues all during the women's senior year.

December 7

The Japanese bomb Pearl Harbor in Hawaii.

December 8

The United States, Great Britain, and Canada declare war on Japan.

December 11

Germany and Italy declare war on the United States.

1942
February 1

For quite some time, the British have been breaking Enigma ciphers. Now, however, German U-boats change the design of their Enigma machines, adding a fourth rotor. Allies largely lose the ability to read that system, endangering convoys at exactly the time that the United States is sending men and materiel across the Atlantic Ocean. Americans will get more and more involved in the Atlantic code-breaking effort, building new, larger, faster bombe machines.

February 15

Japan captures Singapore.

May 12

After several months of hard combat, Japan captures the Philippines.

May 14

President Roosevelt signs the Women's Army Auxiliary Corps (WAAC) bill into law.

May 26

Japan defeats the British in Burma and takes over that country.

June

Female college graduates begin arriving in Washington, D.C., to take up their code-breaking responsibilities.

June 7

The Allies defeat Japan in the Battle of Midway, thanks in large part to Allied code-breaking efforts. The battle marked a turning point in the war. The break was made by male naval officers in the Pacific, all of whom had been trained by one woman: former Texas schoolteacher Agnes Driscoll, who diagnosed how the Japanese naval fleet code worked.

July 21

President Roosevelt creates the women's naval reserve—the WAVES.

1943

January–February

Ann Caracristi and Wilma Berryman break the Japanese Army address code system and begin to excavate code groups revealing the place names of where Japanese Army units are located, and enabling the U.S. military to build order of battle reports.

February 2

German troops surrender in the Soviet Union. It is the first major defeat of Hitler's army.

February 7

After months of hard fighting, the last of the Japanese soldiers evacuate Guadalcanal in the western Pacific.

April 6–7

At Arlington Hall, working through the night, a male-female team of U.S. Army code breakers breaks 2468, the Japanese Army water-transport code. The vital breakthrough leads to the U.S. Army's massive recruitment of civilian schoolteachers, including Dot Braden, to exploit this break. Along with the Battle of Midway and the breaking of the Enigma cipher, the exploitation of 2468 will be one of the most important code-breaking triumphs of World War II and will lead to the sinking of thousands of Japanese supply ships, devastating the Japanese Army.

April 18

Thanks to the work of U.S. code breakers, the Allies are able to carry out Operation Vengeance over Bougainville Island in Papua New Guinea. Japan's admiral Isoroku Yamamoto, the mastermind behind the attacks on Pearl Harbor, is shot down and killed.

May 3

The first U.S. bombe machines, designed to break the four-rotor Enigma cipher, go into testing in Dayton, Ohio. The rotors have been wired by WAVES. Louise Pearsall is one of a group of female Navy mathematicians who travel to Dayton to help troubleshoot the machines.

May 13

The United States and Great Britain achieve a major victory over Germany in North Africa.

August 31

Bombe machines begin to arrive in Washington, where they are run by women, their messages translated by women, and the intelligence reports they generate compiled largely by women. Louise Pearsall and other top women mathematicians return from Dayton to Washington, D.C., to help get the machines up and running. These female mathematicians are held in a naval facility for hours, made to wash windows, and ridiculed by men thinking they are being sent down for bad behavior.

September 8

Italy surrenders to the Allies.

1944

June 6

D-Day. After months of planning and false radio traffic to mislead the Germans, the invasion of France begins. Female code breakers at the Naval Annex experience the landing from the point of view of Germans sending Enigma messages as the landing occurs.

October 26

Japan's navy is defeated by the Allies in the Battle of Leyte Gulf, near the Philippines. It is the first battle that includes organized attacks by kamikaze airmen and is the largest naval battle of the war.

December 16

In a last desperate attempt to avoid defeat, Hitler's forces attack Allied troops in the Ardennes forest in Belgium. The Battle of the Bulge, the largest land battle of World War II, begins.

1945

January 25

The Battle of the Bulge ends with an Allied victory.

March 26

After a monthlong battle, Allied troops capture the Pacific island of Iwo Jima.

April 30

Adolf Hitler commits suicide.

May 7

Germany surrenders to the Allies, ending the war in Europe. The next day becomes an official holiday—V-E Day, or Victory in Europe Day.

August 6

The United States drops an atomic bomb on the Japanese city of Hiroshima. Immediately, between 60,000 and 80,000 people die. The death toll will rise to 135,000 in the coming months.

August 9

A second atomic bomb is dropped on the city of Nagasaki. The death count reaches as high as 80,000 by the end of the year.

August 14

Japan surrenders. Code breakers in Washington, D.C., anxiously await the formal messages. The first person to read the surrender message is a female code breaker at Arlington Hall who is an expert in the lower-level diplomatic cipher system used to transmit it. She knows the war is over even before U.S. president Harry Truman does. The news spreads through Arlington Hall.

September 2

Japan formally surrenders to the Allies in Tokyo Bay on the USS *Missouri*.

Reading Group Guide

Discussion Questions

1. What particular skills and characteristics did the Army and Navy look for in the women recruited to their code-breaking programs? How were stereotypes about women employed or challenged in the recruitment effort?

2. How did World War II affect personal and romantic relationships? What were Americans' attitudes toward marriage then—and did those attitudes change at all for the code girls' generation?

3. Why do you think Dot Braden and Ruth "Crow" Weston became such great friends? If they had met in other circumstances or in peacetime, do you think they would have gotten along just as well?

4. Consider the various motivations Mundy cites for the women who signed up as code breakers. Do you think they differed from those of the men serving in America's military then?

5. Some of the code girls were affected by the extended secrecy of their work. How might keeping secrets, however necessary, affect a person's relationships or her identity in the world?

6. What were the particular successes and struggles of Agnes Driscoll? Why might she have eventually resorted "to extreme measures to retain her authority"?

7. What does it mean that the organizational hierarchy of Arlington Hall was relatively "flat"? How was this beneficial to the code girls?

8. Frank Raven, while acknowledging the skills of the "damn good gals," also concluded that many of the code girls were "damn pretty gals." What effect might this statement and the perspective of people like Raven have had on the women and their work?

9. Barnard's Virginia Gildersleeve noticed in the marching WAVES "a remarkable cross section of the women of the United States of America, from all our economic and social classes...and from all our multitude of racial origins and religions." What might have caused such diversity and cooperation, and how do you think this changed after the war, if at all?

10. What were the challenges for many of the women after the war?

11. Why do you think these women's contributions to cryptanalysis remained a secret for so long?

12. Mundy suggests that "many of the code-breaking women... advance[d] the feminist movement." Do you agree?

13. In January 2016, the American armed services finally lifted a ban on women serving in positions of direct combat. What challenges do you think women still face in the military today?

Notes

Three file collections from the National Archives at College Park have been consulted and are frequently cited. The full citations are:

RG 38, Entry 1030 (A1), Records of the Naval Security Group Central Depository, Crane Indiana, CNSG Library.

RG 0457, Entry 9002 (A1), National Security Agency/Central Security Service, Studies on Cryptology, 1917–1977.

RG 0457, Entry 9032 (A1), National Security Agency/Central Security Service, Historic Cryptographic Collection, Pre–World War I Through World War II.

Transcripts of oral history interviews with the "NSA-OH" ID are from Oral History Interviews, National Security Agency, https://www.nsa.gov/news-features/declassified-documents/oral-history-interviews/index.shtml.

Transcripts of oral history interviews and associated personal materials with the "WV" ID are from the Betty H. Carter Women Veterans Historical Project, Martha Blakeney Hodges Special Collections and University Archives, the University of North Carolina at Greensboro, NC. http://libcdm1.uncg.edu/cdm/landingpage/collection/WVHP/.

The Secret Letters

"Get that fellow's number," he told his junior officer: Gordon W. Prange, *At Dawn We Slept* (New York: Penguin, 1982), 517.

Ann White, a senior at Wellesley College: Ann White Kurtz, "An Alumna Remembers," *Wellesley Wegweiser*, no. 10 (Spring 2003): 3, https://www.wellesley.edu/sites/default/files/assets/departments/german/files/weg03.pdf; Ann White Kurtz, "From Women at War to Foreign Affairs Scholar," *American Diplomacy* (June 2006), http://www.unc.edu/depts/diplomat/item/2006/0406/kurt/kurtz_women.html; Mary Carpenter and

Betty Paul Dowse, "The Code Breakers of 1942," *Wellesley* (Winter 2000):26–30, as well as the underlying notes to that article, which Mary Carpenter shared with the author.

Elizabeth Colby, a Wellesley math major: Carpenter and Dowse, "Code Breakers of 1942."

Anne Barus received her own letter: Anne Barus Seeley, naval code breaker, interview at her Cape Cod home on July 12, 2015.

At Bryn Mawr, Mount Holyoke, Barnard: Craig Bauer and John Ulrich, "The Cryptologic Contributions of Dr. Donald Menzel," *Cryptologia* 30, no. 4 (2006): 306–339. RG 38, Box 113, "CNSG-A History of OP-20-3-GR, 7 Dec 1941–2 Sep 1945," says that the first year, cooperating schools were Barnard, Bryn Mawr, Mount Holyoke, Radcliffe, Smith, Wellesley, and Goucher. Vassar was "under consideration" but does not seem to have cooperated that year. The second year, Vassar and Wheaton were added.

If pressed, they could say: Bauer and Ulrich, "Cryptologic Contributions of Dr. Donald Menzel," 310.

In their introductory meetings: Kurtz, "An Alumna Remembers"; Kurtz, "From Women at War to Foreign Affairs Scholar"; Bauer and Ulrich, "Cryptologic Contributions of Dr. Donald Menzel," 310.

They hid homework under desk blotters: Carpenter and Dowse, "Code Breakers of 1942."

At Goucher, it was an English professor—Ola Winslow: Frederic O. Musser, "Ultra vs Enigma: Goucher's Top Secret Contribution to Victory in Europe in World War II," *Goucher Quarterly* 70, no. 2 (1992): 4–7; Janice M. Benario, "Top Secret Ultra," *Classical Bulletin* 74, no. 1 (1998): 31–33; Robert Edward Lewand, "Secret Keeping 101: Dr. Janice Martin Benario and the Women's College Connection to ULTRA," *Cryptologia* 35, no. 1 (2010): 42–46; Frederic O. Musser, *The History of Goucher College, 1930–1985* (London: Johns Hopkins University Press, 1990), 40, https://archive.org/details/historyofgoucher00muss.

One of the most well-liked students in the Goucher class: Ida Jane Meadows Gallagher, "The Secret Life of Frances Steen Suddeth Josephson," *The Key* (Fall 1996): 26–30; Fran Josephson, uncut interview with South Carolina Educational Television conducted for a DVD called *South Carolina's Greatest Generation.*

At Vassar, nestled in the hills: Edith Reynolds White, naval code breaker, interview at her home in Williamsburg, Virginia, on February 8, 2016.

At first, the Army approached some of the same colleges: RG 38, Box 113, "CNSG-A History of OP-20-3-GR, 7 Dec 1941–2 Sep 1945."

At Indiana State Teachers College: Dorothy Ramale, Arlington Hall and naval code breaker, interviews at her home in Springfield, Virginia, on May 29 and July 12, 2015.

The Army dispatched handsome officers: Dr. Solomon Kullback, oral history interview on August 26, 1982, NSA-OH-17-82, 72; Ann Caracristi, Arlington Hall code breaker, interviews at her home in Washington, D.C., between November 2014 and November 2015.

And so it was that on a Saturday: Dorothy Braden Bruce, Arlington Hall code breaker, interviews at her home near Richmond, Virginia, between June 2014 and April 2017; Personnel Record Folder for War Department Civilian Employee (201) file: "Bruce, Dorothy B., 11 June 1920 Also: Braden, Dorothy V., B-720," National Personnel Records Center, National Archives, St. Louis, MO.

Introduction: "Your Country Needs You, Young Ladies"

Listening in on enemy conversations: David Kahn talks about the intelligence uses of code breaking in many of his writings, including "Pearl Harbor and the Inadequacy of Cryptanalysis," *Cryptologia* 15, no. 4 (1991): 273–294, DOI: 10.1080/0161-119191865948.

The chain of events that led to the women's recruitment: The naval recruiting program, the meetings between the college leaders, and the letters from Comstock, Safford, Noyes, and Menzel are described in Craig Bauer and John Ulrich, "The Cryptologic Contributions of Dr. Donald Menzel," *Cryptologia* 30 (2006): 306–339. The repository of many of these letters, which I also consulted, is Radcliffe's Schlesinger Library, "Office of the President Correspondence and Papers: 1941–42, Harvard-NA, II, Ser. 2," Box 57: 520–529, "National Broadcasting—Naval Communications."

Even before Comstock received: Virginia C. Gildersleeve, "We Need Trained Brains," *New York Times*, March 29, 1942; "Women's College Speed Up," *New York Herald Tribune*, January 24, 1942.

The women's college leaders met at Mount Holyoke: Bauer and Ulrich, "Cryptologic Contributions of Dr. Donald Menzel," 306. Radcliffe, "National Broadcasting—Naval Communications."

Most were in the top 10 percent: Bauer and Ulrich, 313.

(A memo from a Radcliffe administrator explaining: Radcliffe, "National Broadcasting—Naval Communications."

The chosen women not only were cautioned: Bauer and Ulrich, 310.

Pembroke, the women's college affiliated with Brown University: RG 38, Box 113, "CNSG-A History of OP-20-3-GR, 7 Dec 1941–2 Sep 1945."

There was controversy over whether the course: Bauer and Ulrich, "Cryptologic Contributions of Dr. Donald Menzel," 312. Radcliffe, "National Broadcasting—Naval Communications."

He added that, in the Navy's view: Bauer and Ulrich, "Cryptologic Contributions of Dr. Donald Menzel," 311.

By mid-April 1942, Donald Menzel: Ibid., 312.

The women were told that just: Ann White describes this in Mary Carpenter and Betty Paul Dowse, "The Code Breakers of 1942," *Wellesley* (Winter 2000): 26–30. Many other women described it as well.

"Whether women can take it over successfully": Bauer and Ulrich, "Cryptologic Contributions of Dr. Donald Menzel," 310. Radcliffe, "National Broadcasting—Naval Communications."

The women recruits were entering an environment: Robert Louis Benson, *A History of U.S. Communications Intelligence During World War II: Policy and Administration* (Washington, DC: Center for Cryptologic History, National Security Agency, 1997), provides invaluable background on the wartime competition between the Army, Navy, and any number of other federal agencies. Also see RG 38, Box 109, "Resume of Development of American COMINT Organization, 15 Jan 1945." There is also an excellent discussion of the Army-Navy competition in Stephen Budiansky, *Battle of Wits: The Complete Story of Codebreaking in World War II* (New York: Touchstone, 2000), 87.

"Nobody cooperated with the Army, under pain of death": Prescott Currier, oral history interview on November 14, 1980, NSA-OH-38-80, 37.

One British liaison described: Budiansky, *Battle of Wits*, 296.

In the field of astronomy, women long had been employed: Patricia Clark Kenschaft, *Change Is Possible: Stories of Women and Minorities in Mathematics* (Providence: American Mathematical Society, 2005), 32–38.

"It was generally believed": Ann Caracristi, "Women in Cryptology" speech presented at NSA, April 6, 1998.

"The women who gathered together in our world": Ann Caracristi, interview, undated, Library of Congress Veterans History Project, https://memory.loc.gov/diglib/vhp-stories/loc.natlib.afc2001001.30844/transcript?ID=mv0001.

"Don't worry, you'll always have enough": Jeanne Hammond, interview at her home in Scarborough, Maine, on September 30, 2015.

Edith Reynolds, recruited out of Vassar: Edith Reynolds White, naval code breaker, interview with the author.

Suzanne Harpole, a Wellesley code breaker: Suzanne Harpole Embree, naval code breaker, interview at the Cosmos Club in Washington, D.C., on August 11, 2015.

"Come at once; we could use you in Washington": Jeuel Bannister Esmacher, Arlington Hall code breaker, interview at her home in Anderson, South Carolina, on November 21, 2015.

"Nothing had been filed": Jaenn Coz Bailey, oral history interview on January 13, 2000, WV0141.

Women were considered more polite: Kenneth Lipartito, "When Women Were Switches: Technology, Work, and Gender in the Telephone Industry, 1890–1920," *American Historical Review* 99, no 4 (October 1994): 1084.

Soon after Pearl Harbor, however, companies like Hercules Powder: Betty Dowse, telephone interview with the author. In the *Bryn Mawr Alumnae Bulletin*, the president of Bryn Mawr described the dilemma facing educators. Under normal circumstances, they discouraged women from going into math and science, especially fields like physics: "It would have been hard to urge them when there was little promise of a job and a good salary." Now, she said, "there is a new situation for women here, a demand that has never existed for them before." She worried that "the problem of the prospects in science after the war is a serious one for the women's colleges." Katharine E. McBride, "The College Answers the Challenge of War and Peace," *Bryn Mawr Alumnae Bulletin* 23, no. 2 (March 1943): 1–7.

Youth, the memo noted, is "a time": RG 0457, 9032 (A1), Box 778, "Signal Intelligence Service, General Files, 1932–1939."

Lever Brothers and Armstrong Cork also needed chemists: These are all real examples, cited in surveys of members of the class of 1943 at Wellesley College, conducted by Betty Dowse, in which class members were asked what work they did during the war and whether it had been held by a man before they took it on. 6C/1942, Betty Paul Dowse, A01-078a, Wellesley College Archives.

One electrical company asked for: Beatrice Fairfax, "Does Industry Want Glamour or Brains?" *Long Island Star Journal*, March 19, 1943.

On the eve of Pearl Harbor: Army personnel numbers are in RG 0457, 9002 (A1), Box 92, SRH 349, "The Achievements of the Signal Security Agency in World War II." Navy numbers are in RG 0457, 9002 (A1), Box 63, SRH 197, "U.S. Navy Communication Intelligence Organization Collaboration."

Many of the program's major successes: RG 38, Box 4, "COMNAVSECGRU Commendations Received by OP-20-G."

Chapter One: Twenty-Eight Acres of Girls

Unfortunately for Dot Braden, neither: Here and throughout, the details of Dot Braden's application, hiring, employment, and life in Washington are taken from approximately twenty author interviews conducted with Dorothy Braden Bruce, in person at her home near Richmond, Virginia, and over the phone, between June 2014 and April 2017. They also are from her Personnel Record Folder for War Department Civilian Employee (201) file: "Bruce, Dorothy B., 11 June 1920 Also: Braden, Dorothy V., B-720," National Personnel Records Center, National Archives, St. Louis, MO.

Lynchburg was not a big city: A good description of Lynchburg and Randolph-Macon is in Writers' Program of the Work Projects Administration in the State of Virginia, *Virginia: A Guide to the Old Dominion* (New York: Oxford University Press, 1940), 264–266.

As Dot stepped out of the train: An invaluable description of D.C.'s Union Station in wartime is in William M. Wright, "White City to White Elephant: Washington's Union Station Since World War II," *Washington History: Magazine of the Historical Society of Washington, D.C.* 10, no. 2 (Fall/Winter 1998–99): 25–31.

Other women were arriving: In addition to Dorothy's recollection and her personnel file, which includes the signed loyalty oath, this description of the first day at Arlington Hall comes from the author's interviews with Josephine Palumbo Fannon, at her home in Maryland, on April 9 and July 17, 2015. Fannon swore in new hires, and likely was the self-possessed young woman Dot remembers.

Like Arlington Hall, this cluster of female-only dormitories: An excellent history and description of Arlington Farms is in Joseph M. Guyton, "Girl Town: Temporary World War II Housing at Arlington Farms," *The Arlington Historical Magazine* 14, no. 3 (2011): 5–13.

"There's a new army on the Potomac": The *Good Housekeeping* and *Reader's Digest* articles are cited in Megan Rosenfeld, "'Government Girls': World War II's Army of the Potomac," *Washington Post*, May 10, 1999, A1.

Dot didn't know this, but she had found her way into: That Arlington Hall was the biggest message center in the world is in Ann Caracristi, interview, undated, Library of Congress Veterans History Project, https://memory.loc.gov/diglib/vhp-stories/loc.natlib .afc2001001.30844/transcript?ID=mv0001, in comments by Jack Ingram, curator of the National Cryptologic Museum.

Around the time Dot was hired, a harried: RG 0457, 9032 (A1), Box 1016, "Signals Communications Systems."

In America in the 1940s, three-quarters: Claudia Goldin, "Marriage Bars: Discrimination Against Married Women Workers, 1920s to 1950s" (NBER Working Paper 2747, National Bureau of Economic Research, October 1988).

To say that Dot was "trained" would be an overstatement: There are many Arlington Hall training documents, including RG 0457, 9032 (A1), Box 1007, "Training Branch Annual Report," and RG 0457, 9032 (A1), Box 1114, "History of Training in Signal Security Agency and Training Branch."

Specifically, Dot Braden was assigned to: RG 0457, 9032 (A1), Box 1114, "SSA, Intelligence Div, B-II Semi-Monthly Reports, Sept 1942–Dec 1943" shows Dorothy Braden's name on the roster and that she had been assigned to Department K, as part of the sixth group undergoing orientation that fall and winter. It shows that Ruth Weston was in the same orientation group, assigned to research. The mission of Department K is described in RG 0457, 9032 (A1), Box 115, "Organization of Military Cryptanalysis Branch," 12.

Chapter Two: "This Is a Man's Size Job, but I Seem to Be Getting Away with It"

All of these qualities were characteristic of Elizebeth Smith: This description of Elizebeth Friedman's life and background is drawn primarily from "Elizebeth Friedman Autobiography at Riverbank Laboratories, Geneva, Illinois" in the National Cryptologic Museum Library's David Kahn Collection, DK 9-6, in Fort Meade, MD; "Elizebeth Smith Friedman Memoirs—Complete," at the George C. Marshall Foundation, in Lexington, VA, in the Elizebeth Smith Friedman Collection, http://marshallfoundation.org/library/digital-archive/elizebeth-smith-friedman-memoir-complete; "Interview with Mrs. William F. Friedman conducted by Dr. Forrest C. Pogue at the Marshall Research Library, Lexington, Virginia, May 16–17, 1973," http://marshallfoundation.org/library/wp-content/uploads/sites/16/2015/06/Friedman_Mrs-William_144.pdf; and oral history interviews with Elizebeth Friedman on November 11, 1976, NSA-OH-1976-16, NSA-OH-1976-17, and NSA-OH-1976-18.

Bacon, an English statesman and philosopher: An excellent description of Bacon, the biliteral cipher, Riverbank, and the photo are in William H. Sherman, "How to Make Anything Signify Anything," *Cabinet*, no. 40 (Winter 2010–2011), www.cabinetmagazine.org/issues/40/sherman.php. My discussion is also indebted to a talk William Sherman gave at the George C. Marshall Foundation, "From the Cipher Disk to the Enigma Machine: 500 Years of Cryptography" (George C. Marshall Legacy Series sequence on Codebreaking, Lexington, VA, April 23, 2015), and an exhibit he curated at the Folger Shakespeare Library titled *Decoding the Renaissance: 500 Years of Codes and Ciphers* (Washington, DC, November 11, 2014, to February 26, 2015).

Toward that end, Fabyan also had hired: Many sources and biographies of William Friedman have been consulted. In addition to the Elizebeth Friedman memoirs cited in the first note to this chapter, some of the most useful are Rose Mary Sheldon's "The Friedman Collection: An Analytical Guide" to the George C. Marshall

Foundation's extensive William F. Friedman collection, http://marshallfoundation
.org/library/wp-content/uploads/sites/16/2014/09/Friedman_Collection_Guide
_September_2014.pdf; the foundation's brief introduction to its series on code
breaking, "Marshall Legacy Series: Codebreaking," George C. Marshall Foundation,
http://marshallfoundation.org/newsroom/marshall-legacy-series/codebreaking/;
and *The Friedman Legacy: A Tribute to William and Elizebeth Friedman* (Washington,
DC: Center for Cryptologic History, National Security Agency, 2006), https://www
.nsa.gov/resources/everyone/digital-media-center/video-audio/historical
-audio/friedman-legacy/assets/files/friedman-legacy-transcript.pdf.

Codes have been around for as long as civilization, maybe longer: William Friedman
discusses the history of codes and ciphers in *Friedman Legacy*. Also see David Kahn,
The Codebreakers (New York: Scribner, 1967); and Stephen Budiansky, *Battle of Wits:
The Complete Story of Codebreaking in World War II* (New York: Free Press, 2000), 62–68.

Armchair philosophers amused themselves pursuing the "perfect cipher": Robert
Edward Lewand, "The Perfect Cipher," *Mathematical Gazette* 94, no. 531 (November 2010): 401–411, points out that the Vigenère square, invented in 1586, enjoyed
a "good long run" as the *le chiffre indéchiffrable* in that no attacker could divine the
keyword, until it was solved almost three hundred years later by two "Victorian
polymaths," the English mathematician Charles Babbage and a Prussian Army officer, Friedrich Wilhelm Kasiski, within about ten years of each other.

During the American Civil War: A good discussion of Civil War cryptography is in RG
0457, 9032 (A1), Box 1019, "Notes on History of Signal Intelligence Service."

Among those who visited Riverbank during this time: Betsy Rohaly Smoot, "An
Accidental Cryptologist: The Brief Career of Genevieve Young Hitt," *Cryptologia* 35,
no. 2 (2011): 164–175, DOI: 10.1080/01611194.2011.558982.

Known as the "Hello Girls," these were the first American women: Jill Frahm,
"Advance to the 'Fighting Lines': The Changing Role of Women Telephone Operators
in France During the First World War," *Federal History Journal*, no. 8 (2016): 95–108.

The interwar period was not an auspicious time for American code breaking: RG 38,
Box 109, "Resume of Development of American COMINT Organization, 15 Jan 1943."

Yardley was a genial and charismatic man: Frank Rowlett, oral history interview in
1976, NSA-OH-1976-1-10, 87–89. That his employees were mostly women is in RG
0457, 9032 (A1), Box 1019, "Notes on History of Signal Intelligence Service," 44.

Stimson in 1929 shuttered the operation: This comment has been cited far and wide,
including in David Kahn, "Why Weren't We Warned?" *MHQ: Quarterly Journal of
Military History* 4, no. 1 (Autumn 1991): 50–59, and *Friedman Legacy*, 200.

"Our impression—and I think it was a mistaken one—was that": Solomon Kullback,
oral history interview on August 26, 1982, NSA-OH-17-82, 9–11.

Even this influx wasn't enough, however, and it occurred: Susan M. Lujan, "Agnes
Meyer Driscoll," *NCVA Cryptolog*, special issue (August 1988): 4–6.

Born in 1889 in Illinois, Meyer attended Otterbein College: Kevin Wade Johnson, *The
Neglected Giant: Agnes Meyer Driscoll* (Washington, DC: Center for Cryptologic History,
National Security Agency, 2015), https://www.nsa.gov/about/cryptologic-heritage

/historical-figures-publications/publications/assets/files/the-neglected-giant/the
_neglected_giant_agnes_meyer_driscoll.pdf.

In the War Department's "general address and signature" code: RG 0457, 9002 (A1),
Box 91, SRH 344, "General Address and Signature Code No. 2."

She hacked the nut jobs, broke enemy devices and machines that inventors: Colin
Burke, "Agnes Meyer Driscoll vs the Enigma and the Bombe," monograph, http://
userpages.umbc.edu/~burke/driscoll1-2011.pdf.

Impressed, Hebern lured Agnes away to help him develop a better one: That Agnes was
dissatisfied with her advancement prospects is suggested in Johnson, *Neglected Giant*, 9.

"Friedman was always two, three [pay] grades ahead of her": Captain Thomas Dyer's
pay grade comment quoted in Johnson, *Neglected Giant*, 21. That Dyer felt she was
"fully his equal" is in Steven E. Maffeo, *U.S. Navy Codebreakers, Linguists, and Intelligence Officers Against Japan, 1910–1941* (Lanham: Rowman & Littlefield, 2016), 68.

In 1920 George Fabyan wrote a complimentary letter: RG 38, Box 93, "COMNAVSEC-
GRU Letters Between Col Fabyan of Riverbank Laboratories and US Navy Oct
1918–Feb 1932."

In her own civilian post, Agnes would go on to train: Edwin T. Layton, Roger Pineau,
and John Costello, *And I Was There: Pearl Harbor and Midway—Breaking the Secrets*
(New York: Morrow, 1985), 33.

Japan had defeated Russia in 1905 and it clearly wanted: David Kahn, "Pearl Harbor and the Inadequacy of Cryptanalysis," *Cryptologia* 15, no. 4 (1991): 275, DOI:
10.1080/0161-119191865948.

By then, a "research desk" had been set up: Kahn, "Why Weren't We Warned?" 51.

She cursed like a—well, like a sailor: Layton et al., *And I Was There*, 58.

The tiny Navy team—Driscoll, one or two officers: Elliot Carlson, *Joe Rochefort's War:
The Odyssey of the Codebreaker Who Outwitted Yamamoto at Midway* (Annapolis, MD:
Naval Institute Press, 2011), 40.

"The reason you're not getting anywhere": Layton et al., *And I Was There*, 46.

"Mrs. Driscoll got the first break": Robert J. Hanyok, "Still Desperately Seeking 'Miss
Agnes': A Pioneer Cryptologist's Life Remains an Enigma," *NCVA Cryptolog* (Fall 1997): 3.

**Her success "was the most difficult cryptanalytic task ever performed up to that
date":** RG 0457, 9002 (A1), Box 36, SRH 149, "A Brief History of Communications
Intelligence in the United States," by Laurance F. Safford, 11.

"There is . . . only one fully trained individual among the permanent force": Johnson, *Neglected Giant*, 20.

On June 1, 1939, the Japanese fleet began using: Descriptions of how JN-25 worked are
in RG 38, Box 116, "CNSG-OP-20-GYP History for WWII Era (3 of 3)" and "CNSG
History of OP-20-GYP-1 WWII (1 of 2)."

"First break [was] made by Mrs. Driscoll. Solution progressing satisfactorily": RG 38,
Box 115, "CNSG OP-20-GY History." Also Kahn, "Pearl Harbor and the Inadequacy
of Cryptanalysis."

Years after World War II ended: "I saw not all that long ago maybe eight or ten years
ago, two people who had served in FRUPAC and one in Washington, who were still

arguing about the value of a code group." Captain Prescott Currier, oral history interview on April 14, 1972, NSA-OH-02-72, 32.

"In the navy she was without peer as a cryptanalyst": Layton et al., *And I Was There*, 58.

"If the Japanese Navy had changed the code-book along with the cipher keys": RG 0457, 9002 (A1), Box 36, SRH 149, "A Brief History of Communications Intelligence in the United States" by Laurance F. Safford, 15.

Chapter Three: The Most Difficult Problem

On an upper floor, occupying: David Kahn, "Pearl Harbor and the Inadequacy of Cryptanalysis," *Cryptologia* 15, no. 4 (1991): 282, DOI: 10.1080/0161-119191865948.

Sometimes in private they called him "Uncle Willie": RG 0457, 9002 (A1), Box 17, SRH 58, "The Legendary William F. Friedman."

As the staff expanded, Friedman had done something else: RG 0457, 9032, Box 751, "SIS Organization and Duties/SIS Personnel," and Box 779, "Signal Intelligence Service (SIS) General Correspondence Files."

The Coast Guard's mission was enforcing "neutrality": R. Louis Benson Interview of Mrs. E. S. Friedman, January 9, 1976, Washington, D.C., https://www.nsa.gov/news-features/declassified-documents/oral-history-interviews/assets/files/nsa-OH-1976-22-efriedman.pdf.

In October 1939, following the outbreak of war in Europe: RG 0457, 9002 (A1), SRH 361, "History of the Signal Security Agency," vol. 1, "Organization," part 1, "1939–45." NSA Cryptologic Histories, https://www.nsa.gov/news-features/declassified-documents/cryptologic-histories/assets/files/history_of_the_signal_security_agency_vol_1SRH364.pdf.

Berryman hailed from Beech Bottom, West Virginia: Wilma Berryman Davis, oral history interview on December 3, 1982, NSA-OH-25-82, 2–8.

Asking around, Wilma Berryman found out: A history of the naval correspondence course is in Chris Christensen and David Agard, "William Dean Wray (1910–1962) the Evolution of a Cryptanalyst," *Cryptologia* 35, no. 1 (2010): 73–96, DOI: 10.1080/01611194.2010.485410.

It was a teach-yourself kind of place: Frank Rowlett, oral history interview in 1976 (otherwise undated), NSA-OH-1976-(1-10), 380.

"What four things were thought by Captain Hitt to be essential": RG 0457, 9032, Box 751, "Army Extension Course in Military Cryptanalysis."

Since the United States was not, strictly speaking, at war: Rowlett in his oral history NSA-OH-1976-(1-10), 350, says, "We knew it was illegal" but "we figured that as long as we didn't let it be openly published that we were still legal if we intercepted and if we cryptanalyzed...that we sort of had a little bit of an island to stand on."

The Communications Act of 1934: RG 0457, 9032 (A1), Box 1019, "Notes on History of Signal Intelligence Service," 76.

And there was twenty-seven-year-old Genevieve Marie Grotjan: RG 0457, 9032, Box 780, "Signal Intelligence Service—General Correspondence file 1941"; Genevieve Grotjan, interview with David Kahn, May 12, 1991, National Cryptologic Museum

Library, Fort Meade, MD, David Kahn Collection, DK 35–44, notes; Personnel Record Folder for War Department Civilian Employee (201) file: "Grotjan, G.," National Personnel Records Center, National Archives, St. Louis, MO.

She was known for her thoroughness, powers of observation: Frank B. Rowlett, *The Story of Magic: Memoirs of an American Cryptologic Pioneer* (Laguna Hills, CA: Aegean Park Press, 1998), 128.

Friedman's office kept its own "nut file," recording the outlandish systems: Abraham Sinkov, oral history interviews, NSA-OH-02-79 through NSA-OH-04-79.

The first message in the new machine cipher: RG 0457, 9002 (A1), SRH 361, "History of the Signal Security Agency," vol. 2, "The General Cryptanalytic Problems," 31–32. Also available online from NSA "Cryptologic Histories."

The British had tried to break: RG 0457, 9002 (A1), Box 92, SRH 349, "The Achievements of the Signal Security Agency in World War II," 17.

The U.S. Navy, in the wing next door, worked on Purple: Frank Rowlett, oral history interview on June 26, 1974, NSA-OH-01-74 to NSA-OH-12-74, available in notebooks (not online) at National Cryptologic Museum Library, 457.

Friedman's team had figured out that the old Red: Kahn, "Pearl Harbor and the Inadequacy of Cryptanalysis," 280–281; "History of the Signal Security Agency," vol. 2, 32; William Friedman, "Preliminary Historical Report on the Solution of the 'B' Machine," October 14, 1940, https://www.nsa.gov/news-features/declassified-documents/friedman-documents/assets/files/reports-research/FOLDER_211/41760789079992.pdf; Rowlett, *Story of Magic*, 145.

Frank Rowlett liked to go to bed early: Kahn, "Pearl Harbor and the Inadequacy of Cryptanalysis," 282.

William Friedman often thought of solutions: Davis, oral history, 13.

He studied how French letters behaved: RG 0457, 9032 (A1), Box 751, "Army Extension Course in Military Cryptanalysis."

They had mastered the behavior of romanized Japanese: Rowlett, *Story of Magic*, 117.

They reviewed the workings of known machines: Ibid., 138.

When the Purple machine was being installed in Japanese embassies: Ibid., 139.

The code breakers talked to radio intercept operators: Ibid., 141.

Friedman liked his team to do their own pen work: Grotjan interview with Kahn.

Cribs are educated guesses about what: Kahn, "Why Weren't We Warned?" 56. Stephen Budiansky, *Battle of Wits: The Complete Story of Codebreaking in World War II* (New York: Free Press, 2000) also describes the importance of cribs.

The fact that they had broken the sixes meant: "History of the Signal Security Agency," vol. 2, 33–34.

They theorized that the Purple machine was using some kind of switching device: Rowlett, *Story of Magic*, 150.

Their hopes renewed by this latest theory, Frank Rowlett and his Purple team: "History of the Signal Security Agency," vol. 2, 41.

Mary Louise Prather—keeping her meticulous files: RG 0457, 9002 (A1), Box 37, SRH 159, William F. Friedman, "The Solution of the Japanese Purple Machine."

"We were looking for this phenomenon," he would later say: Frank Rowlett, oral history, NSA-OH-01-74 to NSA-OH-12-74, 283.

Sitting there engrossed in what Rowlett later rather sheepishly: Ibid.

She was "obviously excited": Rowlett, *Story of Magic*, 151.

Then, at the end of a long stream of letters, she circled a third: Some accounts have her circling three spots, but Rowlett, *Story of Magic*, 152, says four.

While she stood quietly, they erupted in cheers: Ibid.

"I was just doing what Mr. Rowlett told me to do": Grotjan interview with Kahn.

"When Gene . . . brought in those worksheets and pointed out": Frank Rowlett, oral history, NSA-OH-01-74 to NSA-OH-12-74, 284.

Three years later, Friedman wrote a top secret: William Friedman, "Recommendations for Legion of Merit and Medal of Merit Awards," September 27, 1943, https://www.nsa.gov/news-features/declassified-documents/friedman-documents/assets/files/correspondence/FOLDER_529/41771309081039.pdf.

His private announcement was made on September 27, 1940: "History of the Signal Security Agency," vol. 2, 44.

"What was really mysterious was the fact": RG 0457, 9002 (A1), Box 81, SRH 280, "An Exhibit of the Important Types of Intelligence Recovered Through Reading Japanese Cryptograms."

"England and America are jingling money": RG 0457, 9032 (A1), Box 833, "Diplomatic Translations of White House Interest, 1942–1943."

Each day, messages like these were deciphered: RG 0457, 9002 (A1), Box 78, SRH 269, Robert L. Benson, "US Army Comint Policy: Pearl Harbor to Summer 1942," notes that at first raw message transcripts were delivered, but intelligence officials later drew up summaries. Kahn, "Why Weren't We Warned?" says that fifty to seventy-five intercepts were solved and translated each day; the most important were sent on, carried in locked briefcases. The messenger then retrieved the papers and burned them.

"Through their almost naïve confidence": RG 0457, 9032 (A1), Box 1115, "History of the Language Branch, Army Security Agency."

So eager were the two services for credit: The odd-even compromise is discussed in a number of places, including Budiansky, *Battle of Wits*, 168. That Frank Raven, with the U.S. Navy, found a way to predict the keys is in John Prados, *Combined Fleet Decoded* (New York: Random House, 1995), 165.

The Purple machine could not predict the attack: A good description of what could and could not be read on December 7, 1941, is in Robert J. Hanyok, "How the Japanese Did It," *Naval History* (December 2009): 48–49. Kahn points out that the Imperial Japanese Navy did not clue in the Japanese diplomats in "Why Weren't We Warned?" 59.

Driving back from a tour of the proposed installation: These descriptions of Arlington Hall are taken from a sheaf of internal histories and briefing papers provided to the author by Michael Bigelow, historian at the U.S. Army Intelligence and Security Command.

Delia Taylor and Wilma Berryman had rented rooms one summer: Wilma Berryman Davis, oral history interview, December 3, 1982, NSA-OH-25-82, 10.

"The finishing school atmosphere was shattered": RG 0457, 9032 (A1), Box 1125, "Signal Security Agency, History of the Cryptographic Branch."

The Purple machine was installed: Budiansky, *Battle of Wits*, 226.

There was a French code they called Jellyfish: RG 0457, 9032 (A1), Box 1114, "Signal Security Agency Weekly Reports, Jan to Oct 1943," Weekly Report for Section B-III, July 9, 1943.

"The outstanding solution of the week was that of the SAUDI cipher": RG 0457, 9032 (A1), Box 1114, "Signal Security Agency, B-III Weekly Reports Oct–Dec 1943," Weekly Report for October 9, 1943.

Sometimes, when she was riding the bus: Grotjan interview with Kahn.

Chapter Four: "So Many Girls in One Place"

She had grown up in the crossroads of Bourbon, Mississippi: Kitty Weston, interview at her niece's home in Oakton, Virginia, on April 10, 2015, and Clyde Weston, telephone interview on October 9, 2015. Also Personnel Record Folder for War Department Civilian Employee (201) file: "Weston, Carolyn Cable," National Personnel Records Center, National Archives, St. Louis, MO.

The goal of the lecture series, This Is Our War: RG 0457, 9002 (A1), Box 17, SRH 057, Lecture Series This Is Our War, Autumn 1943.

Early in December 1943: Curtis Paris to Dot Braden, December 4, 1943.

Chapter Five: "It Was Heart-Rending"

The two diplomats were hatching plans to sow discord: RG 0457, 9032 (A1), Box 606, "Items of Propaganda Value."

Japan had a brilliant top commander: RG 0457, 9002 (A1), Box 71, SRH 230, Henry F. Schorreck, "The Role of COMINT in the Battle of Midway," *Cryptologic Spectrum* (Summer 1975): 3–11.

The U.S. Navy had a small cryptanalytic team: Captain Rudolph T. Fabian, oral history interview on May 4, 1983, NSA-OH-09-83, 8–30.

In the Atlantic, things were going equally badly: RG 0457, 9002 (A1), Box 95, SRH 367, "A Preliminary Analysis of the Role of Decryption Intelligence in the Operational Phase of the Battle of the Atlantic."

The Battle of the Atlantic began: David Kahn, *Seizing the Enigma* (New York: Barnes and Noble, 2001), vii.

To be sure, the Allies for some of this time: "The British were using an enciphered code for the convoy thing and we were convinced that the Germans were reading it. And we told them that and it was hard to persuade them that it was true....I think we did some monitoring to see if we could prove our point, which we couldn't." Dr. Howard Campaigne, oral history interview on June 29, 1983, NSA-OH-14-83, 49.

Even if it was theoretically possible: Lieutenant Howard Campaigne said that a fact-finding team went to Germany after the war and "we found that the Germans were well aware of the way the Enigma could be broken, but they had concluded that it would take a whole building full of equipment to do it. And that's what we had. A building full of equipment. Which they hadn't pictured as really feasible." Campaigne, oral history, 15–16.

The Poles broke the Enigma: Chris Christensen, "Review of IEEE Milestone Award to the Polish Cipher Bureau for the 'First Breaking of Enigma Code,'" *Cryptologia* 39, no. 2 (2015): 178–193, DOI: 10.1080/01611194.2015.1009751; Ann Caracristi, interview, undated, Library of Congress Veterans History Project, https://memory .loc.gov/diglib/vhp-stories/loc.natlib.afc2001001.30844/transcript?ID=mv0001, comments by National Cryptologic Museum curator Jack Ingram.

In January 1941, naval code breaking consisted: RG 38, Box 110, "Historical Review of OP-20-G."

According to a November 1941 proposed salary memo: RG 38, Box 1, "CNSG Officer/ Civilian Personnel Procurement 1929–1941 (1 of 2)."

The early women came from a variety: Francis Raven, oral history interview on March 28, 1972, NSA-OH-03-72, 3, provided the Puffed Rice anecdote.

"Could you start within a week or two after": Personnel Record Folder for War Department Civilian Employee (201) file: "Beatrice A. Norton," National Personnel Records Center, National Archives, St. Louis, MO.

By now the women's ranks had winnowed: RG 38, Box 113, "CNSG-A History of OP-20-3-GR, 7 Dec 1941–2 Sep 1945."

Goucher graduates Constance McCready: RG 38, Box 1, "CNSG, General Personnel, 4 Dec 1941–31 Jan 1944 (3 of 3)."

Fearful the women might quit if they couldn't find housing: Radcliffe, Schlesinger Library, "Office of the President Correspondence and Papers: 1941–42, Harvard-NA, II, Sec. 2," Box 57: 520–529, "National Broadcasting—Naval Communications."

The women would start each day: Ann Ellicott Madeira, interview, undated, Library of Congress Veterans History Project, https://memory.loc.gov/diglib/vhp-stories/ loc.natlib.afc2001001.07563/transcript?ID=sr0001.

Vi Moore, a French major from Bryn Mawr: Viola Moore Blount, email correspondence between April 22 and April 30, 2016.

Margaret Gilman, who had majored in biochemistry: Margaret Gilman McKenna, Skype interview on April 18, 2016.

"German submarines were literally controlling": Ibid. The work of the cribbing group can be found in RG 38, Box 63, "Crib Study of Message Beginnings, Signature, etc-German Weather Msgs (1941–1943)."

Ann White also was assigned to the Enigma unit: Ann White Kurtz, "An Alumna Remembers," *Wellesley Wegweiser*, no. 10 (Spring 2003), https://www.wellesley.edu /sites/default/files/assets/departments/german/files/weg03.pdf; Mary Carpenter, underlying notes for Mary Carpenter and Betty Paul Dowse, "The Code Breakers of 1942," *Wellesley* (Winter 2000): 26–30.

Everyone we knew and loved: Ibid.

Erma Hughes, a psychology major: Erma Hughes Kirkpatrick, oral history interview on May 12, 2001, WV0213.

The disaster of Pearl Harbor had called into question: A good analysis of the state of affairs is in Schorreck, "Role of COMINT."

Often, though, Washington had to wait: RG 38, Box 116, "CNSG-OP20-GYP History for WWII Era (3 of 3)."

Things had begun looking up for the code breakers: Frederick Parker, *A Priceless Advantage: U.S. Navy Communications Intelligence and the Battles of Coral Sea, Midway, and the Aleutians* (Washington, DC: Center for Cryptologic History, National Security Agency, 1993).

So Rochefort and Edwin Layton: Edwin T. Layton, Roger Pineau, and John Costello, *And I Was There: Pearl Harbor and Midway—Breaking the Secrets* (New York: Morrow, 1985), 421.

"He knew the targets; the dates; the debarkation points": Schorreck, "Role of COMINT."

The Japanese duly showed up on June 4: Laurance E. Safford, "The Inside Story of the Battle of Midway and the Ousting of Commander Rochefort," 1944, in Naval Cryptologic Veterans Association, *Echoes of Our Past* (Pace, FL: Patmos Publishing, 2008), 26.

Even so, "the Battle of Midway gave the Navy confidence": RG 38, Box 116, "CNSG-OP20-GYP History for WWII Era (3 of 3)."

The Midway victory also set in motion: Safford, "Inside Story," 27; Stephen Budiansky, *Battle of Wits: The Complete Story of Codebreaking in World War II* (New York: Free Press, 2000), 23; John Prados, *Combined Fleet Decoded* (New York: Random House, 1995), 410–411.

"I never felt that I should go tell her that the world had fallen": Captain (ret.) Prescott Currier, oral history interview on November 14, 1988, NSA-OH-38-80, 44.

"She became fearful that she wouldn't:" Campaigne, oral history, 33–34.

Before her accident, "she was a very strikingly beautiful woman": Francis Raven, oral history interview on January 24, 1983, NSA-OH-1980-03, 11.

"You can't visualize the climate around Aggie": Ibid., 86.

"There wasn't a regular Navy officer except Safford": Ibid., 18.

So Raven decided to pillage Agnes Driscoll's safe: Ibid., 13–25.

Driscoll was the "curse of the Enigma effort": Ibid., 34.

One of the officers had a hair-raising collection: Ibid., 42–43.

As the war went on, Agnes tended to be given projects: Campaigne, oral history, "And so there was a period there when she was given assignments which were very difficult assignments, and everybody else had given up on them. And they were given to her more or less to keep her busy. They figured they were hopeless anyhow and there wasn't anything bad she could do," 34.

During a lull in his own work, Raven returned: Francis Raven, oral history, NSA-OH-1980-03, 56–63.

"In retrospect I am convinced that Aggie Driscoll": Ibid., 37–38.

Room assignments were made and remade: RG 38, Box 1, "CNSG-General Personnel, 5 Dec 1940–31 Jan 1944"; RG 38, Box 2, "CNSG-Civilian Personnel, 18 Feb 1942–31 Dec 1943 (3 of 3)"; RG 38, Box 116, "CNSG History of OP-20-GYP-1 WWII Era (3 of 3)."

Vi Moore found herself doing additive: July 1942 room assignments are in RG 38, Box 117, "CNSG-OP-20-GY-A/GY-A-1."

The volume of messages grew: RG 38, Box 115, "CNSG OP-20-GY History."

The women rose to the challenge: Ann Barus Seeley described her additive work, in detail, in interviews with the author. Her memories are supported by many archival files that confirm "tailing," additives, and *shoo-goichi* messages, including in RG 38, Box 116, "CNSG-OP20-GYP History for WWII Era (3 of 3)."

Looking at her work sheet: Anne Barus Seeley, interview. Also, Elizabeth Corrin, from Smith College, recalled: "They lined up the messages horizontally. So, what you tried to do was to get the additive that would work with the whole column....The code had to be divisible by three. So, if you added up all the digits—the five digits—it had to be divisible by three...the people who worked in the priority rooms saw the more interesting things....If we broke a message, we'd pass it to the priority room and they had the codebook." Elizabeth Corrin, oral history interview on February 8, 2002, NSA-OH-2002-06.

Elizabeth Bigelow, an aspiring architect: Elizabeth Bigelow Stewart, essay of reminiscence, shared with the author by her daughter Cam Weber.

The operation developed the swiftness: RG 38, Box 116, "CNSG-OP20-GYP History for WWII Era (3 of 3)" and "CNSG History of OP-20-GYP-1 WWII Era (1 of 2)."

By the fourth quarter of 1943: RG 38, Box 115, "CNSG OP-20-GY History."

And they shared the outrage: The chain of events around the news stories is detailed in Safford, "Inside Story," 27–30.

The Japanese periodically changed JN-25 books: Parker, in *Priceless Advantage*, 66, notes, "Whether the Japanese ever discovered that U.S. cryptologists had successfully penetrated their most secret operational code...remains a matter of conjecture to this day," but at the time, officials within OP-20-G were convinced of it.

During the battle, U.S. Marines came: RG 38, Box 116, "CNSG History of OP-20-GYP-1 WWII (1 of 2)."

Among these were Bea Norton and Bets Colby: The decision to implement a minor-cipher unit, and the fact that JN-20 was a substitution-transposition cipher, is in RG 38, Box 116, "CNSG-OP20-GYP History for WWII Era (3 of 3)." The makeup of the minor-cipher unit is in RG 38, Box 117, "CNSG-Op-20-GY-A/GY-A-1," and RG 38, Box 115, "CNSG OP-20-GY History (1,2,3,4,5)." Bea Norton Binns described the minor ciphers in Carpenter and Dowse, "Code Breakers of 1942," and Carpenter, underlying notes.

"Whenever the main code was not being read": RG 38, Box 4, "COMNAVSECGRU Commendations Received by OP-20G."

When U.S. Marines hit the beaches of Guadalcanal: RG 38, Box 117, "CNSG History of OP-20-GYP-1 (Rough), 1945 (1 of 2)."

"I have not seen the sea for two weeks": Ibid.

Her college training course did come in handy: Carpenter, underlying notes for Carpenter and Dowse, "Code Breakers of 1942."

"JN-20 ciphers were broken with increasing speed": RG 38, Box 116, "CNSG-OP20-GYP History for WWII Era (3 of 3)."

Bets Colby, a math major from Wellesley, was a favorite of Raven: Raven, oral history, NSA-OH-1980-03, 55.

"I felt so lucky to be in this small interesting unit": Carpenter, underlying notes for Carpenter and Dowse, "Code Breakers of 1942."

"Never in my life since have I felt": Carpenter and Dowse, "Code Breakers of 1942."

The only hitch was the heat: Carpenter, underlying notes for Carpenter and Dowse, "Code Breakers of 1942."

"The women are arriving in great numbers": Graig Bauer and John Ulrich, "The Cryptologic Contributions of Dr. Donald Menzel," *Cryptologia* 30, no. 4, (2006): 313.

Chapter Six: "Q for Communications"

Women were proving so useful to the war effort: A good summary of the creation of the WAVES and their training is in an unpublished history by Jacqueline Van Voris, "Wilde and Collins Project," Folder "Women in the Military, Box 6," Ready Reference Section, Naval History and Heritage Command in Washington, D.C., as well as articles cited below. Also Jennifer Wilcox, *Sharing the Burden: Women in Cryptology During World War II* (Washington, DC: Center for Cryptologic History, National Security Agency, 2013), and underlying files from the Wilcox archives that she provided.

The WAACS, coming first, bore the brunt: Harriet F. Parker, "In the Waves," *Bryn Mawr Alumnae Bulletin* 23, no. 2 (March 1943): 8–11.

10,000 WOMEN IN U.S. RUSH TO JOIN NEW ARMY CORPS: Lucy Greenbaum, *New York Times*, May 28, 1942, A1.

Despite fears that women would become hysterical: Mattie E. Treadwell, *United States Army in World War II: Special Studies; The Women's Army Corps* (Washington, DC: Center of Military History, United States Army, 1995), 290–91.

When Nimitz polled the naval bureaus—the branches of the Navy: D'Ann Campbell, "Fighting with the Navy: The WAVES in World War II," in Sweetman, Jack, ed., *New Interpretations in Naval History: Selected Papers from the Tenth Naval History Symposium Held at the United States Naval Academy, 11–13 September 1991* (Annapolis, MD: Naval Institute Press, 1993), 344.

"If the Navy could possibly have used dogs or ducks or monkeys": Virginia Crocheron Gildersleeve, *Many a Good Crusade* (New York: MacMillan, 1954), 267.

"Volunteer" assured the public that women: Ibid., 273.

Virginia Gildersleeve would later recall that at one: Ibid., 271.

McAfee, appalled, thought this so gaudy: D'Ann Campbell, "Fighting with the Navy," 346.

"Utility was sacrificed to looks": Gildersleeve, *Many a Good Crusade*, 272.

Others chose the Navy over the Army: Myrtle O. Hanke, oral history interview, on February 11, 2000, WV0147 Myrtle Otto Hanke Papers.

"The work the women are now doing is too important": Wilcox, *Sharing the Burden*, 5.

"Hide your disappointment," he urged his students: Smith College Archives, 1939–45 WAVES, Box 1, "Broadcasts by Mr. Davis 1942–1943."

The heels presented a problem: The description of officer training is drawn from Fran Steen interviews; Nancy Dobson Titcomb, interview with the author at her home in Maine on October 1, 2015; Nancy Gilman McKenna, interview with the author; Edith Reynolds White, interview with the author; Anne Barus Seeley, interview with the author; and Viola Moore Blount, correspondence with the author.

Ordered to salute the first lady: Erma Hughes Kirkpatrick, oral history interview on May 12, 2001, WV0213.

Erma Hughes, the bricklayer's daughter, came to Smith: Ibid.

The women retaliated by singing, with spirit: Frances Lynd Scott, *Saga of Myself* (San Francisco: Ithuriel's Spear, 2007).

During services the men would sing the original and the women: Ibid.

They were missed so badly that most were snatched: Ann White Kurtz, "From Women at War to Foreign Affairs Scholar," *American Diplomacy* (June 2006), http://www.unc.edu/depts/diplomat/item/2006/0406/kurt/kurtz_women.html.

Cars would have fender benders: Mary Carpenter and Betty Paul Dowse, "The Code Breakers of 1942," *Wellesley* (Winter 2000): 28.

Their bosses were glad to see them: Ibid.

Blanche DePuy sensed veiled resentment: Carpenter, underlying notes for Carpenter and Dowse, "Code Breakers of 1942."

Nancy Dobson from Wellesley was asked: Nancy Dobson Titcomb, interview with the author.

There was nobody who outranked her: Fran Steen Suddeth Josephson, interview with South Carolina Educational Television.

Jaenn Coz, a bored librarian: Jaenn Coz Bailey, oral history interview on January 13, 2000, and papers, WV0141.

At the outset there was a cap on the number: Van Voris, "Wilde and Collins Project," 17.

Georgia O'Connor joined the WAVES out of curiosity: Georgia O'Connor Ludington, oral history interview on September 5, 1996, NSA-OH-1996-09, 4.

Ava Caudle joined because: Ava Caudle Honeycutt, naval code breaker, oral history interview on November 22, 2008, and papers, WV0438.

"I had such a yearning to do something": Hanke, oral history.

Ida Mae Olson: Ida Mae Olson Bruske, naval code breaker, telephone interview with the author on May 8, 2015.

On the train to Cedar Falls, Betty Hyatt: Betty Hyatt Caccavale, naval code breaker, oral history interview on June 18, 1999, and papers, WV0095.

Since southerners were considered slow: Betty Hyatt Caccavale, *Sing On Mama, Sing On*, self-published memoir, shared with the author.

The women officers continued to train: A description of the Hunter College boot camp is in Campbell, "Fighting with the Navy," 349.

Jaenn Magdalene Coz, the librarian from California: WV0141 Jaenn Coz Bailey papers.

Other women were in for shocks of a different nature: Veronica Mackey Hulick, telephone interview with the author, undated.

The WAVES by mid-1943 were a big deal: Gildersleeve, *Many a Good Crusade*, recalls the mayor's last-minute calls, 285, and the splendid marching, 278.

It was not only women from rural families: Jane Case Tuttle, interview with the author at her home in Scarborough, Maine, on September 30, 2015.

The Navy began to cast around for a bigger facility: A good history of Mount Vernon Seminary, the naval takeover, and the transformation of the campus is in Nina Mikhalevsky, *Dear Daughters: A History of Mount Vernon Seminary and College* (Washington, DC: Mount Vernon Seminary and College Alumnae Association, 2001), 63–135.

Cover names were proposed, such as "Naval Research Station": RG 38, Box 81, "CNSG Staff Conference Notes-Oct-Dec 1942."

It was located near: Elizabeth Allen Butler, *Navy Waves* (Charlottesville, VA: Wayside Press, 1988), 38.

When Elizabeth Bigelow showed up: Elizabeth Bigelow Stewart, essay of reminiscence, shared with the author by her daughter Cam Weber.

On rainy days, the women code breakers took off: Tuttle interview.

When she went in and out: Seeley interview.

One day, Jane Case, the former debutante: Tuttle interview.

After arriving in Washington: Ruth Rather Vaden reminiscence; Jennifer Wilcox archives.

At Mount Vernon the work remained the same: A description of the modifications of the Mount Vernon campus and the increasingly female makeup of most units is in RG 38, Box 110, "A Historical Review of OP-20-G."

As it happened, Elizebeth Friedman also moved: RG 0457, 9002 (A1), Box 79, SRH 270, Robert L. Benson, "The Army-Navy-FBI Comint Agreements of 1942," explains the complex system by which the Coast Guard since 1940 had been intercepting and processing German intelligence covert traffic to and from Germany and the Western Hemisphere, passing it to the FBI as well as other entities, including the British. In March 1942 this operation was merged into OP-20-G.

The publishing world was represented: Stewart, essay of reminiscence.

The women called it the Booby Hatch: Kirkpatrick, oral history.

Suzanne Harpole, recruited from Wellesley: Suzanne Harpole Embree, interview with the author.

When Ensign Marjorie Faeder reported for duty: Marjorie E. Faeder, "A Wave on Nebraska Avenue," *Naval Intelligence Professionals Quarterly* 8, no. 4 (October 1992): 7–10.

Yeoman Ruth Schoen was put to work: Ruth Schoen Mirsky, interviews with the author.

Jaenn Coz one day was whistling: WV0141 Jaenn Coz Bailey papers.

He liked to publish a little mimeographed broadsheet: "Packard, Wyman (Capt, USN), Naval Code Room (OP-19C) Watch 3," in folder "Packard, Wyman Papers of Capt USN 1944–1945," Ready Reference Section, Naval History and Heritage Command in Washington, D.C.

Jane Case—told, growing up, that she was: Tuttle interview.

On May 25, 1943, the Naval Annex additive recovery room: RG 38, Box 119, "Daily Log of Room 1219 (Additive Recovery) 1 Apr–23 June 1943."

They had not only exceeded 2,500; they had broken: Ibid.

"Please convey to all hands my congratulations:" Ibid.

Before long, she had clearance to go in almost any room: Caccavale, oral history.

Betty Hyatt was on duty in 1944 when a naval: Caccavale, *Sing On Mama, Sing On*, and WV0095 Betty Hyatt Caccavale papers.

Ruth Schoen was the only Jewish member: Descriptions of Ruth's background, and the friendship group, are from Mirsky interviews and from Butler, *Navy Waves*.

Georgia O'Connor was next in the group to marry: This description is from Butler, *Navy Waves*, and Bill Ludington (Georgia's son), telephone interview with the author.

"The history of the Navy in the Pacific is the history": Francis Raven, oral history interview on January 24, 1983, NSA-OH-1980-03, 79–81.

One enlisted member of the WAVES "had such a knack for running": RG 38, Box 116, "CNSG-OP20-GYP History for WWII Era (3 of 3)."

Every year, Suzanne Harpole got a standard form: Embree interview.

Might women officers be taught to shoot?: Staff meetings regarding teaching WAVES to shoot, women not being saluted, and the need to come up with a consistent cover story for the Q rating are in RG 38, Box 81, "CSNG Staff Conference Notes, Jul–Dec 1943."

Jelleff's department store had a fashion show: Mirsky interview.

Vi Moore heard the Budapest String Quartet: Many of the details about life in Washington are from Viola Moore Blount, correspondence with the author.

The women visited a roadside joint: Kirkpatrick, oral history.

Jaenn Coz, whose mother had been a flapper: WV0141 Jaenn Coz Bailey papers.

A group of WAVES officers lived in a house: Titcomb interview.

Ida Mae Olson invited her friend Mary Lou: Bruske interview.

When Jane Case learned that her father was dying: Tuttle interview.

The war, for her, was "this period of very": Embree interview.

In 1944, Eleanor Roosevelt and: Campbell, "Fighting with the Navy," 351.

In a June 1945 memo: J. N. Wenger, "Memorandum for Op-20-1," June 26, 1945, Wilcox archives.

As hard as the women worked, there were lighthearted moments: Tuttle interview.

Edith Reynolds, from Vassar, found herself: White interview.

One code breaker was standing in a movie line: Lyn Ramsdell Stewart (who was not the WAVES member in question), telephone interview with the author on October 27, 2015.

"I hesitate to write this letter": RG 38, Box 1, "CNSG-General Personnel, 5 Dec 1940–31 Jan 1944."

In late 1943, Wellesley's Bea Norton, now married: Carpenter, underlying notes for Carpenter and Dowse, "Code Breakers of 1942." Her resignation is recorded in RG 38, Box 1, "COMNAVSECGRU-OP-20G Headquarters Personnel Rosters & Statistics" (3 of 4).

She conceived on her honeymoon: Scott, *Saga of Myself*, 167, and Library of Congress oral history interview.

Dorothy Ramale, the would-be math teacher: Dorothy Ramale, interviews with the author.

It was a WAVES officer, Ensign Janet Burchell: RG 38, Box 91, "CNSG-COMNAVSECGRU Joint Army-Navy Liaison."

The Navy women had just missed taking part: Details about the Yamamoto message breaking are in RG 38, Box 138, files marked "Yamamoto Shootdown, 1–4." That the minor cipher JN-20 "carried further details" of Admiral Yamamoto's last tour of inspection is in RG 38, Box 116, "CNSG-OP20-GYP History for WWII Era (3 of 3)" and "CNSG History of OP-20-GYP-1 WWII (1 of 2)." Bea Norton Binns, in a September 27, 1998, letter to her class secretary, said that the inter-island cipher led "to many opportunities for our forces, including the shooting down of Admiral Yamamoto's plane taking him on an inspection tour." Carpenter, underlying notes for Carpenter and Dowse, "Code Breakers of 1942." Fran Steen talks about taking part in the Yamamoto effort in her interview with South Carolina ETV. In her Library of Congress oral history, Ann Ellicott Madeira mentions being excited "when the work we did ensured that our flyers were able to shoot down Yamamoto."

"The day his plane went down": Hanke, oral history.

Chapter Eight: "Hell's Half-Acre"

Young Annie Caracristi washed her hair with laundry soap: Wilma Berryman Davis, oral history interview, December 3, 1982, NSA-OH-25-82, 39.

One of the bookish men, a New York editor: Robert L. Benson, former NSA historian, interview with the author in The Plains, Virginia, in June 2015.

At Arlington Hall there also were "BIJs": Ann Caracristi, interview, undated, Library of Congress Veterans History Project, https://memory.loc.gov/diglib/vhp-stories/loc.natlib.afc2001001.30844/transcript?ID=mv0001; Stuart H. Buck, "The Way It Was: Arlington Hall in the 1950s," *Phoenician* (Summer 88): 3–11.

Josephine Palumbo at eighteen was virtually running: Josephine Palumbo Fannon, interviews with the author on April 9 and July 17, 2015.

Unlike the Navy, Arlington Hall also had an: Jeannette Williams with Yolande Dickerson, *The Invisible Cryptologists: African-Americans, WWII to 1956* (Washington, DC: Center for Cryptologic History, National Security Agency, 2001), https://www.nsa.gov/about/cryptologic-heritage/historical-figures-publications/publications/wwii/assets/files/invisible_cryptologists.pdf.

When Juanita Morris, a college student fresh from North Carolina: Juanita Moody, oral history interview on June 12, 2003, NSA-OH-2003-12.

When the code breakers figured out how to rig a Coke machine: Solomon Kullback, oral history interview on August 26, 1982, NSA-OH-17-82, 119.

Designed to hold 2,200 people, it quickly proved inadequate: Descriptions of the grounds and physical plant are in RG 0457, 9032 (A1), Box 1370, "Signal Security Agency Summary Annual Report for the Fiscal Year 1944."

"You didn't go by rank," said Solomon Kullback: Kullback, oral history, 117.

During 1942 the U.S. Army and Navy had hammered out: Robert Louis Benson, *A History of U.S. Communications Intelligence During World War II: Policy and Administration* (Washington, DC: Center for Cryptologic History, National Security Agency, 1997).

Part of the problem, at first, had been a lack of message traffic: Kullback, oral history, 34–37; David Alvarez, *Secret Messages: Codebreaking and American Diplomacy, 1930–1945* (Lawrence: University Press of Kansas, 2000), 150.

The job was just too big: Davis, oral history, 11; RG 0457, 9032 (A1), Box 1016, "Signals Communications Systems."

Each unit remained tied to its home base in Japan: Kullback, oral history, 38.

Lewis, the Utah-bred son of an Englishman turned cowboy: Douglas Martin, "Frank W. Lewis, Master of the Cryptic Crossword, Dies at 98," *New York Times*, December 3, 2010, http://www.nytimes.com/2010/12/03/arts/03lewis.html.

It was not unusual to find an exhausted code breaker napping in a tub: Davis, oral history, 24.

"Visualize, if you will, the entire communications set-up": RG 0457, 9002 (A1), Box 95, SRH 362, "History of the Signal Security Agency," vol. 3, "The Japan Army Problems: Cryptanalysis, 1942–1945."

The Japanese devised a host of minor codes: RG 0457, 9002 (A1), Box 92, SRH 349, "The Achievements of the Signal Security Agency in World War II," 23.

She had been recruited out of Russell Sage: Ann Caracristi, interviews with the author.

Thinking it a bit of a lark, the three friends: Ibid.; Ann Caracristi, oral history interview on July 16, 1982, NSA-OH-15-82, 2.

Soon enough, Ann too found herself laboring: Caracristi, interview, Library of Congress Veterans History Project.

The suggestion—it came to be known as de-duping: Caracristi, oral history, 7.

To the naked eye, the major Japanese Army code systems: RG 0457, 9032 (A1), Box 831, "Japanese Army Codes Solution Section."

Wilma Berryman was assigned to the address problem in April 1942: RG 0457, 9032 (A1), Box 1016, "Signals Communications Systems."

At the suggestion of one military officer: Caracristi, oral history, 10.

They had caught a small break, though: RG 0457, 9032 (A1), Box 1016, "Signals Communications Systems."

They agreed that chaining differences was "silly": Caracristi, oral history, 10.

At times, the Japanese Army was obliged to send messages over Navy radio circuits: RG 0457, 9032 (A1), Box 831, "Japanese Army Codes Solution Section."

"I sort of remembered having seen something in that file": Davis, oral history, 51. The usefulness of the Navy cribs is described in RG 0457, 9032 (A1), Box 831, "Japanese Army Codes Solution Section"; RG 0457, 9032 (A1), Box 827, "Monthly Report No. 5, 15 February 1943."

Ann Caracristi dove in, blissfully at home: Caracristi, oral history, 11.

Arlington Hall began producing weekly memos: RG 0457, 9032 (A1), Box 1114, "SSA, Intelligence Div, B-II Semi-Monthly Reports, Sept 1942–Dec 1943."

On March 15, 1943, a memo: RG 0457, 9032 (A1), Box 827, "Monthly Report No. 6, 15 March 1943."

If a code group was 0987, for example, 098 was the actual code group: RG 0457, 9032 (A1), Box 831, "Japanese Army Codes Solution Section"; RG 0457, 9032 (A1), Box 1016, "Signals Communications Systems."

The address codes carried a bounty: RG 0457, 9002 (A1), SRH 349, Box 92, "The Achievements of the Signal Security Agency in World War II," 25.

"That outfit was 100 percent female": Kullback, oral history, 113–115.

Wilma's team worked alongside a unit: Davis, oral history, 41.

"I think that's one of the things that made it": Ibid., 43–53.

"We were in an awful pickle, because it was war": Ibid., 48–49. The odds and evens problem is also discussed in RG 0457, 9032 (A1), Box 831, "Japanese Army Codes Solution Section," and RG 0457, 9032 (A1), Box 827, "Monthly Report May 15, 1943."

"It was fascinating, actually, to work in the world": Caracristi, interview with the author.

When Solomon Kullback received visitors: Kullback, oral history, 39.

Wilma Berryman would give Annie: Caracristi, interviews with the author.

Years later, when Solomon Kullback was asked whom: Caracristi, interview, Library of Congress Veterans History Project.

The address unit also did its bit: Davis, oral history, 37.

In the long-running beef: Solomon Kullback said that "the attitude of the Navy was such that they didn't want to tell the Army people too much because the Army was practically all civilian . . . they didn't trust the security." Oral history, 121.

Wilma liked to say nobody working on: Davis, oral history, 26.

Ann and Wilma and a few others even: Ibid., 22.

The code breakers formed a glee club: Caracristi, interview, Library of Congress Veterans History Project.

Gert, Ann, and Wilma would save up: Davis, oral history, 43.

Tooth and nail they worked: Ibid., 38–43.

One civilian woman complained often: Caracristi, oral history, 15.

There was competition: Ibid., 22.

"The mere statement of facts and figures": RG 0457, 9002 (A1), Box 95, SRH 362, "History of the Signal Security Agency," vol. 3.

Then in April 1943, a couple of things happened: The breaking of 2468 is described in RG 0457, 9032 (A1), Box 827, "Monthly Report No. 7, 15 April 1943," and RG 0457, 9032 (A1), Box 1016, "Signals Communications Systems," 393; Joseph E. Richard, "The Breaking of the Japanese Army's Codes," *Cryptologia* 28, no. 4 (2004): 289–308, DOI: 10.1080/0161-110491892944; Peter W. Donovan, "The Indicators of Japanese Ciphers 2468, 7890, and JN-25A1," *Cryptologia* 30, no. 3 (2006): 212–235, DOI: 10.1080/01611190 500544695.

"New life has been given to the entire section": RG 0457, 9032 (A1), Box 827, "Monthly Report No. 7, 15 April 1943."

In July 1943, one of the first 2468 messages: Kullback, oral history, 81.

The break into 2468 was one of the most important: David Kahn, *The Codebreakers* (New York: Scribner, 1967), 594.

"What nicer bit of information": Kullback, oral history, 80.

Buoyed and elated, Arlington Hall became: RG 0457, 9032 (A1), Box 831, "Japanese Army Codes Solution Section."

They attacked a major administrative code: RG 0457, 9032 (A1), Box 1016, "Signals Communications Systems," 244–246.

"There wasn't a damn thing that the Japanese": Kullback, oral history, 87–89.

In January 1944, Australian soldiers: Donovan, "Indicators of Japanese Ciphers"; Kullback discusses Japanese code security in his NSA oral history, 40.

Chapter Nine: "It Was Only Human to Complain"

That's when Arlington Hall decided to lure: For a history of recruiting in 1943 and 1944, see RG 0457, 9032 (A1), Box 1115, "Signal Security Agency Annual Report Fiscal Year 1944."

The void of information led recruiting officers: RG 0457, 9002 (A1), Box 95, "History of the Signal Security Agency," vol. 1, "Organization," part 2, "1942–1945" (also online at NSA Cryptologic Histories site).

"Young Army boys we used": Solomon Kullback, oral history interview on August 26, 1982, NSA-OH-17-82, 72. Elsewhere in the same interview, Kullback admitted that they lied: "I think unfortunately, some of the recruiting officers may have lied a little bit in order to get those girls to come to work in Washington by maybe implying that there would be more younger officers available," 112.

"I think the northern members of our community": Ann Caracristi, interview with the author.

By 1944, recruiters were allowed to expand: RG 0457, 9002 (A1), Box 96, "History of the Signal Security Agency," vol. 1, "Organization," part 2, "1942–1945."

An article in a Minnesota newspaper: Jennifer Wilcox archives.

The instructor, Ruth W. Stokes: Letter provided to the author by Winthrop University archivist Susanna O. Lee. Information about the Winthrop program can be found in the Winthrop University Louise Pettus Archives, http://digitalcommons.winthrop .edu/winthroptowashington/.

Titled *Private Smith Goes to Washington*: Recruiting pamphlet provided to the author by Josephine Palumbo Fannon.

The "stinkinest jobs that there were to have": Davis, oral history, 41.

"I don't care if you're a colonel; you can't go in there!": Ibid.

Norma Martell, one of the WACs assigned to Vint Hill: Norma Martell, oral history interview, WV0072 Norma Martell Papers.

In May 1945, two WACs working at the Vint Hill: RG 0457, 9002 (A1), Box 95, "History of the Signal Security Agency, vol. 1, "Organization," part 1, "1939–1945."

A report conducted in 1943 concluded that: RG 0457, 9032 (A1), Box 991, "Report on Progress and Improvements in Section BII, 1943."

In the early fall of 1943, a "morale survey": RG 0457, 9032 (A1), Box 1027, "Survey of Morale, Signal Security Agency, 1943."

In April 1944, two code breakers identified only as M. Miller: Document ID A69346, "A Poem for a Birthday Celebration on April 6th," William F. Friedman

Collection of Official Papers, National Security Agency, https://www.nsa.gov/news-features/declassified-documents/friedman-documents/assets/files/reports-research/FOLDER_060/41709519074880.pdf. The identity of the authors was suggested by NSA historian Elizabeth Smoot.

Chapter Ten: Pencil-Pushing Mamas Sink the Shipping of Japan

Ambon. Canton. Davao. Haiphong: RG 0457, 9032 (A1), Box 844, "Japanese Army Transport Codes."

"This material is extremely secret and must be treated with the utmost care": Ibid.

A few dealt with transportation of the wounded: Ibid.

For example, one station transmitting from Singapore: These and the subsequent examples of stereotypes are found in RG 0457, 9032 (A1), Box 877, "Stereotypes in Japanese Army Cryptographic Systems (Vol III)."

Dot's workday consisted of messages: This example is given in RG 0457, 9032 (A1), Box 844, "Japanese Army Transport Codes."

The administrators at Arlington Hall concluded: RG 0457, 9032 (A1), Box 1115, "Signal Security Agency Annual Report Fiscal Year 1944."

"The history of the department during the past year": Ibid.

"This is a business organization," said one memo: RG 0457, 9032 (A1), Box 1016, "Signals Communications Systems."

A pneumatic tube: RG 0457, 9032 (A1), Box 1380, "History of the Distribution of Intercept Traffic in SSA."

The women in Department K—Dot's unit—were a "very fine group": Ibid.

"The great value of the intelligence derived": RG 0457, 9032 (A1), Box 1115, "Signal Security Agency Annual Report Fiscal Year 1944."

Another memo pointed out that the *New York Times*: RG 0457, 9002 (A1), Box 92, SRH 349, "The Achievements of the Signal Security Agency in World War II."

November 1943, one month after Dot's arrival: RG 0457, 9002 (A1), Box 82, SRH 284, "Radio Intelligence in WWII Submarine Operations in the Pacific Ocean Areas November 1943."

"The success of undersea warfare is to a certain extent": Ibid.

After the war, a census would be taken: RG 0457, 9002 (A1), Box 36, SRH 156, "Weekly Listing of Merchant Vessels Sunk in Far East Waters 14 Dec–March 1945."

At the end of the war, a U.S. naval report found: RG 0457, 9002 (A1), Box 84, SRH 306, "OP-20-G Exploits and Communications World War II."

Citing a few of the biggest achievements: RG 0457, 9032 (A1), Box 878, "Capt. Fuld's Reports, 'Intelligence Derived from Ultra.'"

"When they were planning some major moves against": Solomon Kullback, oral history, interview on August 26, 1982, NSH-OH-17-82, 89.

"By resorting to chewing it raw": RG 0457, 9002 (A1), Box 18, SRH-66, "Examples of Intelligence Obtained from Cryptanalysis 1 August 1946."

"If the latter," the document noted: RG 0457, 9032 (A1), Box 876, "Stereotypes in Japanese Army Cryptographic Systems."

Chapter Eleven: Sugar Camp

They boarded the train at midnight: The memory of arriving at Sugar Camp is from Iris Bryant Castle, "Our White Gloves," letter of reminiscence, Deborah Anderson private archives. Throughout this chapter, I have also drawn from detailed letters of reminiscence written by Jimmie Lee Hutchison Powers Long, Dot Firor, and Esther Hottenstein, in the Deborah Anderson private archives, which Debbie Anderson shared. I have also drawn from Curt Dalton, *Keeping the Secret: The Waves & NCR Dayton, Ohio 1943–1946* (Dayton, OH: Curt Dalton, 1997), and from transcripts of the underlying interviews, which Dalton kindly provided. And I have drawn from interviews I conducted with Millie Weatherly Jones, Veronica Mackey Hulick, and Betty Bemis Robarts.

"The WAVES will take courses in the operation": "Waves to be Occupants of Sugar Camp This Summer," *NCR News*, May 5, 1943.

Jimmie Lee Hutchison was another: Jimmie Lee Hutchison Powers Long, oral history interview on June 30, 2010, NSA-OH-2010-46.

He flirted and had the hearty, slightly false air: Millie Weatherly Jones, in an interview with the author, recalled Meader's flirtatiousness and the women's reaction. His demeanor also was mentioned by Howard Campaigne: "He was quite a politician. He would be all 'hail fellow well met' with everybody." Howard Campaigne, oral history interview on June 29, 1983, NSA-OH-14-83, 38.

For the Allies, 1942 marked the low point: The discussion of the Allied merchant shipping losses, the Battle of the Atlantic, and the role the bombes played in it are taken from a number of sources: David Kahn, *Seizing the Enigma: The Race to Break the German U-Boat Codes, 1939–1943* (New York: Barnes and Noble, 1998); Jim DeBrosse and Colin Burke, *The Secret in Building 26: The Untold Story of America's Ultra War Against the U-Boat Enigma Codes* (New York: Random House, 2004); John A. N. Lee, Colin Burke, and Deborah Anderson, "The US Bombes, NCR, Joseph Desch, and 600 WAVES: The First Reunion of the US Naval Computing Machine Laboratory," *IEEE Annals of the History of Computing* (July–September 2000): 1–15; RG 0457, 9032 (A1), Box 705, "History of the Bombe Project"; and Jennifer Wilcox, *Solving the Enigma: History of the Cryptanalytic Bombe* (Washington, DC: Center for Cryptologic History, National Security Agency, 2015).

The submarine, which had surfaced, began to sink: Wilcox, *Solving the Enigma*, 21–22.

Raised and schooled in Dayton: A good description of Desch's background is in DeBrosse and Burke, *Secret in Building 26*, 6–9.

On January 31, 1943, a unit diary: The transfer of Agnes Driscoll's team, and the trips taken by John Howard between D.C. and Dayton, are in RG 38, Box 113, "CNSG-OP-20-GM-6/GM-1-C-3/GM-1/GE-1/GY-A-1 Daily War Diary."

One such woman was Louise Pearsall: The details about Louise Pearsall's life, enlistment, life in Washington, and work on the bombe project are from an oral history: "Interview with Louise Pearsall Canby," taken by her daughter, Sarah Jackson, May 17, 1997, University of North Texas Oral History Collection Number 1163, and from an author interview with her daughter, Sarah Jackson, and her brother, William Pearsall.

If they suspected that a line of cipher such as: This example is offered in Chris Christensen, "Review of IEEE Milestone Award to the Polish Cipher Bureau for 'The First Breaking of Enigma Code,'" *Cryptologia* 39, no. 2 (2015): 188.

She always remembered one terrible night: Ann White Kurtz, from Mary Carpenter, underlying notes for Mary Carpenter and Betty Paul Dowse, "The Code Breakers of 1942," *Wellesley* (Winter 2000): 26–30.

He was under terrible pressure: DeBrosse and Burke, *Secret in Building 26*, 86, describe Meader as a hard taskmaster, as did Deborah Anderson in an interview with the author.

"The design of the Bombe eventually required": RG 38, Box 109, "CNSG Report of Supplementary Research Operations in WWII."

"The first two experimental bombes were under preliminary tests": The saga of the bombes' first summer is in RG 38, Boxes 38 and 39, "Watch Officer's Log, 26 June–9 August 1943."

Commander Meader had ejected a number of: RG 38, Box 2, "CNSG, Assignment/Transfers, (Enlisted Pers), (1 of 5)."

Back in 1942, when the WAVES were formed: The policy of what to do about pregnancy and abortion is discussed in "Women in the Military Box 7," in the folder marked "Bureau of Naval Personnel Women's Reserve, First Draft Narrative Prepared by the Historical Section, Bureau of Naval Personnel" in the Ready Reference Section of the Naval History and Heritage Command in Washington, D.C.

Once, when a sloppy (or tired): Jennifer Wilcox, *Sharing the Burden: Women in Cryptology During World War II* (Washington, DC: Center for Cryptologic History, National Security Agency, 2013), 10.

"He had nightmares for years about men dying": Deborah Anderson, daughter of Joseph Desch, interview with the author.

The people working on the Enigma project: DeBrosse and Burke, *Secret in Building 26*, make this point very well.

One of the women in charge of maintaining: Graham Cameron, son of Charlotte McLeod Cameron, interview with the author.

A daily log on February 25, 1944: RG 38, Box 40, "Watch Officers Log, 2 Feb–4 March 1944."

Another was reprimanded for coming in: RG 38, Box 39, "Watch Officers Log 26 September–26 November 1943."

After two months at Smith: Pearsall's return date as an officer is in RG 38, Box 1, "COMNAVSECGRU-OP-20G Headquarters Personnel Rosters & Statistics (3 of 4)."

Jimmie Lee by now had married: Jimmie Lee's reflections here and elsewhere are taken from transcripts of her interviews with Curt Dalton, author of *Keeping the Secret*; letters of reminiscence she wrote Deborah Anderson; and her NSA oral history interview, Jimmie Lee Hutchison Power Long, on June 30, 2010, OH-2010-46.

As Jimmie Lee and the other women: Jennifer Wilcox pointed out the all-female nature of the operation in an interview with the author.

Promoted to watch officer, Fran had access: Jed Suddeth, son of Fran Steen Suddeth Josephson, interview with the author.

The effect of the U.S. bombes on solving the Atlantic U-boat cipher: RG 38, Box 141, "Brief Resume of Op-20-G and British Activities vis-à-vis German Machine Ciphers," in folder marked "Photograph of Bombe Machine, about 1943."

Once they were broken, the messages would pass to: Janice Martin Benario, interviews with the author. She also describes the tracking room in Janice M. Benario, "Top Secret Ultra," *Classical Bulletin* 74, no. 1 (1998): 31–33; and Robert Edward Lewand, "Secret Keeping 101: Dr. Janice Martin Benario and the Women's College Connection to ULTRA," *Cryptologia* 35, no. 1 (2010): 42–46.

There, a commander named Kenneth Knowles: David Kohnen, *Commanders Winn and Knowles: Winning the U-Boat War with Intelligence, 1939–1943* (Krakow: Enigma Press, 1999), describes the tracking room. The working together of submarine tracking rooms in the UK and United States is described in Kahn, *Seizing the Enigma*, 191.

One male officer said the WAVES did a better job: Kahn, *Seizing the Enigma*, 242–244.

After the carnage of 1942 and early 1943: Good descriptions of the innovations in the first six months of 1943—HF/DF, hunter-killers, and so on—are in DeBrosse and Burke, *Secret in Building 26*, 117; and Richard Overy, *Why the Allies Won* (New York: Norton, 1996), 45–62.

These refuelers were known as milch cows: Kahn, *Seizing the Enigma*, 274–275.

In October 1943, the U-Boats reappeared: RG 0457, 9002 (A1), Box 95, SRH 367, "A Preliminary Analysis of the Role of Decryption Intelligence in the Operational Phase of the Battle of the Atlantic."

"Congratulations from Hut six": RG 38, Box 4, "COMNAVSECGRU Commendations Received by the COMINT Organization, Jan 1942–8 July 1948."

These were long and desperate sea journeys: RG 0457, 9002 (A1), Box 84, SRH 306, "OP-20-G Exploits and Communications World War II."

"The attack against the U-boat cipher has been so successful": RG 38, Box 141, "Brief Resume of Op-20-G and British Activities vis-à-vis German Machine Ciphers," in folder marked "Photograph of Bombe Machine, about 1943."

Chapter Twelve: "All My Love, Jim"

"I don't have anything exciting to write you, Dot": Jim Bruce to Dorothy Braden, April 21, 1944.

"I get treated worse than anyone I know": Jim Bruce to Dorothy Braden, April 30, 1944.

"The plane went every way but the right way": Jim Bruce to Dorothy Braden, May 26, 1944.

"I was just sitting here in bed, waiting for Carolyn": Dot Braden, April 3, 1944.

"I enjoy very much reading your letters, Dot": Jim Bruce to Dorothy Braden, August 7, 1944.

"I guess you are still having a good time with your friends": Jim Bruce to Dorothy Braden, October 28, 1944.

"The weather situations are rather interesting here": Jim Bruce to Dorothy Braden, November 28, 1944.

"That is a long ways, in fact it is six thousand miles": Jim Bruce to Dorothy Braden, December 1, 1944.

"In a letter from you that I received yesterday": Jim Bruce to Dorothy Braden, December 19, 1944.

"I love you and am looking forward with great anxiety": Jim Bruce to Dorothy Braden, January 9, 1945.

"It stopped raining and cleared up when I predicted": Jim Bruce to Dorothy Braden, January 14, 1945.

"The letter that I received from you today is the one": Jim Bruce to Dorothy Braden, February 21, 1945.

Chapter Thirteen: "Enemy Landing at the Mouth of the Seine"

In November 1943, the Purple machine: RG 0457, 9002 (A1), Box 17, "Achievements of U.S. Signal Intelligence During WWII."

Raven and his crew called him Honest Abe: The Coral team's monitoring of Abe, and his message about coastal fortifications, are in RG 38, Box 116, "CNSG-OP20-GYP History for WWII Era (3 of 3)."

At Bletchley, code breakers broke a long message: Arthur J. Levenson, oral history interview on November 25, 1980, NSA-OH-40-80, https://www.nsa.gov/news-features /declassified-documents/oral-history-interviews/assets/files/nsa-oh-40-08-levenson.pdf.

"So great were the chances of all the traffic": RG 0457, 9032 (A1), Box 763, "Cover Plan in Operation Overlord." A good description of Operation Fortitude North and South is in Thaddeus Holt, *The Deceivers: Allied Military Deception in the Second World War* (New York: Scribner, 2004), 510–584.

"The ploughman homeward plods his weary way": RG 0457, 9032 (A1), Box 833, "Security Posters and Miscellaneous Documents."

A whole section of Arlington Hall was devoted: The role of women in the protective security branch, and their involvement in planning and implementing a number of deception programs, including at Yalta and Normandy, is in RG 0457, 9032 (A1), Box 980, "Pictorial History of the SSA Security Division Protective Security Branch Communications Security Branch."

"No invasion tonight," thought Wellesley's Ann White: Ann White Kurtz, in "From WomenatWartoForeignAffairsScholar,"*AmericanDiplomacy*(June2006),http://www .unc.edu/depts/diplomat/item/2006/0406/kurt/kurtz_women.html, describes the receipt of the D-Day messages, bolting up and down the stairs, the first and second messages, and then "sporadic bulletins followed."

"At 0130, messages on coastal circuits": RG 38, Box 113, "CNSG-OP-20-GM-6/ GM-1-C-3/GM-1/GE-1/GY-A-1 Daily War Diary."

At 1:40 in the morning they were warned: RG 38, Box 30, "OP-20-GM Watch Office Logs, 22 June 1943–31 Dec 1943."

Going to church was the only way: Carpenter and Dowse, "Wellesley Codebreakers," 30, and Mary Carpenter, underlying notes for Mary Carpenter and Betty Paul Dowse, "The Code Breakers of 1942," *Wellesley* (Winter 2000): 26–30.

Ann would remember the Normandy invasion: Ibid.

"A great quantity of administrative traffic": RG 38, Box 113, "CNSG-OP-20-GM-6/ GM-1-C-3/GM-1/GE-1/GY-A-1 Daily War Diary."

At the Naval Annex, Georgia O'Connor: Georgia O'Connor Ludington, oral history interview on September 5, 1996, NSA-OH-1996-09. The extensive nature of the code rooms devoted to communications intelligence coming in from the Atlantic and Pacific theaters is in RG 38, Box 111, "CNSG-OP-20GC War Diary, 1941–1943."

"There were many signs the enemy was disintegrating": Elizabeth Bigelow Stewart, essay of reminiscence, shared with the author by her daughter Cam Weber.

Others were not so lucky: Stewart, essay of reminiscence.

Donna Doe Southall was one of: Donna Doe Southall, interview, undated, Library of Congress Veterans History Project.

Chapter Fourteen: Teedy

Teedy Braden finished high school on a Friday: Here and throughout this chapter reminiscences are from John "Teedy" Braden, interview with the author in Good Hope, Georgia, on December 1, 2015.

"I sure do hope that you won't [be] too busy": Teedy Braden to Dot Braden, June 26, 1944.

"How's everything, gal?": Teedy Braden to Dot Braden, July 20, 1944.

"If I do go it'll mean that it's the first step": Teedy Braden to Dot Braden, July 31, 1944.

He was one of thousands of very young men: Antony Beevor, *Ardennes 1944: The Battle of the Bulge* (New York: Viking, 2015), 50–53.

Teedy's unit, the 112th Infantry, suffered: Ibid., 68, 151–156.

And it was one of the war's worst: Arthur J. Levenson, a U.S. Army cryptanalyst sent to work at Bletchley Park, said, "Battle of the Bulge took us a little by surprise and we were a little ashamed of the intelligence dearth because they had put on a silence and I remember just before there was no traffic.... They had imposed a silence and that should have been a real indicator." Oral history interview on November 25, 1980, NSA-OH-40-80, 38.

"I suppose that you've been kinda worried:" Teedy Braden to Dot Braden, January 1, 1945.

Admiral Dönitz—the new head of state in Germany—ordered: Richard Overy, *Why the Allies Won* (New York: Norton, 1996), 62.

"You have fought like lions:" RG 0457, 9032 (A1), Box 623, "COMINCH File of Memoranda Concerning U-Boat Tracking Room Operations."

"My Dear Teedy," she wrote, "Hope everything": Virginia Braden to Teedy Braden, June 7, 1945.

Chapter Fifteen: The Surrender Message

Not long after, Alethea Chamberlain: Karen Kovach, "Breaking Codes, Breaking Barriers: The WACs of the Signal Security Agency, World War II" (Fort Belvoir, VA: History Office, U.S. Army Intelligence and Security Command, 2001), 41.

The minute Ann Caracristi set foot in Arlington Hall: Ann Caracristi, interview, undated, Library of Congress Veterans History Project, https://memory.loc.gov/ diglib/vhp-stories/loc.natlib.afc2001001.30844/transcript?ID=mv0001.

JAH "theoretically was restricted to low grade traffic": RG 0457, 9032 (A1), Box 1115, "History of the Language Branch, Army Security Agency."

She owned that code: That Virginia Aderholdt attended Bethany College is in RG 0457, 9032 (A1), Box 1007, "Personnel Organization." That she "scanned and translated JAH and related texts" and that the Japanese translators followed the war by monitoring the diplomatic messages is in RG 0457, 9032 (A1), Box 1115, "History of the Language Branch, Army Security Agency."

She "had worked on that code and loved": Frank Rowlett, oral history interview in 1976, NSA-OH-1976-1-10, 189–192.

At Arlington Hall, the rule was: Solomon Kullback recalls that though Arlington Hall knew the Japanese surrender message was coming twenty-four hours in advance, "no word leaked out." Oral history interview on August 26, 1982, NSA-OH-17-82.

Chapter Sixteen: Good-Bye to Crow

"I went down town yesterday to do some shopping": Virginia Braden to Dot Braden, December 10, 1945.

"I think I enjoyed them more than any letters I received": Jim Bruce to Dot Braden Bruce, January 16, 1946.

Epilogue: The Mitten

Hugh Erskine, a younger relative: Hugh Erskine, interview with the author.

In an interview before her death: Ann Caracristi, interviews with the author.

Polly Budenbach, a Smith College graduate: Mary H. "Polly" Budenbach, oral history interview on June 19, 2001, NSA-OH-2001-27. Budenbach also discusses the difficulty of having an NSA career and having a spouse.

In the very early days of 1943, an ex-schoolteacher: Robert L. Benson, "The Venona Story," https://www.nsa.gov/about/cryptologic-heritage/historical-figures-publications/publications/coldwar/assets/files/venona_story.pdf.

"Gene was just an independent person": Eleanor Grabeel, interview with the author.

"In every case the response was the same": Elizabeth Bigelow Stewart, essay of reminiscence, shared with the author by her daughter Cam Weber.

Janice Martin Benario, the Goucher Latin major: Janice Martin Benario, interviews with the author.

Dorothy Ramale, the aspiring math teacher: Dorothy Ramale, interviews with the author.

Betty Bemis, the champion swimmer: Betty Bemis Roberts, naval code breaker, interview with the author in Georgia on December 2, 2015.

Louise Pearsall, who worked on Enigma: "Interview with Louise Pearsall Canby," oral history taken by her daughter, Sarah Jackson, May 17, 1997, University of North Texas Oral History Collection Number 1163; Sarah Jackson, interviews with the author; William Pearsall, interview with the author.

Betty Allen, one of the group of friends: Elizabeth Allen Butler, *Navy Waves* (Charlottesville, VA: Wayside Press, 1988).

Here is how the round-robin letter worked: Ruth Schoen Mirsky, interviews with the author.

"We hadn't won any battles and didn't feel": Edith Reynolds White, interviews with the author.

Fran Steen, the Goucher biology major: Fran Josephson, SCETV interview; Jed Suddeth, interview with the author; David Shimp, interview with the author.

"There were Japanese that went down with that ship": Jeuel Bannister Esmacher, interview with the author.

"I had always done everything I was told": Jane Case Tuttle, interview with the author.

"Oh golly, did I miss it": Mary Carpenter and Betty Paul Dowse, "The Code Breakers of 1942," *Wellesley* (Winter 2000): 26–30.

"A lot of people don't bother to learn their names": Dorothy Braden Bruce, interviews with the author.

Bibliography

Selected Interviews

Deborah Anderson in Dayton, Ohio, and by telephone, numerous between May 2015–May 2017

Janice Martin Benario, in Atlanta, Georgia, December 2, 2015

Viola Moore Blount (email correspondence) numerous between April 27–30, 2016

John "Teedy" Braden, in Good Hope, Georgia, December 1, 2015

Dorothy "Dot" Braden Bruce, in Midlothian, Virginia, between June 2014 and April 2017

Ida Mae Olson Bruske, by telephone May 8, 2015

Ann Caracristi, at her home in Washington, D.C., numerous between November 2014 and November 2015

Suzanne Harpole Embree in Washington, D.C., August 11, 2015

Jeuel Bannister Esmacher, in Anderson, South Carolina, November 21, 2015

Josephine Palumbo Fannon in Maryland, April 9 and July 17, 2015

Jeanne Hammond in Scarborough, Maine, September 30, 2015

Veronica "Ronnie" Mackey Hulick, by telephone, undated

Millie Weatherly Jones in Dayton, Ohio, May 1, 2015

Margaret Gilman McKenna by Skype, April 19, 2016

Ruth Schoen Mirsky, in Belle Harbor, New York, June 8, June 16, and October 20, 2015

Helen Nibouar, by telephone, July 3, 2015

Dorothy Ramale in Springfield, Virginia, May 29 and July 12, 2015

Betty Bemis Robarts in Georgia, December 2, 2015

Anne Barus Seeley in Yarmouth, Massachusetts, June 12, 2015

Lyn Ramsdell Stewart, by telephone, October 27, 2015

Nancy Thompson Tipton, by telephone, January 27, 2016

Nancy Dobson Titcomb in Springvale, Maine, October 1, 2015

Jane Case Tuttle, in Scarborough, Maine, September 30, 2015

Clyde Weston, by telephone, October 9, 2015

Kitty Weston, in Oakton, Virginia, April 10, 2015

Edith Reynolds White, in Williamsburg, Virginia, February 8, 2016

Jean Zapple, by telephone, undated

Manuscript and Archival Sources

National Archives and Records Administration II in College Park, Maryland

National Archives and Records Administration Personnel Records Center in St. Louis, Missouri

Library of Congress Veterans History Project in Washington, D.C.

Betty H. Carter Women Veterans Historical Project, Martha Blakeney Hodges Special Collections and University Archives, The University of North Carolina at Greensboro, North Carolina.

William F. Friedman Papers and Elizebeth Smith Friedman Collection at George C. Marshall Foundation Library in Lexington, Virginia

National Cryptologic Museum Library in Fort Meade, Maryland

Naval History and Heritage Command Ready Reference Room in Washington, D.C.

Personal Archives of Deborah Anderson in Dayton, Ohio

Dayton History in Dayton, Ohio

Winthrop University Louise Pettus Archives

Center for Local History, Arlington Public Library, Arlington, Virginia

Historical Society of Washington, D.C.

Wellesley College Archives in Wellesley, Massachusetts

Schlesinger Library at Radcliffe Institute for Advanced Study in Cambridge, Massachusetts

Smith College Archives in Northampton, Massachusetts

Randolph College Archives in Lynchburg, Virginia

Jones Memorial Library in Lynchburg, Virginia

Pittsylvania County History Research Center and Library in Chatham, Virginia

Women Veterans Oral History Project, University of North Texas

Oral Histories

National Security Agency (available online):

Hildegarde Bearg-Hopt, NSA-OH-2013-30, March 15, 2013

Mary H. (Polly) Budenbach, NSA-OH-2001-27, June 19, 2001

Benson Buffham, NSA-OH-51-99, June 15, 1999

Lambros D. Callimahos, NSA-OH-2013-86, Summer 1966

Howard Campaigne, NSA-OH-20-83, June 29, 1983

Ann Caracristi, NSA-OH-15-82, July 16, 1982

Gloria Chiles, NSA-OH-32-80, September 15, 1980

Elizabeth Corrin, NSA-OH-2002-06, February 8, 2002

Prescott Currier, NSA-OH-02-72, April 14, 1972

Prescott Currier, NSA-OH-03-80, November 14, 1980

Wilma (Berryman) Davis, NSA-OH-25-82 December 3, 1982

Elizebeth Friedman, NSA-OH-1976-16, November 11, 1976

Elizebeth Friedman, NSA-OH-1976-17, November 11, 1976

Elizebeth Friedman, NSA-OH-1976-18, November 11, 1976

Theresa G. Knapp, NSA-OH-1999-67, November 8, 1999

Solomon Kullback, NSA-OH-17-82, August 26, 1982

Jimmie Lee Hutchison Powers Long, NSA-OH-2010-46, June 30, 2010
Arthur J. Levenson, NSA-OH-40-80, November 25, 1980.
Georgia Ludington, NSA-OH-1996-09, September 5, 1996
Dorothy Madsen, NSA-OH-2010-24, April 27, 2010
Georgette McGarrah, NSA-OH-2013-06, January 22, 2013
Juanita (Morris) Moody, NSA-OH-1994-32, June 16, 1994
Juanita Moody, NSA-OH-2001-28, June 20, 2001
Juanita Moody, NSA-OH-2003-12, June 12, 2003
Helen Nibouar, NSA-OH-2012-39, June 7, 2012
Francis Raven, NSA-OH-03-72, March 28, 1972
Francis Raven, NSA-OH-1980-03, January 24, 1980
Frank Rowlett, NSA-OH-1976-(1-10), undated 1976
Abraham Sinkov, NSA-OH-02-79, May 1979
Sally Speer, NSA-OH-18-84, August 28, 1984
Margueritte Wampler, NSA-OH-2002-03, April 29, 2003

Betty H. Carter Women Veterans Historical Project, Martha Blakeney Hodges Special Collections and University Archives, The University of North Carolina at Greensboro, NC (available online):

Helen R. Allegrone, April 21, 1999, WV0062
Jaenn Coz Bailey, January 13, 2000, WV0141
Betty Hyatt Caccavale, June 18, 1999, WV0095
Myrtle O. Hanke, February 11, 2000 WV0147
Ava Caudle Honeycutt, November 22, 2008, WV0438
Erma Hughes Kirkpatrick, May 12, 2001, WV0213

University of North Texas Oral History Collection:

Louise Pearsall Canby, OH Collection No. 1163, March 17, 1997
William W. Pearsall, OH Collection No. 1185, June 18, 1997

Library of Congress Veterans History Project (available online, most undated):

Ann Caracristi
Ann Ellicott Madeira
Donna Doe Southall
Ethel Louise Wilson Poland
Elizabeth Bigelow Stewart
Frances Lynd Scott

Books

Alvarez, David. *Secret Messages: Codebreaking and American Diplomacy, 1930–1945*. Lawrence: University Press of Kansas, 2000.
Atkinson, Rick. *The Guns at Last Light: The War in Western Europe, 1944–1945*. New York: Henry Holt, 2013.
Beevor, Antony. *Ardennes 1944: The Battle of the Bulge*. New York: Viking, 2015.

———. *D-Day: The Battle for Normandy.* New York: Penguin, 2010.

Benson, Robert Louis. *A History of U.S. Communications Intelligence During World War II: Policy and Administration.* Washington, DC: Center for Cryptologic History, National Security Agency, 1997.

Browne, Jay, and C. Carlson, eds. *Echoes of Our Past: Special Publication.* Pace, FL: Naval Cryptologic Veterans Association; Patmos, 2008.

Budiansky, Stephen. *Battle of Wits: The Complete Story of Codebreaking in World War II.* New York: Free Press, 2000.

Butler, Elizabeth Allen. *Navy Waves.* Charlottesville, VA: Wayside Press, 1988.

Carlson, Elliot. *Joe Rochefort's War: The Odyssey of the Codebreaker Who Outwitted Yamamoto at Midway.* Annapolis: Naval Institute Press, 2011.

Center for Cryptologic History. *The Friedman Legacy: A Tribute to William and Elizebeth Friedman.* National Security Agency, 2006.

Clark, Ronald. *The Man Who Broke Purple: The Life of Colonel William F. Friedman, Who Deciphered the Japanese Code in World War II.* Boston: Little Brown, 1977.

Dalton, Curt. *Keeping the Secret: The Waves & NCR Dayton, Ohio 1943–1946.* Dayton: Curt Dalton, 1997.

DeBrosse, Jim, and Colin Burke. *The Secret in Building 26: The Untold Story of America's Ultra War Against the U-Boat Enigma Codes.* New York: Random House, 2004.

De Leeuw, Karl, and Jan Bergstra, eds. *The History of Information Security: A Comprehensive Handbook.* Amsterdam: Elsevier, 2007.

Drea, Edward J. *MacArthur's ULTRA: Codebreaking and the War Against Japan, 1942–1945.* Lawrence: University Press of Kansas, 1992.

Ebbert, Jean, and Marie-Beth Hall. *Crossed Currents: Navy Women in a Century of Change.* Washington, DC: Brassey's, 1999.

Friedman, William. *Elementary Military Cryptography.* Laguna Hills, CA: Aegean Park, 1976.

———. *Elements of Cryptanalysis.* Laguna Hills, CA: Aegean Park, 1976.

———. *Six Lectures Concerning Cryptography and Cryptanalysis.* Laguna Hills, CA: Aegean Park, 1996.

Gilbert, James L., and John P. Finnegan, eds. *U.S. Army Signals Intelligence in World War II.* Washington, DC: Center of Military History, United States Army, 1993.

Gildersleeve, Virginia Crocheron. *Many a Good Crusade.* New York: Macmillan, 1954.

Godson, Susan H. *Serving Proudly: A History of Women in the U.S. Navy.* Annapolis: Naval Institute Press, 2002.

Hanyok, Robert J. *Eavesdropping on Hell: Historical Guide to Western Communications Intelligence and the Holocaust, 1939–1945.* Mineola, NY: Dover Publications, 2012.

Hart, Scott. *Washington at War: 1941–1945.* Englewood Cliffs, NJ: Prentice Hall, 1970.

Harwood, Jeremy. *World War II at Sea: A Naval View of the Global Conflict: 1939 to 1945.* Minneapolis: Zenith, 2015.

Hinsley, F. H., and Alan Stripp, eds., *Code Breakers: The Inside Story of Bletchley Park.* Oxford: Oxford University Press, 2001.

Holt, Thaddeus. *The Deceivers: Allied Military Deception in the Second World War.* New York: Scribner, 2004.

Isaacson, Walter. *The Innovators: How a Group of Hackers, Geniuses, and Geeks Created the Digital Revolution.* New York: Simon & Schuster, 2014.

Johnson, Kevin Wade. *The Neglected Giant: Agnes Meyer Driscoll.* Washington, DC: National Security Agency Center for Cryptologic History, 2015.

Kahn, David. *The Codebreakers.* New York: Scribner, 1967.

———. *Seizing the Enigma: The Race to Break the German U-Boat Codes, 1939–1943.* New York: Houghton Mifflin, 1991.

Keegan, John. *The Second World War.* New York: Viking Penguin, 1990.

Kenschaft, Patricia Clark. *Change Is Possible: Stories of Women and Minorities in Mathematics.* American Mathematical Society, 2005.

Kessler-Harris, Alice. *Out to Work: A History of Wage-Earning Women in the United States.* Oxford: Oxford University Press, 1982.

Kohnen, David. *Commanders Winn and Knowles: Winning the U-Boat War with Intelligence, 1939–1943.* Krakow: Enigma Press, 1999.

Kovach, Karen. *Breaking Codes, Breaking Barriers: The WACs of the Signal Security Agency, World War II.* History Office, U.S. Army Intelligence and Security Command, 2001.

Layton, Edwin T., Roger Pineau, and John Costello. *And I Was There: Pearl Harbor and Midway—Breaking the Secrets.* New York: Morrow, 1985.

Lewin, Ronald. *The American Magic: Codes, Ciphers and the Defeat of Japan.* New York: Farrar Straus & Giroux, 1982.

———. *Ultra Goes to War: The Secret Story.* London: Hutchinson, 1978.

Maffeo, Steven E. *US Navy Codebreakers, Linguists, and Intelligence Officers Against Japan, 1910–1941.* Lanham, MD: Rowman & Littlefield, 2015.

Marston, Daniel, ed. *The Pacific War: From Pearl Harbor to Hiroshima.* Oxford: Osprey, 2005.

McGinnis, George P., ed. *U.S. Naval Cryptologic Veterans Association.* Paducah, KY: Turner, 1996.

McKay, Sinclair. *The Secret Lives of Codebreakers: The Men and Women Who Cracked the Enigma Code at Bletchley Park.* New York: Plume, 2012.

Mikhalevsky, Nina. *Dear Daughters: A History of Mount Vernon Seminary and College.* Washington, DC: Mount Vernon Seminary and College Alumnae Association, 2001.

Musser, Frederic O. *The History of Goucher College, 1930–1985.* Baltimore: Johns Hopkins University Press, 1990.

Overy, Richard. *Why the Allies Won.* New York: Norton, 1996.

Parker, Frederick. *A Priceless Advantage: U.S. Navy Communications Intelligence and the Battles of Coral Sea, Midway, and the Aleutians.* Washington, DC: Center for Cryptologic History, National Security Agency (1993).

Pimlott, John. *The Historical Atlas of World War II.* New York: Henry Holt, 1995.

Prados, John. *Combined Fleet Decoded.* New York: Random House, 1995.

Prange, Gordon W. *At Dawn We Slept: The Untold Story of Pearl Harbor.* New York: McGraw-Hill, 1981.

Rowlett, Frank B. *The Story of Magic: Memoirs of an American Cryptologic Pioneer.* Laguna Hills, CA: Aegean Park, 1998.

Scott, Frances Lynd. *Saga of Myself.* San Francisco: Ithuriel's Spear, 2007.

Smith, Michael. *The Debs of Bletchley Park and Other Stories.* London: Aurum, 2015.

Treadwell, Mattie, *United States Army in World War II: Special Studies; The Women's Army Corps.* Washington, DC: Center of Military History, United States Army, 1991.

Weatherford, Doris. *American Women During World War II: An Encyclopedia.* New York: Routledge, 2010.

Wilcox, Jennifer. *Sharing the Burden: Women in Cryptology During World War II.* Fort Meade, MD: Center for Cryptologic History, National Security Agency, 1998.

———. *Solving the Enigma: History of the Cryptanalytic Bombe.* Fort Meade, MD: Center for Cryptologic History, National Security Agency, 2006.

Williams, Jeannette, with Yolande Dickerson. *The Invisible Cryptologists: African-Americans, WWII to 1956.* Fort Meade, MD: Center for Cryptologic History, National Security Agency, 2001. https://www.nsa.gov/about/cryptologic-heritage/historical -figures-publications/publications/wwii/assets/files/invisible_cryptologists.pdf.

Writers' Program of the Work Projects Administration in the State of Virginia. *Virginia: A Guide to the Old Dominion.* New York: Oxford University Press, 1940.

Articles and Pamphlets

Bauer, Craig, and John Ulrich. "The Cryptologic Contributions of Dr. Donald Menzel," *Cryptologia* 30.4: 306–339.

Benario, Janice M. "Top Secret Ultra," *The Classical Bulletin* 74.1 (1998): 31–33.

Benson, Robert L. "The Venona Story," https://www.nsa.gov/about/cryptologic-her itage/historical-figures-publications/publications/coldwar/assets/files/venona _story.pdf.

Buck, Stuart H. "The Way It Was: Arlington Hall in the 1950s," *The Phoenician* (Summer 1988).

Burke, Colin. "Agnes Meyer Driscoll vs the Enigma and the Bombe," monograph, http://userpages.umbc.edu/~burke/driscoll1-2011.pdf.

Campbell, D'Ann. "Fighting with the Navy: The WAVES in World War II," in Sweetman, Jack, ed., *New Interpretations in Naval History: Selected Papers from the Tenth Naval History Symposium Held at the United States Naval Academy, 11–13 September 1991.* Annapolis: Naval Institute Press, 1993.

Carpenter, Mary, and Betty Paul Dowse. "The Code Breakers of 1942," *Wellesley* (Winter 2000): 26–30.

Christensen, Chris. "Review of IEEE Milestone Award to the Polish Cipher Bureau for 'The First Breaking of Enigma Code,'" *Cryptologia* 39.2: 178–193.

———. "US Navy Cryptologic Mathematicians During World War II," *Cryptologia* 35.3: 267–276.

Christensen, Chris, and David Agard. "William Dean Wray (1910–1962): The Evolution of a Cryptanalyst," *Cryptologia* 35.1: 73–96.

Donovan, Peter W. "The Indicators of Japanese Ciphers 2468, 7890, and JN-25A1," *Cryptologia* 30.3: 212–235.

Faeder, Marjorie E. "A Wave on Nebraska Avenue," *Naval Intelligence Professionals Quarterly*, 8.4 (October 1992): 7–10.

Fairfax, Beatrice. "Does Industry Want Glamour or Brains?" *Long Island Star Journal*, March 19, 1943.

Frahm, Jill. "Advance to the 'Fighting Lines': The Changing Role of Women Telephone Operators in France During the First World War," *Federal History Journal* Issue 8 (2016): 95–108.

Gallagher, Ida Jane Meadows. "The Secret Life of Frances Steen Suddeth Josephson," *The Key* (Fall 1996): 26–30.

Gildersleeve, Virginia C. "We Need Trained Brains," *New York Times Magazine*, March 29, 1942.

Goldin, Claudia D. "Marriage Bars: Discrimination Against Married Women Workers, 1920s to 1950s," NBER Working Paper 2747 (October 1988).

———. "The Quiet Revolution That Transformed Women's Employment, Education, and Family," *AEA Papers and Proceedings* (May 2006): 1–21.

———. "The Role of World War II in the Rise of Women's Employment," *The American Economic Review* 81.4 (September 1991): 741–756.

Greenbaum, Lucy. "10,000 Women in U.S. Rush to Join New Army Corps," *New York Times*, May 28, 1942, A1.

Guton, Joseph M. "Girl Town: Temporary World War II Housing at Arlington Farms," *Arlington Historical Magazine* 14.3 (2011): 5–13.

Kahn, David. "Pearl Harbor and the Inadequacy of Cryptanalysis," *Cryptologia* 15.4: 273–294.

———. "Why Weren't We Warned?" *MHQ: Quarterly Journal of Military History* 4.1 (Autumn 1991): 50–59.

Kurtz, Ann White. "An Alumna Remembers," *Wellesley Wegweiser*, Issue 10 (Spring 2003).

———. "From Women at War to Foreign Affairs Scholar," *American Diplomacy: Foreign Service Dispatches and Periodic Reports on U.S. Foreign Policy* (June 2006).

Lee, John A. N., Colin Burke, and Deborah Anderson. "The US Bombes, NCR, Joseph Desch, and 600 WAVES: The First Reunion of the US Naval Computing Machine Laboratory," *IEEE Annals of the History of Computing* (July–September 2000): 1–15.

Lewand, Robert Edward. "The Perfect Cipher," *The Mathematical Gazette* 94.531 (November 2010): 401–411.

———. "Secret Keeping 101—Dr. Janice Martin Benario and the Women's College Connection to ULTRA," *Cryptologia* 35.1: 42–46.

Lipartito, Kenneth. "When Women Were Switches: Technology, Work, and Gender in the Telephone Industry, 1890–1920," *American Historical Review* 99.4 (October 1994): 1075–1111.

Lujan, Susan M. "Agnes Meyer Driscoll," *NCA Cryptolog* (August Special 1988): 4–6.

Martin, Douglas, "Frank W. Lewis, Master of the Cryptic Crossword, Dies at 98," *New York Times*, December 3, 2010.

McBride, Katharine E. "The College Answers the Challenge of War and Peace," *Bryn Mawr Alumnae Bulletin* 23.2 (March 1943).

Musser, Frederic O. "Ultra vs Enigma: Goucher's Top Secret Contribution to Victory in Europe in World War II," *Goucher Quarterly* 70.2 (1992): 4–7.

Parker, Harriet F. "In the Waves," *Bryn Mawr Alumnae Bulletin* 23.2 (March 1943).

Richard, Joseph E. "The Breaking of the Japanese Army's Codes," *Cryptologia* 28.4: 289–308.

Rosenfeld, Megan. "'Government Girls:' World War II's Army of the Potomac," *Washington Post*, May 10, 1999. A1.

Safford, Captain Laurance F. "The Inside Story of the Battle of Midway and the Ousting of Commander Rochefort," essay written in 1944, published in *Echoes of Our Past*, Naval Cryptologic Veterans Association (Pace, FL: Patmos, 2008).

Sheldon, Rose Mary. "The Friedman Collection: An Analytical Guide," http://marshall foundation.org/library/wp-content/uploads/sites/16/2014/09/Friedman_Collection _Guide_September_2014.pdf.

Sherman, William H. "How To Make Anything Signify Anything," *Cabinet* Issue 40 (Winter 2010/11). www.cabinetmagazine.org/issues/40/sherman.php.

Smoot, Betsy Rohaly. "An Accidental Cryptologist: The Brief Career of Genevieve Young Hitt," *Cryptologia* 35.2: 164–175.

Stickney, Zephorene. "Code Breakers: The Secret Service," *Wheaton Quarterly* (Summer 2015).

Wright, William M. "White City to White Elephant: Washington's Union Station Since World War II," *Washington History* 10.2 (Fall/Winter 1998–99): 25–31.

Websites, DVDs, Speeches, Essays

Dayton Code Breakers: http://daytoncodebreakers.org/

Undated television interview with Nancy Dobson Titcomb

Fran Steen Suddeth Josephson, *South Carolina's Greatest Generation* DVD, interview with South Carolina ETV, uncut version, undated

Margaret Gilman McKenna, videotape interview provided by family

Elizabeth Bigelow Stewart, essay of reminiscence

Ann Caracristi speech, "Women in Cryptology," presented at the NSA on April 6, 1998

Larry Gray essay of reminiscence about his mother, Virginia Caroline Wiley, "Nobody Special, She Said"

Nancy Tipton letter of reminiscence, "Memoirs of a Cryptographer 1944–1946," February 2, 2006

Betty Dowse publication of wartime reminiscences by the class of 1942 at Wellesley, "The World of Wellesley '42"

Index